Advances in Experimental Medicine and Biology

Volume 1104

More information about this series at http://www.springer.com/series/5584

Yoshiki Yamaguchi • Koichi Kato
Editors

Glycobiophysics

Springer

Editors
Yoshiki Yamaguchi
Synthetic Cellular Chemistry Laboratory
RIKEN Cluster for Pioneering Research
Wako, Saitama, Japan

Koichi Kato
Exploratory Research Center on Life and Living Systems
National Institutes of Natural Sciences
Okazaki, Aichi, Japan

ISSN 0065-2598 ISSN 2214-8019 (electronic)
Advances in Experimental Medicine and Biology
ISBN 978-981-13-2157-3 ISBN 978-981-13-2158-0 (eBook)
https://doi.org/10.1007/978-981-13-2158-0

Library of Congress Control Number: 2018958878

This Springer imprint is published by the registered company Springer Nature Singapore Pte Ltd.
The registered company address is: 152 Beach Road, #21-01/04 Gateway East, Singapore 189721, Singapore

Preface

The glycan chains modifying proteins and lipids are critically involved in various physiological and pathological events occurring both in the cells and on their surfaces. This major posttranslational modification is crucial not only for the physical properties of proteins but also for their functional regulation and fate determination. Despite the importance of glycosylation, biophysicists have often hesitated to study glycoproteins because of glycan complexity, heterogeneity, and flexibility. Moreover, although evidence indicates that glycolipids form dynamic clusters on cell membranes and provide platforms of molecules involved in cell recognition and consequent signal transduction, their structural complexity and transient assembling properties hamper the biophysical approaches used to characterize these molecular systems. Significant advances in biophysical approaches to the complex glycoconjugates have recently been achieved by using calorimetry, interferometry, mass spectrometry, solution scattering, and computational simulation, along with various crystallographic, microscopic, and spectroscopic techniques. These methods have been conjugated with biochemical and synthetic approaches. This book comprises 13 chapters, each of which will introduce the current state-of-the-art biophysical approaches used to solve glycobiological questions with their cutting-edge applications. Therefore, the publication of this book would be timely and useful for graduate students and postdoctoral researchers who are potentially interested in the frontier topics in glycosciences, as well as for senior bioresearchers in academic and industrial fields aiming to get acquainted with glyco-issues.

Wako, Saitama, Japan — Yoshiki Yamaguchi
Okazaki, Aichi, Japan — Koichi Kato
June 15, 2018

Acknowledgments

We wish to express our sincere thanks to all authors who took the time to contribute to the chapters. We particularly thank Tadashi Satoh at Nagoya City University, who kindly helped us to handle the manuscripts, and Kripa Guruprasad at Springer, who greatly contributed to manuscript development process. We are very grateful to all staff members of Springer who generously edited this Glycobiophysics book. This book would not have been possible without all their contributions.

Contents

Contributors

Kathirvel Alagesan Institute for Glycomics, Griffith University, Southport, QLD, Australia

Yoko Amazaki Department of Chemistry, Osaka University, Toyonaka, Japan

Hiromune Ando Center for Highly Advanced Integration of Nano and Life Sciences (G-CHAIN), Gifu University, Gifu, Japan

Institute for Integrated Cell-Material Sciences (WPI-iCeMS), Kyoto University, Kyoto, Japan

Peggi M. Angel Department of Cell and Molecular Pharmacology and Experimental Therapeutics, Medical University of South Carolina, Charleston, SC, USA

Cristina De Castro Department of Agricultural Sciences, University of Napoli, Portici, NA, Italy

Richard R. Drake Department of Cell and Molecular Pharmacology and Experimental Therapeutics, Medical University of South Carolina, Charleston, SC, USA

Garry A. Duncan Department of Biology, Nebraska Wesleyan University, Lincoln, NE, USA

James L. Van Etten Department of Plant Pathology and Nebraska Center for Virology, University of Nebraska, Lincoln, NE, USA

Arun Everest-Dass Institute for Glycomics, Griffith University, Southport, QLD, Australia

Takahiro Fujiwara Institute for Integrated Cell-Material Sciences (WPI-iCeMS), Kyoto University, Kyoto, Japan

Domenico Garozzo CNR, Institute for Polymers, Composites and Biomaterials, Catania, Italy

Kunihiko Gekko Hiroshima Synchrotron Radiation Center, Hiroshima University, Higashi-Hiroshima, Hiroshima, Japan

Teruaki Hasegawa Faculty of Life Sciences, Toyo University, Itakura-machi, Ora-gun, Gumma, Japan

Bio-Nano Electronics Research Centre, Toyo University, Kawagoe, Saitama, Japan

Hien Minh Nguyen School of Medicine, Vietnam National University, Ho Chi Minh City, Vietnam

Ryota Iino Okazaki Institute for Integrative Bioscience, National Institutes of Natural Sciences, Aichi, Japan

Department of Functional Molecular Science, School of Physical Sciences, Kanagawa, Japan

Institute for Molecular Science, National Institutes of Natural Sciences, Aichi, Japan

Ye Ji Complex Carbohydrate Research Center, University of Georgia, Athens, GA, USA

Yasuhiro Kajihara Department of Chemistry, Osaka University, Toyonaka, Japan

Koichi Kato Exploratory Research Center on Life and Living Systems, National Institutes of Natural Sciences, Okazaki, Aichi, Japan

Makoto Kiso Institute for Integrated Cell-Material Sciences (WPI-iCeMS), Kyoto University, Kyoto, Japan

Daisuke Kohda Division of Structural Biology, Medical Institute of Bioregulation, Kyushu University, Fukuoka, Japan

Daniel Kolarich Institute for Glycomics, Griffith University, Southport, QLD, Australia

Naoko Komura Center for Highly Advanced Integration of Nano and Life Sciences (G-CHAIN), Gifu University, Gifu, Japan

Institute for Integrated Cell-Material Sciences (WPI-iCeMS), Kyoto University, Kyoto, Japan

Akihiro Kusumi Institute for Integrated Cell-Material Sciences (WPI-iCeMS), Kyoto University, Kyoto, Japan

Membrane Cooperativity Unit, Okinawa Institute of Science and Technology (OIST), Onna-son, Okinawa, Japan

Yuta Maki Department of Chemistry, Osaka University, Toyonaka, Japan

Koichi Matsuo Hiroshima Synchrotron Radiation Center, Hiroshima University, Higashi-Hiroshima, Hiroshima, Japan

Anand S. Mehta Department of Cell and Molecular Pharmacology and Experimental Therapeutics, Medical University of South Carolina, Charleston, SC, USA

Antonio Molinaro Department of Chemical Sciences, University of Napoli, Napoli, Italy

Masamichi Nagae Graduate School of Pharmaceutical Sciences, The University of Tokyo, Tokyo, Japan

Akihiko Nakamura Okazaki Institute for Integrative Bioscience, National Institutes of Natural Sciences, Aichi, Japan

Department of Functional Molecular Science, School of Physical Sciences, Kanagawa, Japan

Ryo Okamoto Department of Chemistry, Osaka University, Toyonaka, Japan

Tadashi Satoh Graduate School of Pharmaceutical Sciences, Nagoya City University, Nagoya, Aichi, Japan

Luisa Sturiale CNR, Institute for Polymers, Composites and Biomaterials, Catania, Italy

Kenichi G. N. Suzuki Center for Highly Advanced Integration of Nano and Life Sciences (G-CHAIN), Gifu University, Gifu, Japan

Institute for Integrated Cell-Material Sciences (WPI-iCeMS), Kyoto University, Kyoto, Japan

Michela Tonetti Department of Experimental Medicine and Center of Excellence for Biomedical Research, University of Genova, Genova, Italy

Yoshiki Yamaguchi Synthetic Cellular Chemistry Laboratory, RIKEN Cluster for Pioneering Research, Wako, Saitama, Japan

Hirokazu Yagi Graduate School of Pharmaceutical Sciences, Nagoya City University, Nagoya, Japan

Saeko Yanaka Exploratory Research Center on Life and Living Systems (ExCELLS) and Institute for Molecular Science, National Institutes of Natural Sciences, Okazaki, Aichi, Japan

Connor A. West Department of Cell and Molecular Pharmacology and Experimental Therapeutics, Medical University of South Carolina, Charleston, SC, USA

Robert J. Woods Complex Carbohydrate Research Center, University of Georgia, Athens, GA, USA

Chapter 1
Effects of N-Glycans on Glycoprotein Folding and Protein Dynamics

Yoko Amazaki, Hien Minh Nguyen, Ryo Okamoto, Yuta Maki, and Yasuhiro Kajihara

Abstract This chapter describes the folding of synthetic homogeneous glycosyl-polypeptides into glycoproteins depending on the position and number of glycosylation sites and oligosaccharide structures. To evaluate the role of oligosaccharides in protein folding, we synthesized small glycoprotein models, homogeneous misfolded glycoproteins, and erythropoietins. In addition to these chemical syntheses, this chapter introduces a unique method for ^{15}N-labeling of synthetic glycoproteins to enable structural analysis. Based on experimental results, it can be suggested that N-glycans stabilize the structure of glycoproteins.

Keywords Glycoprotein · Glycoprotein folding · Erythropoietin · Crambin · Ovomucoid · Fractalkine

1.1 Introduction

Protein oligosaccharides are essential for a variety of biological processes such as acceleration of inflammation, cell-cell recognition, cytokine/antibody activation, and glycoprotein quality control in the endoplasmic reticulum (ER) (Aebi et al. 2010; Varki 2017; Hebert et al. 2010). Oligosaccharides can be found on the side chains of asparagine (N-linked type) and serine/threonine side chains (O-linked type). There are three types of N-linked type oligosaccharides: high-mannose content, hybrid, and complex-type oligosaccharides. High-mannose and complex-type sialyloligosaccharides, shown in Fig. 1.1A, have major effects on the glycoprotein-folding process and on the bioactivity of secreted cytokines, respectively (Varki 2017).

Y. Amazaki · R. Okamoto · Y. Maki · Y. Kajihara (✉)
Department of Chemistry, Osaka University, Toyonaka, Japan
e-mail: kajihara@chem.sci.osaka-u.ac.jp

H. M. Nguyen
School of Medicine, Vietnam National University, Ho Chi Minh City, Vietnam

Y. Yamaguchi, K. Kato (eds.), *Glycobiophysics*, Advances in Experimental Medicine and Biology 1104, https://doi.org/10.1007/978-981-13-2158-0_1

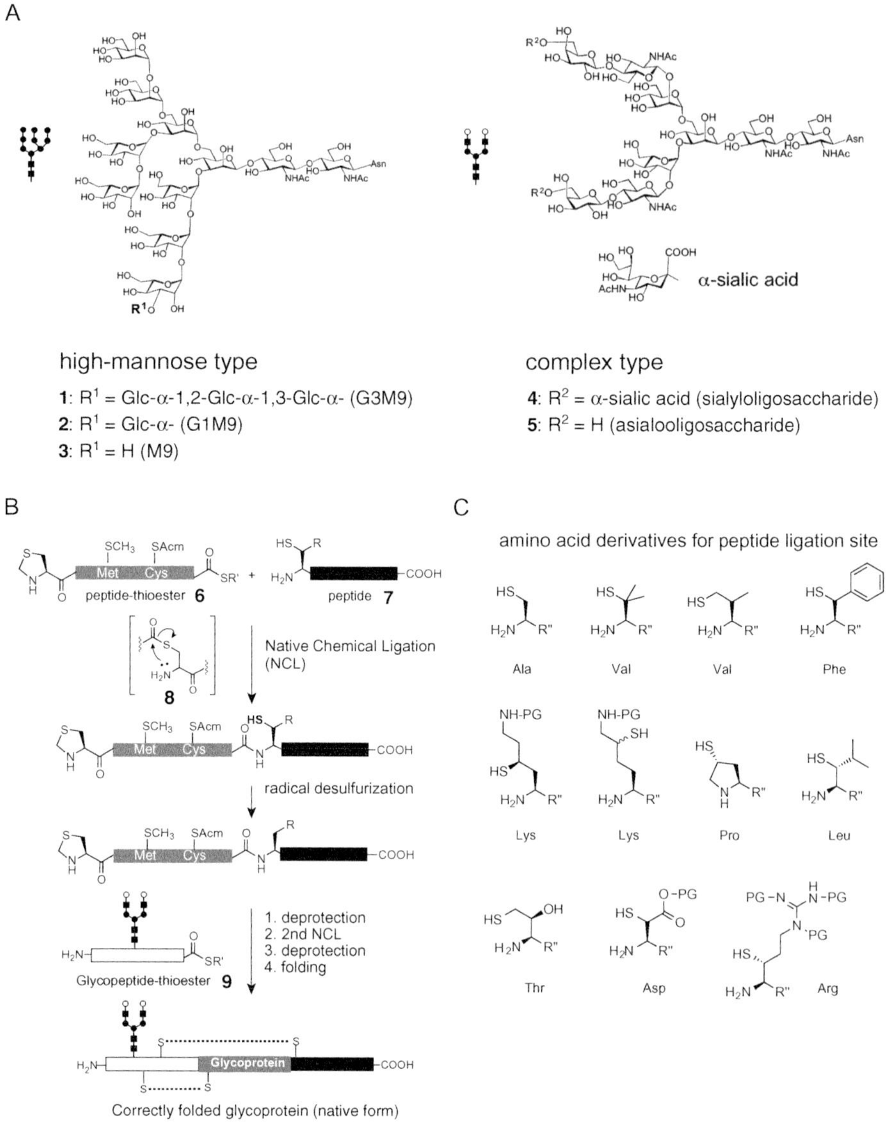

Fig. 1.1 Glycan structure and synthetic strategy of glycoproteins. (**A**) Glycan structure. (**B**) Native chemical ligation (NCL) and synthetic strategy of glycoproteins (**C**) Cysteine surrogates for NCL

The synthesis of glycoproteins in mammalian cells commences with the production of glycosylpolypeptides having high-mannose-type oligosaccharides **1**–**3** (Fig. 1.1A). The oligosaccharide (G3 M9) **1**, having three glucoses and nine mannoses as a result of co-translational modification in the ER, attaches to polypeptides. Subsequent folding process is regulated by a glycoprotein quality control (GQC) system (Hebert et al. 2010; Aebi et al. 2010). After undergoing successful folding processes in the ER, glycoproteins having native protein structures are transported into the Golgi apparatus for modification of their mature oligosaccharide structures. In the Golgi apparatus, the glycoproteins with high-mannose-type oligosaccharides undergo further modification by several mannosidases and glycosyltransferases to form glycoproteins with complex-type sialyloligosaccharides **4** (Fig. 1.1A). However, misfolded glycoproteins accumulate in the ER and are released into the cytosol to be digested by the ER-associated degradation (ERAD) system (Hebert et al. 2010).

Because glycoproteins having native form are known to exhibit important bioactivities, they are recognized as useful drugs (biologics). To produce biologics, it is essential to obtain glycoproteins with homogeneous oligosaccharides (Sinclair and Elliott 2005). Recent studies have demonstrated that glycoprotein drugs are more effective when prepared in mammalian cell-based expression systems than when prepared in bacterial protein expression systems, owing to the lack of glycosylation in the latter. A biosynthetic platform for pharmaceutical glycoprotein drugs bearing complex-type acidic sialyloligosaccharides **4** has been developed using Chinese hamster ovary (CHO) cells, which are potent in producing recombinant glycoproteins bearing mammalian-type oligosaccharides.

However, it is known that natural sialyloligosaccharides **4** exhibit considerable heterogeneity in terms of oligosaccharide branching, including di-, tri-, and tetra-antennary structures (Park et al. 2009). Moreover, oligosaccharides lacking sialic acid can be found frequently. Thus, the functions of oligosaccharides remain unclear owing to differences in its structure and the insufficient purity of the glycoprotein drugs. To overcome these issues, it is essential to establish an effective methodology for synthesizing homogeneous glycoproteins in varying amounts.

Over the past two decades, numerous improvements have been made to peptide-coupling reactions, allowing larger polypeptides to be obtained (Dawson et al. 1994). This development has enabled us to chemically synthesize glycoproteins more effectively, as shown in Fig. 1.1B. While the folding processes of glycoproteins after chemical synthesis are also interesting for illustrating protein behavior, the processes and yields are unpredictable when polypeptides have large oligosaccharides.

This chapter describes the difference in the folding processes of synthetic homogeneous glycosylpolypeptides into glycoproteins depending on the glycosylation position and oligosaccharide structures.

1.2 Protocol for Synthesis of Glycoproteins

Figure 1.1B illustrates a general scheme for glycoprotein synthesis (Unverzagt and Kajihara 2013). Native chemical ligation (NCL) is a representative coupling reaction between two peptides, one of which is a peptide-α-thioester **6** and the other is a peptide **7** bearing a cysteine at the N-terminal. The coupling reaction proceeds through a thiol-exchange reaction between the thioester and the cysteine thiol. Subsequent N-S intramolecular migration (**8**: Fig. 1.1B) yields a native amide bond.

This coupling reaction enables us to synthesize large polypeptides containing more than 50 amino acids, overcoming the limited lengths achievable when synthesized by solid-phase peptide synthesis (SPPS).

However, this NCL method requires a cysteine at the ligation site. Therefore, it cannot be used on proteins without cysteine in their peptide backbones. To overcome this problem, an alanine site was chosen as a ligation site because cysteine residues can be transformed into alanines by a desulfurization reaction after NCL (Fig. 1.1B). This ligation-desulfurization concept can be applied to other amino acids having alkyl side chains. Recent studies have shown that many amino acid derivatives can have a thiol at the β-position as a cysteine surrogate (Fig. 1.1C) (Unverzagt and Kajihara 2013).

For glycoprotein synthesis, the NCL method should be applied to a glycopeptide thioester **9** (Fig. 1.1B) and a polypeptide bearing a cysteine at the N-terminus. For the preparation of asparaginyl oligosaccharides, asparaginyl complex-type oligosaccharides (Kajihara et al. 2004) and high-mannose-type oligosaccharides (Makimura et al. 2012) were isolated from egg yolk and used for SPPS. The folding processes of the glycosylpolypeptide thereby synthesized occur subsequently, forming disulfide bonds under thermodynamically controlled conditions to acquire the correctly folded protein.

1.3 Folding Protocol

The conventional protocol for glycoprotein folding demonstrated by Anfinsen is shown in Fig. 1.2 (Anfinsen 1973). This method is achieved by stepwise dialysis in the presence of redox reagents. The glycosylpolypeptides (0.3–0.01 mg/ml) are dissolved in a Tris-HCl buffer (100 mM, pH 7.5) containing 6–8 M guanidine-HCl (Gn-HCl), and the solution is incubated for 1 h. This solution is poured into a dialysis tube (molecular weight cutoff at 8000 kDa, Spectra/Por) for dialysis against the first folding buffer (3.0 M Gn-HCl, 100 mM tris-HCl, pH 8.5) containing 4 mM cysteine and 0.5 mM cystine to create the redox system and incubated for 24 h at 4 °C. This is followed by another 16 h of dialysis by replacing the external buffer with a second folding buffer solution (1.0 M Gn-HCl, 100 mM tris-HCl, pH 8.0). During this process, undesired disulfide bonds are cleaved by the reaction of cysteine thiol. This process accelerates the re-formation of other disulfide bond patterns along with the refolding of the protein backbone. This reversible process finally yields native protein structures with correct disulfide bond patterns under thermodynamic control (Fig. 1.2B). Next, the external buffer solution is replaced with a third folding buffer solution (10 mM tris-HCl, pH 7.0) to perform the final dialysis, which is conducted for 16 h.

Next, the folded glycoproteins are monitored and purified by RP-HPLC (reverse phase column with C4–16 types, 0.1% TFA: 0.1% TFA in 90% MeCN for gradient elution, 60:40 to 25:75 for 15–30 min). The yield and purity of the folded glycoproteins can be confirmed by high-performance liquid chromatography (HPLC) and electrospray ionization mass spectrometry (ESI-MS).

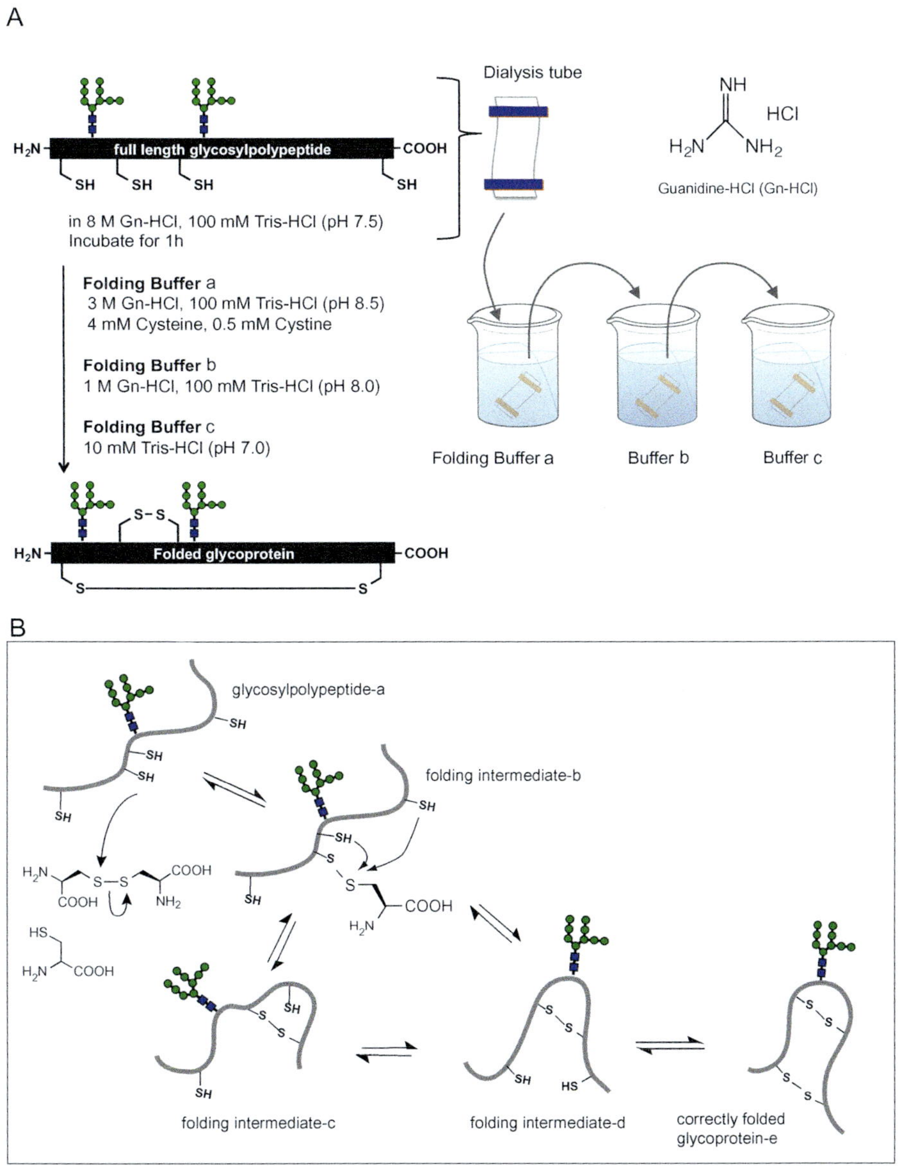

Fig. 1.2 Glycoprotein folding protocol (**A**) Reagents and conditions for glycoprotein folding (**B**) Disulfide bond formation under the thermodynamically controlled conditions

To confirm the disulfide bond positions, the folded glycoprotein is first digested with trypsin or other peptidases. Then, the resulting peptide fragments are analyzed by HPLC-MS. To identify the inclusion of disulfide bonds in each peptide fragment, tris(2-carboxyethyl)phosphine (TCEP) treatment is performed for the reduction of disulfide bonds, and all peptide fragments are analyzed by HPLC-MS to compare their HPLC profiles before and after TCEP treatment.

This chapter also describes several experiments pursuing our interests in how oligosaccharide fluctuations on protein surfaces exert an influence on protein conformation and protein folding processes. From this prospect, our laboratory carried out research into the synthesis of several different glycoproteins having various glycosylation positions. Our laboratory also investigated the effects of oligosaccharide fluctuation on β-sheet and α-helical structures.

1.4 Folding Experiments with Small Glycoproteins

To evaluate oligosaccharide functions on protein folding, our laboratory selected glycosylated crambins and ovomucoid third domain to serve as small glycoprotein models, as shown in Figs. 1.3 and 1.4, respectively (Kajihara et al. 2012).

Crambin is a plant protein without oligosaccharides in its native form (Teeter et al. 1981). This protein consists of 46 amino acid residues, which is a suitably short length for a model glycoprotein. Therefore, we intentionally inserted asparaginyl complex-type biantennary asialo-oligosaccharides (**5**: Figs. 1.1A and 1.3) into the β-sheet **10** and the α-helix **11**, respectively. These syntheses of these peptides were implemented according to the synthetic protocol depicted in Fig. 1.1B.

Folding experiments involving two types of glycosylpolypeptides corresponding to crambins **10** and **11** were performed under the cysteine-cystine redox conditions employing 3 M guanidine-HCl. Multiple peaks were observed on HPLC profiles during the folding process due to protein denaturation, which merged into one main peak within 1 h (Fig. 1.3B). Mass spectroscopic analysis revealed the formation of three disulfide bonds in glycosylated crambins **10** and **11** (Fig. 1.3B e and h).

Ovomucoid **17** (Fig. 1.4B), which consists of 56 amino acid residues and bears an oligosaccharide at position 28 in its native form (Fig. 1.4C) (Tirado-Rives and Jorgensen 1990), was also chosen as a model glycoprotein. The glycosylpolypeptides were synthesized by segment coupling of **14–16** by NCL as shown in Fig. 1.1B. Although the native glycopeptide has a hybrid-type oligosaccharide at Asn28 (Yamashita et al. 1983) located on the loop between the antiparallel β-sheets, it was substituted with the complex-type **5** in these experiments.

Figure 1.4 shows folding experiments with the non-glycosylated form **17** and glycosylated ovomucoid **18**. The folding experiments were performed under cysteine-cystine redox conditions employing 3 M Gn-HCl. Different to the folding process of crambins (Fig. 1.3B), both glycosylated and non-glycosylated polypeptide chains exhibited multiple peaks on HPLC profiles. Extensive studies performed under various folding conditions, however, resulted in the same pattern of multiple peaks.

While evaluating the outcome of non-glycosylated ovomucoid **17** (Fig. 1.4B a–e), only broad peaks were observed on HPLC profiles after 24 h of folding experiments, although there was still a main peak, which also exhibited a broad profile. This main peak was purified using reverse phase HPLC. The ESI mass data indicated the formation of three disulfide bonds (Fig. 1.4F).

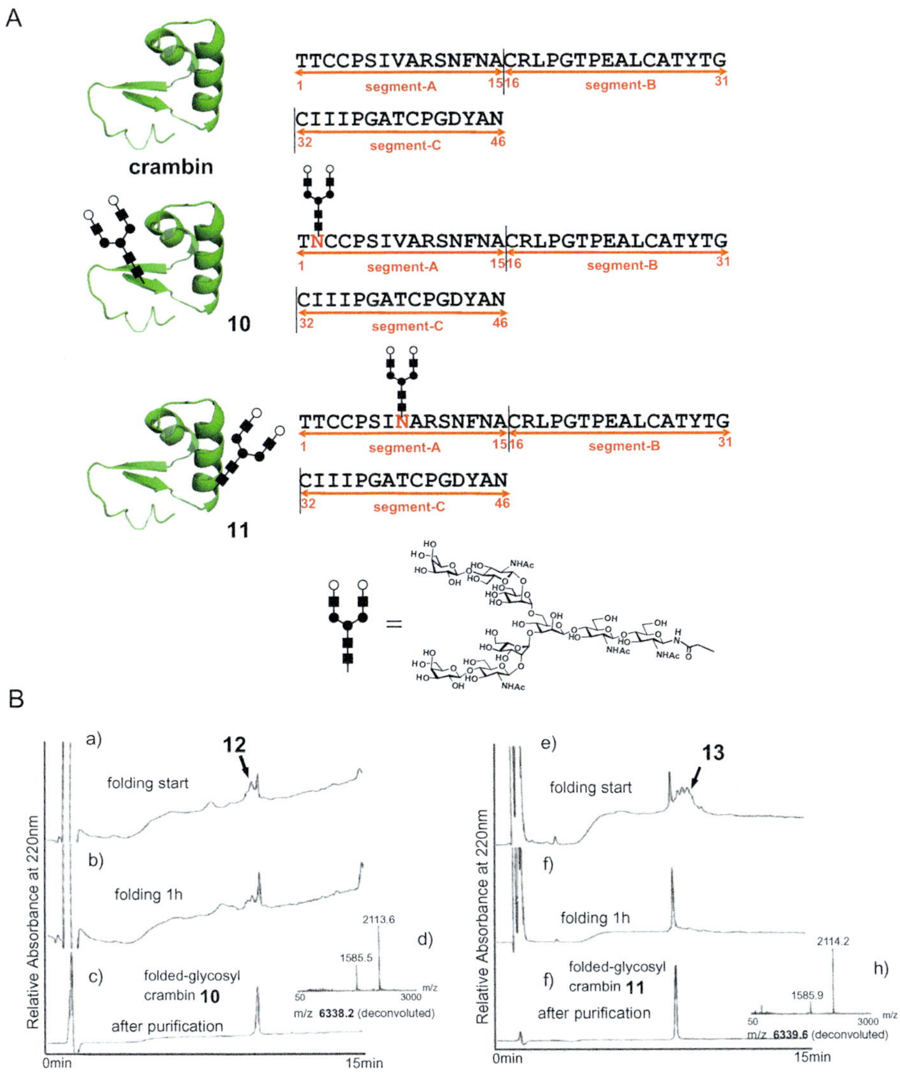

Fig. 1.3 Synthesis and folding of small glycoprotein, crambins. (**A**) Amino acid sequences of crambin glycoforms. (**B**) Monitoring of folding by reverse phasehigh performance liquid chromatography (RP-HPLC)

The folding experiments involving the glycosylated ovomucoid **18** having a complex-type oligosaccharide also showed multiple HPLC peaks (Fig. 1.4C, j). All peaks were isolated and subsequently evaluated to find the correctly folded glycosylated ovomucoid **18**. The main peak was revealed as the target product (Fig. 1.4C, k) from the ESI mass data (Fig. 1.4C, n), disulfide bond mapping, and the protease inhibitory activity (inhibition of chymotrypsin catalytic activity: IC 50 = 10 nM).

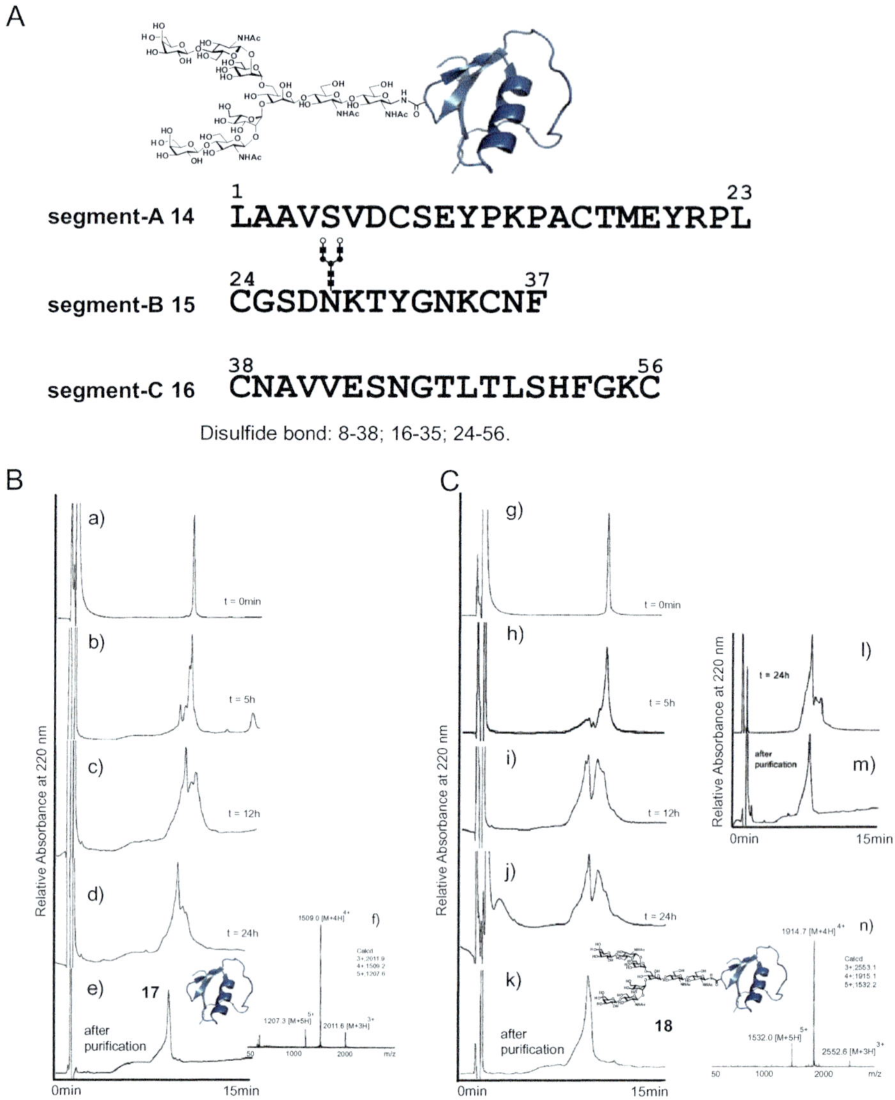

Fig. 1.4 Synthesis and folding of small glycoprotein, ovomucoids. (**A**) Amino acid sequences of ovomucoid glycoforms. (**B**) Monitoring of folding of non-glycosylated ovomucoid by RP-HPLC. (**C**) Monitoring of folding of glycosylated ovomucoid by RP-HPLC

However, the folding results were not reproducible. Extensive studies performed under various folding conditions resulted in the same pattern of multiple peaks. In some experiments, the number of multiple products reduced, and these showed good HPLC profiles (Fig. 1.4C, l, m). Uniform conditions should be rigorously set in order to maintain the reproducibility of the desired folding processes.

As shown in Figs. 1.3 and 1.4, three types of small glycoproteins showed different folding patterns depending on their glycosylation positions. In the usual cases of protein folding, several peaks were observed on HPLC profiles as denaturation occurred initially, but these multiple peaks easily converged into a single main peak within an hour. In the case of glycosylated ovomucoid **18**, a rather low folding yield was obtained when compared with non-glycosylated ovomucoid **17** (Fig. 1.4). The experimental folding results changed, depending on the experimental conditions and due to the formation of antiparallel β-sheets under the influence of the oligosaccharide, although both glycosylated and non-glycosylated ovomucoids were stable in the solution after purification.

1.5 Intentional Synthesis of Homogeneous Misfolded Glycoproteins

Next, our laboratory focused on glycoproteins that have stable misfolded structures in solution. In the biosynthesis of glycoproteins, sequential protein folding occurs in the ER, generating folding intermediates in various conformational states (Aebi et al. 2010). To study the process of folding in the ER, it is necessary to synthesize homogeneous glycoproteins in both correctly folded and misfolded forms. A trial for the successful synthesis of stable misfolded glycoproteins, which have a single protein conformation, was performed for the first time. The substrate specificity of several enzymes used in the GQC system in the ER was successfully demonstrated using the synthesized misfolded glycoproteins **20–23** (Izumi et al. 2012).

For the preparation of misfolded glycoproteins, the scaffold of interleukin 8 (IL-8) was selected as the model glycopeptide. Native IL-8 consists of 72 amino acids, with two disulfide bonds between Cys7-Cys34 and Cys9-Cys50 (Baldwin et al. 1991). These two disulfide bonds were shuffled to prepare intentionally misfolded proteins. In terms of intentional glycosylation, the M9 high-mannose-type oligosaccharide **3** (Fig. 1.1A) was inserted at Asn36 (Fig. 1.5A), which is located on the loop between two antiparallel β-sheets.

Next, folding process of the synthesized glycosylated IL-8 polypeptide **19** was analyzed. To obtain a correctly folded glycosylated IL-8, the oxidative folding process was examined with the full-length IL-8 glycopeptide under redox conditions (1 mM cysteine and 0.05 mM cystine). The progress of the reaction was carefully monitored using RP-HPLC (Fig. 1.5). Several peaks, which were observed at the early stages showing folding intermediates, converged into a main single product after 2 h (Fig. 1.5B). This major product **20** was confirmed to be the native form of IL-8, containing disulfide bonds between Cys7-Cys34 and Cys9-Cys50. The existence of the disulfide bonds was confirmed by disulfide mapping experiments, including peptidase digestion and subsequent MS/MS analysis.

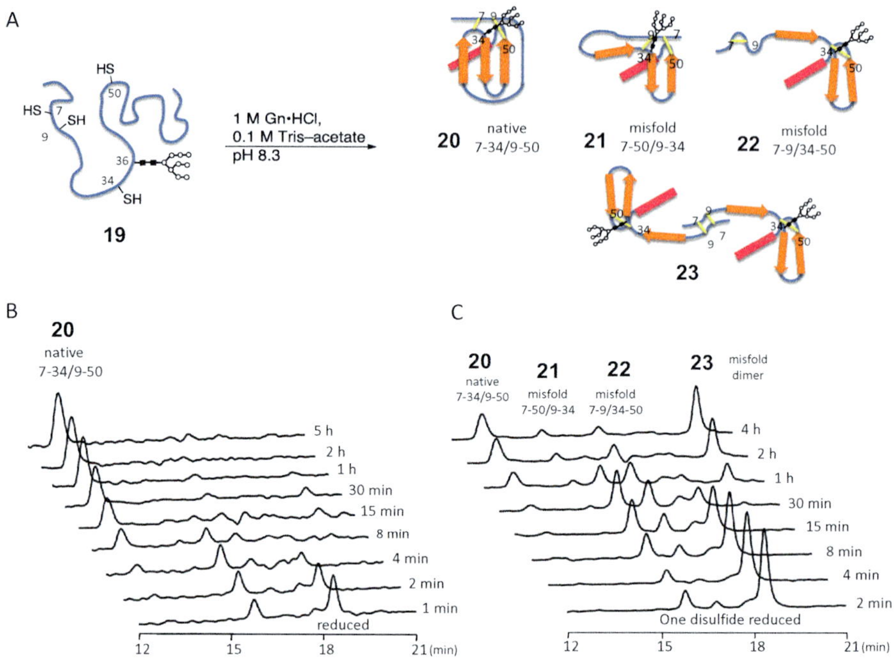

Fig. 1.5 Synthesis of folding of misfolded glycoproteins. (**A**) Conditions for synthesis of misfolded glycoproteins. (**B**) Monitoring of thermodynamically controlled folding process by RP-HPLC. (**C**) Monitoring of kinetically controlled folding process by RP-HPLC

To acquire misfolded glycopeptides, the oxidative folding of **19** was performed in the absence of cysteine and cystine. Because the disulfide bonds were not cleaved in the absence of cysteine, misfolded protein structures with incorrect disulfide bond formation accumulated under the kinetic control processes. After conducting 4 h of folding reactions, four derivatives appeared. Each peak was determined to be **20** (native form), **21** (structure having disulfide bonds between Cys7- Cys50, Cys9-Cys34), **22** (Cys7-Cys9, Cys34-Cys50), and **23**, respectively. The mass spectrum of structure **23** indicated it to be a homodimer. The dimeric misfolded glycoprotein had two intermolecular disulfide linkages between Cys7 and Cys9 of two IL-8 polypeptides, as well as an intramolecular disulfide bond between Cys34 and Cys50. These experimental data demonstrated that the synthesis of homogeneous misfolded glycoproteins could be carried out under kinetic folding conditions (Izumi et al. 2012; Dedola et al. 2014).

1.6 Folding of Larger Glycoproteins

With knowledge of small glycoprotein folding, our research shifted to larger molecules (Murakami et al. 2012). Erythropoietin (EPO), which consists of 166 amino acid residues, is known as a cytokine that stimulates the production of red blood cells. EPO has four sialyloligosaccharides, three of which are N-linked types, located at residues Asn24, Asn38, and Asn83. The structures of these sialyloligosaccharides vary from bi- to tetra-antennary forms. These changes in the structure have great effects on EPO activities, because the structural changes interfere with glomerular filtration and galactose-binding lectin interactions, which extend EPO's half-life in blood, and ultimately result in the enhancement of its biological activity. The fourth sialyloligosaccharide exists at Ser126 and consists of three to four sugars. This short sialyloligosaccharide does not have any effect on the biological activity of EPO.

To understand the effects of oligosaccharides on the folding processes of EPO and bioactivity, five variants of EPO with different glycoforms were synthesized, with each having different glycosylation numbers and positions (Fig. 1.6A) (Murakami et al. 2016). The EPO polypeptide chain was divided into six segments. Using NCL, all segments were coupled to obtain the full-length glycopolypeptides (Fig. 1.1B).

The folding conditions were then optimized, and the folding processes (Fig. 1.6B) were traced using SDS-polyacrylamide gel electrophoresis (SDS-PAGE: Fig. 1.6C), ESI-MS, and combined with trypsin digestion and MS/MS analysis. The Gn-HCl (8 M) solution for denaturing all the sialylglycopolypeptides was gradually diluted in the presence of cysteine-cystine redox reagents.

The disulfide bond formation in EPO glycosylpolypeptide **29** bearing two sialyloligosaccharides at positions Asn24 and Asn83 was monitored using LC-MS/MS analysis after peptidase digestion. Disulfide bond formation **30** was observed between residues Cys29 and Cys33, when the concentration of Gn-HCl in the folding buffer solution was changed from 8 M to 3 M (Fig. 1.6B, C). However, when the Gn-HCl concentration was subsequently reduced to 1 M, both oligomeric misfolded EPO and correctly folded EPO (Fig. 1.6B, C) were generated. Therefore, the diluted condition in which the initial concentration of the glycopolypeptides was set at 0.01 mg/ml was used to avoid the formation of oligomeric misfolded structures (Fig. 1.6D). This condition successfully achieved correctly folded EPO glycoforms in good yields and minimized protein aggregation.

Notably, the yields of EPO folding were found to be different depending on the number of sialyloligosaccharides and glycosylation positions (Fig. 1.6D) (Murakami et al. 2016). EPON38, N83 **25** showed the highest yield (90%), whereas EPO lacking the sialyloligosaccharide at Asn38 had relatively lower folding yields with a considerable quantity of misfolded products. These results suggested that the sialyloligosaccharide at asparagine 38 might be important for generating folding intermediates in an appropriate folding route or for preventing aggregation.

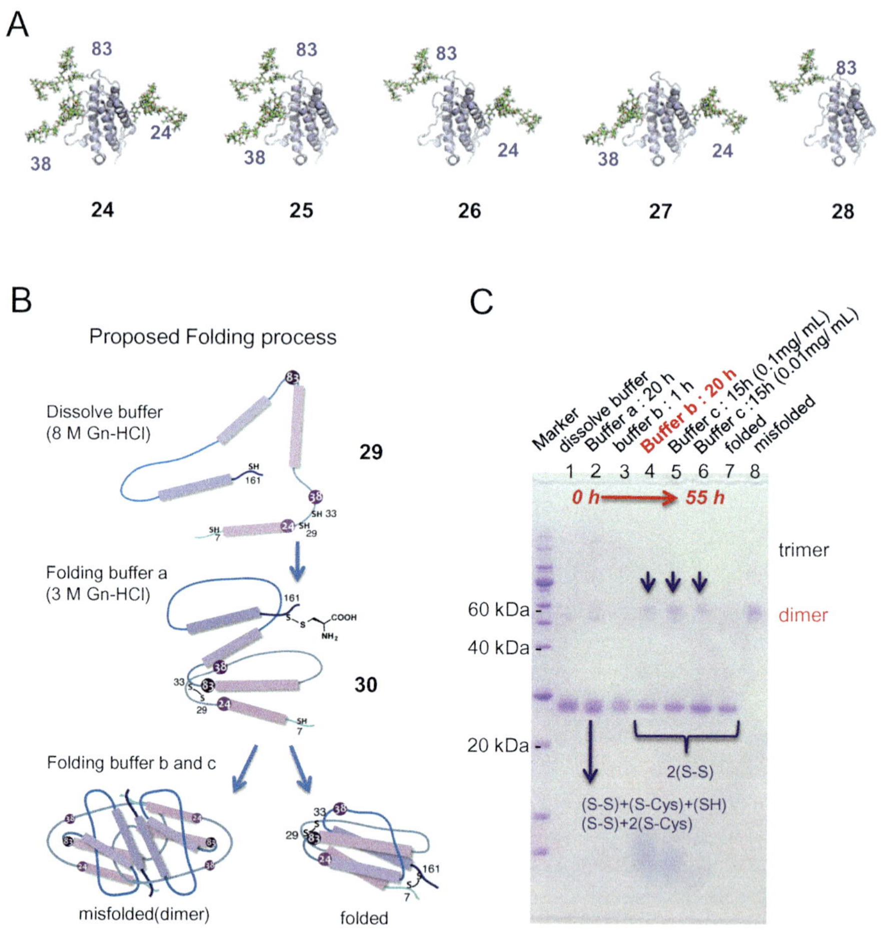

D

	Yields from HPLC peaks (%)				
Concentrations	**EPO$^{24, 38, 83}$**	**EPO$^{38, 83}$**	**EPO$^{24, 83}$**	**EPO$^{24, 38}$**	**EPO^{83}**
0.1 mg/mL	67.1 (± 0.9)	57.6 (± 4.6)	42.3 (± 2.5)	48.8 (± 4.8)	37.0 (± 3.2)
0.01 mg/mL	86.0 (± 5.4)	**89.9** (± 5.7)	65.9 (± 1.6)	76.4 (± 11.2)	**62.5** (± 2.4)

Fig. 1.6 Synthesis of erythropoietin (EPO) glycoforms. (**A**) EPO glycoforms synthesized. (**B**) Proposed folding process. (**C**) Monitoring folding process by SDS-electrophoresis. (**D**) Yields of oxidative folding of EPO glycoforms

1.7 Glycoprotein Dynamics

NMR analysis is useful for studying how glycosylation affects protein conformation and protein dynamics. However, the preparation of homogeneous glycoproteins for NMR studies was difficult, and there was no method to insert ^{15}N atoms into homogeneous synthetic glycoproteins.

Therefore, our laboratory has studied a novel and efficient ^{15}N isotope-labeling method for the synthesis of ^{15}N-labeled glycoproteins as probes for NMR experiments (Nguyen et al. 2017). This method that we established is termed "specific and gradient ^{15}N isotope (SGI) labeling," as it can arrange the insertion of ^{15}N-enriched amino acids at any position in the target glycoprotein during solid-phase peptide synthesis and varying the enrichment of ^{15}N-stable isotope atoms. Using this method enables us to discriminate the same types of amino acids by using the difference in intensities of ^{1}H-^{15}N HSQC signals caused by the varied enrichment of ^{15}N-atoms on the peptide backbone. The relevant concepts have been reported for the analyses of nucleic acid and non-glycosylated proteins (Yabuki et al. 1998; Phan and Patel 2002; Kasai et al. 2015). The resulting assignments can also be verified by means of the classic homonuclear TOCSY and NOESY methods developed by Wuthrich (Wuthrich 1986).

Our laboratory selected the chemical domain of fractalkine (CDF) as a glycoprotein model (Bazan et al. 1997). CDF is a member of the chemokine family, with Asn9 as its *N*-glycosylation site in mammalian cells. CDF also has a unique consensus sequence (Asn-Ile-Thr) between two cysteines (Cys8-Asn9-Ile10-Thr11-Cys12: Fig. 1.7 A) (Mizoue et al. 1999; Hoover et al. 2000). However, the effect of glycosylation on CDF is still unclear.

Before the synthesis of ^{15}N-labeled glycosyl CDF, the synthesis of CDF with a complex-type oligosaccharide and in its non-glycosylated form was performed to evaluate whether glycosylation changes conformational properties of the protein. The synthesis employed the same synthetic protocol shown in Fig. 1.1B.

^{1}H-^{15}N HSQC measurements of synthetic non-glycosylated CDF and glycosylated CDF were performed in order to examine the effects of oligosaccharides on protein conformation (^{15}N natural abundance samples, Fig. 1.7D). The ^{1}H-^{15}N HSQC spectrum of synthetic non-glycosylated CDF showed the same signal patterns when compared with the previous reports of CDF using *E. coli* expression (Mizoue et al. 1999). The comparison of ^{1}H-^{15}N HSQC signal patterns of non-glycosylated and glycosylated CDF showed mostly the same signal patterns (Fig. 1.7D; red, glycosylated, and black, non-glycosylated). These data clearly showed that glycosylation did not have an effect on the formation of the CDF structures.

However, the slight differences in the chemical shifts indicated slight changes in local protein conformation. CDF has a unique amino acid sequence of Cys8-Asn9-Ile10-Thr11-Cys12 around its glycosylation site, which forms a bulged loop sustained by two disulfide bonds between Cys8–34 and Cys12–50 (Fig. 1.7B, C). Analyses regarding how glycosylation affects local protein dynamics and stability

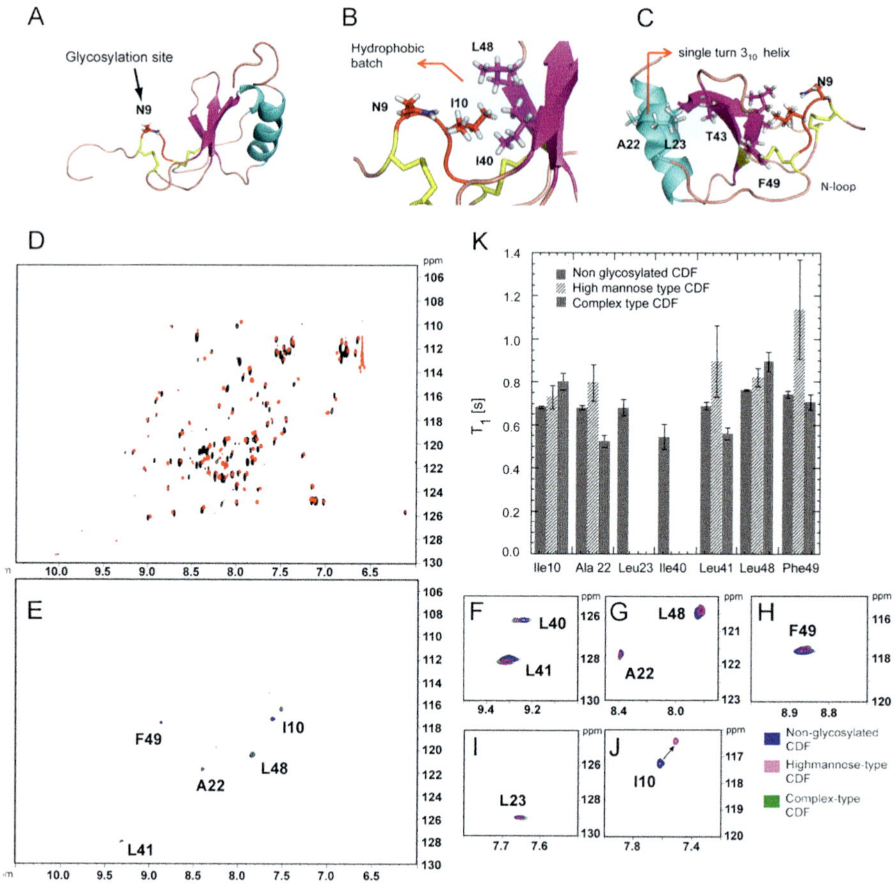

Fig. 1.7 Analyses of glycoprotein dynamics by means of ^{15}N-labeled glycoproteins. (**A**) Structure of CDF. (**B**) (**C**) The positions of amino acids labeled with ^{15}N atom. (**D**) Superimposed HSQC spectra of glycosylated and non-glycosylated CDF. (**E**)-(**J**) Superimposed HSQC spectra of ^{15}N labeled non-glycosylated CDF, CDF having high-mannose type and complex type glycan. (**K**) T_1 values of amino acids of CDFs

were examined using the SGI-labeling method. The ^{15}N-labeled CDF polypeptides that were synthesized had either a high-mannose-type oligosaccharide **3** or a complex-type **5** (Fig. 1.1A). They were synthesized by SPPS and NCL to be used as probes for analyzing the effects of oligosaccharides on protein behavior.

To determine the position at which the glycoprotein amino acids should be labeled with ^{15}N-atoms, the glycoprotein's conformational properties were taken into account. The conformational properties of non-glycosylated CDF indicated that hydrophobic interactions between Ile10 and Leu48 in the bulged loop sustained the local structure (Fig. 1.7B). Therefore, Ile10, Leu48, and Ile40, incorporating ^{15}N atoms, were installed to analyze their dynamics. In this synthesis, Ala22 and Leu23 at the rear side of a hydrophobic loop were also labeled with ^{15}N atoms. Even though

Ala22 and Leu23, which form a short and single-turn 3_{10} helix (Fig. 1.7C), are not close to the glycosylation site, this type of structure may be important for stabilizing the antiparallel sheets (Fig. 1.7 B: purple color) and the bulged loop (Fig. 1.7B red color) due to the hydrogen bond network. In addition to these, the *N*-loop conformation (Fig. 1.7C: residues N9-I19) involving the large bulged loop Cys8-Asn9-Ile10-Thr11-Cys12 was considered to contribute to the specific chemokine structure.

According to those considerations, ^{15}N-labeled amino acids were inserted during the chemical synthesis of glycosylated and non-glycosylated CDFs. The concentrations of ^{15}N atoms were varied as follows: Leu23 (2%), Leu41 (10%), and Leu48 (20%), respectively. Amino acids adjacent to the labeled leucine were also labeled with ^{15}N atoms (Ala22: 10%, Ile40: 5%, and Phe49: 10%). ^{15}N-Labeling of Ile10 in the bulged loop was 20% (Fig. 1.7B, C) (Minh Hien et al. 2017).

After the synthesis and folding experiments involving ^{15}N-labeled non-glycosylated and glycosyl CDFs having complex-type and high-mannose-type, ^{1}H-^{15}N HSQC spectra were measured. The superimposed HSQC spectra of synthetic ^{15}N-labeled CDF (Fig. 1.7F-J) showed that glycosylation with complex-type and high-mannose-type oligosaccharides did not change their chemical shifts, suggesting that glycosylation did not cause conformational changes in the protein.

In addition to these analyses, T_1 values were estimated to evaluate the flexibility of the glycoproteins (Fig. 1.7 K). The measurement of T_1 (^{15}N) in labeled amino acids was performed with a standard inversion recovery method using the pseudo 3D ^{1}H-^{15}N HSQC method (Farrow et al. 1994). The T_1 values observed were almost the same among glycoproteins and non-glycosylated proteins. However, T_1 values for Ile40 and Leu23 of the glycoproteins were weak in comparison to their non-glycosylated counterparts. It was hypothesized that the dynamics of Ile40 and Leu23 may be retarded or hindered by glycosylation.

In addition to NMR analyses, temperature-dependent CD measurements were implemented. The data revealed remarkable differences between non-glycosylated CDF and glycosylated CDF bearing either a complex-type or a high-mannose-type oligosaccharide (Fig. 1.8). CD profiles for the non-glycosylated CDF and its glycoforms at 298 K were similar, although when the temperature increased, the CD spectra of non-glycosylated CDF revealed altered absorption at 200–230 nm, indicating a decrease in α-helical patterns.

Glycosylated CDFs did not display such CD spectra changes, and this result indicated that glycosylated CDFs are more thermally stable than non-glycosylated CDFs. As shown in Fig. 1.8, the CD spectra of glycosylated CDFs depended on the temperature: there was a slight change when the temperature was increased (Fig. 1.8A–C), but this reverted to the original CD spectra when the temperature was decreased (black line in Fig. 1.8B vs 8E; 8C vs 8F). These data clearly indicated that thermal denaturation of glycosylated CDFs is reversible, whereas non-glycosylated CDF did not return to its original CD profile after the temperature was decreased (black line: Fig. 1.8A vs 8D).

These CD experiments (Fig. 1.8A–F) suggest that oligosaccharides may inhibit fluctuations in protein structure. In other words, they may stabilize protein structure. This phenomenon may correspond with the changes in signal intensities observed

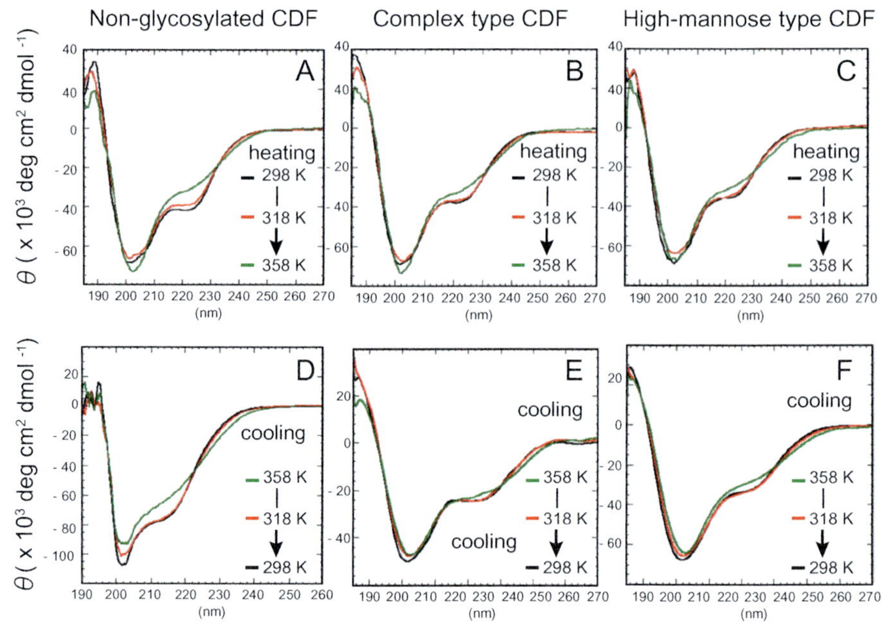

Fig. 1.8 Circular Dichroism (CD) spectra depending on temperature. (**A**)-(**C**) Heating from 298K to 358 K. (**D**)-(**F**) Cooling from 358 K to 298 K

from the measurement of T_1 values involving Leu23 and Ile40 (Fig. 1.7). Because these analyses were implemented at a small number of selected positions in the ^{15}N-labeled peptide backbone, it is important to note that the dynamics of other local positions are unknown. However, the conformational properties of the total protein are unlikely to be affected by glycosylation, based on the ^{1}H-^{15}N HSQC spectra.

1.8 Discussion

Proteins bearing homogeneous oligosaccharides have always been essential in understanding the function of oligosaccharides in protein folding and dynamics. Although biosynthetic methods can introduce an N-glycan at any position in the protein backbone by the insertion of an Asn-X-Ser/Thr sequence, this method also results in heterogeneous oligosaccharides (Martin et al. 1995). However, chemical synthesis can insert N-glycans at desired positions without needing to insert specific sequences, which is useful for experiments seeking to understand the function of N-glycans (Unverzagt and Kajihara 2013).

In the experiments involving small glycoprotein folding, the folding processes were not disturbed by the presence of oligosaccharides. However, EPO showed different folding yields depending on the number and location of glycosylations.

Furthermore, for EPO, which has hydrophobic surfaces that cause protein aggregation by hydrophobic interactions, oligosaccharides are seen to be suppressing such phenomena during protein folding. As mentioned earlier in this chapter, synthetic glycoproteins are able to undergo optimal protein folding processes and finally yield their native protein forms.

According to glycoprotein-folding experiments, the folding of glycoproteins having helical structures seems to be more feasible than the folding of glycoproteins having multiple β-strands. The glycosylpolypeptide backbone forms α-helical structures faster than forming globular structures. The hydrophobic surfaces of individual helices undergo specific interactions to form globular protein structures, and these processes support the formation of correct disulfide bonds.

The results of CD experiments involving CDFs subjected to various temperatures showed that, although oligosaccharides disturbed protein dynamic flexibility, this effect might suppress the formation of undesired misfolding or protein aggregation. This result indicated that oligosaccharides act as obstacles for proteins to behave freely, which can reduce undesired aggregation and irreversible misfolding.

As shown in the CD spectra, the glycosylation of CDF might reduce molecular motion, which may affect the bioactivity of the CDF chemokine group. The reason for this is the lower chemotactic activity that was observed in the case of glycosylated chemokine CCL-1, compared with non-glycosylated CCL-1(Okamoto et al. 2014a, b). The oligosaccharide structure of the active CDF is unclear, but N-glycans may exert effects on the biological activity.

In addition to our experiments, several groups have reported interesting discussions with sophisticated experimental procedures. Although the glycoproteins employed in our experiments had several disulfide bonds, which stabilized protein structures, other experiments used glycoproteins having no disulfide bonds. Therefore, their protein structures were stabilized by protein interactions and hydrogen bond networks. Those experiments reported clearly demonstrated the effects of glycosylation to stabilize protein structures (Price et al. 2011; Culyba et al. 2011; Hackenberger et al. 2005; Huang et al. 2017).

Based on these results and those of our experiments, it can be suggested that N-glycans stabilize the structure of glycoproteins.

References and Note

Aebi M, Bernasconi R, Clerc S, Molinari M (2010) N-glycan structures: recognition and processing in the ER. Trends Biochem Sci 35(2):74–82. https://doi.org/10.1016/j.tibs.2009.10.001

Anfinsen CB (1973) Principles that govern the folding of protein chains. Science 181:223–230

Baldwin ET, Weber IT, St Charles R, Xuan JC, Appella E, Yamada M, Matsushima K, Edwards BF, Clore GM, Gronenborn AM (1991) Crystal structure of interleukin 8: symbiosis of NMR and crystallography. Proc Natl Acad Sci 88(2):502–506

Bazan JF, Bacon KB, Hardiman G, Wang W, Soo K, Rossi D, Greaves DR, Zlotnik A, Schall TJ (1997) CDF finding. Nature 385:640–644

Culyba EK, Price JL, Hanson SR, Dhar A, Wong CH, Gruebele M, Powers ET, Kelly JW (2011) Protein native-state stabilization by placing aromatic side chains in N-glycosylated reverse turns. Science 331(6017):571–575. https://doi.org/10.1126/Science.1198461
Dawson PE, Muir TW, Clark-Lewis I, Kent SBH (1994) Synthesis of proteins by native chemical ligation. Science 266(5186):776–779
Dedola S, Izumi M, Makimura Y, Seko A, Kanamori A, Sakono M, Ito Y, Kajihara Y (2014) Folding of synthetic homogeneous glycoproteins in the presence of a glycoprotein folding sensor enzyme. Angew Chem Int Ed 53(11):2883–2887. https://doi.org/10.1002/anie.201309665
Farrow NA, Muhandiram R, Singer AU, Pascal SM, Kay CM, Gish G, Shoelson SE, Pawson T, Forman-Kay JD, Kay LE (1994) Backbone dynamics of a free and a phosphopeptide-complexed Src homology 2 domain studied by 15N NMR relaxation. Biochemistry 33(19):5984–6003. https://doi.org/10.1021/bi00185a040
Hackenberger CPR, Friel CT, Radford SE, Imperiali B (2005) Semisynthesis of a glycosylated Im7 analogue for protein folding studies. J Am Chem Soc 127:12882–12889
Hebert DN, Bernasconi R, Molinari M (2010) ERAD substrates: which way out? Semin Cell Dev Biol 21(5):526–532
Hien Minh N, Izumi M, Sato H, Okamoto R, Kajihara Y (2017) Chemical synthesis of glycoproteins with the specific installation of gradient-enriched 15N-labeled amino acids for getting insights into glycoprotein behavior. Chem Eur J 23(27):6579–6585. https://doi.org/10.1002/chem.201606049
Hoover DM, Mizoue LS, Handel TM, Lubkowski J (2000) The crystal structure of the chemokine domain of fractalkine shows a novel quaternary arrangement. J Biol Chem 275:23187–23193
Huang YW, Yang HI, Wu YT, Hsu TL, Lin TW, Kelly JW, Wong CH (2017) Residues comprising the enhanced aromatic sequon influence protein N-glycosylation efficiency. J Am Chem Soc 139(37):12947–12955. https://doi.org/10.1021/jacs.7b03868
Izumi M, Makimura Y, Dedola S, Seko A, Kanamori A, Sakono M, Ito Y, Kajihara Y (2012) Chemical synthesis of intentionally misfolded homogeneous glycoprotein: a unique approach for the study of glycoprotein quality control. J Am Chem Soc 134(17):7238–7241. https://doi.org/10.1021/ja3013177
Kajihara Y, Suzuki Y, Yamamoto N, Sasaki K, Sakakibara T, Juneja LR (2004) Prompt chemoenzymatic synthesis of diverse complex-type oligosaccharides and its application to the solid-phase synthesis of a glycopeptide with asn-linked sialyl-undeca- and asialo-nonasaccharides. Chem Eur J 10(4):971–985. https://doi.org/10.1002/chem.200305115
Kajihara Y, Tanabe Y, Sasaoka S, Okamoto R (2012) Homogeneous Human Complex-Type Oligosaccharides in Correctly Folded Intact Glycoproteins: Evaluation of Oligosaccharide Influence On Protein Folding, Stability, and Conformational Properties. Chem Eur J 18(19):5944–5953
Kasai T, Koshiba S, Yokoyama J, Kigawa T (2015) Stable isotope labeling strategy based on coding theory. J Biomol NMR 63(2):213–221
Makimura Y, Kiuchi T, Izumi M, Dedola S, Ito Y, Kajihara Y (2012) Efficient synthesis of glycopeptide-α-thioesters with a high-mannose type oligosaccharide by means of tert-Boc-solid phase peptide synthesis. Carbohydr Res 364:41–48
Martin G, Ulrich V, Manfred N, Roland W, Harald SC (1995) Characterization of changes in the glycosylation pattern of recombinant proteins from BHK-21 cells due to different culture conditions. J Biotechnol 42:117–131
Mizoue LS, Bazan JF, Johnson EC, Handel TM (1999) CDF NMR. Biochemistry 38:1402–1414
Murakami M, Kiuchi T, Nishihara M, Tezuka K, Okamoto R, Izumi M, Kajihara Y (2016) Chemical synthesis of erythropoietin glycoforms for insights into the relationship between glycosylation pattern and bioactivity. Sci Adv 2:e1500678. https://doi.org/10.1126/sciadv.1500678
Murakami M, Okamoto R, Izumi M, Kajihara Y (2012) Chemical Synthesis of an Erythropoietin Glycoform Containing a Complex-type Disialyloligosaccharide. Angew Chem Int Ed 51(15):3567–3572

Nguyen HM, Izumi M, Sato H, Okamoto R, Kajihara Y (2017) Chemical synthesis of glycoproteins with the specific installation of gradient-enriched 15N-labeled amino acids for getting insights into glycoprotein behavior. Chem Eur J 23(27):6579–6585. https://doi.org/10.1002/chem.201606049

Okamoto R, Mandal K, Ling M, Luster AD, Kajihara Y, Kent SBH (2014a) Total chemical synthesis and biological activities of glycosylated and non-glycosylated forms of the chemokines CCL1 and Ser-CCL1. Angew Chem Int Ed 53(20):5188–5193. https://doi.org/10.1002/anie.201310574

Okamoto R, Mandal K, Sawaya MR, Kajihara Y, Yeates TO, Kent SBH (2014b) (Quasi-)racemic X-ray structures of glycosylated and non-glycosylated forms of the chemokine Ser-CCL1 prepared by total chemical synthesis. Angew Chem Int Ed 53(20):5194–5198. https://doi.org/10.1002/anie.201400679

Park SS, Park J, Ko J, Chen L, Meriage D, Crouse-Zeineddini J, Wong W, Kerwin BA (2009) Biochemical assessment of erythropoietin products from Asia versus US Epoetin alfa manufactured by Amgen. J Pharm Sci 98(5):1688–1699. https://doi.org/10.1002/jps.21546

Phan AT, Patel DJ (2002) A site-specific low-enrichment ^{15}N,^{13}C isotope-labeling approach to unambiguous NMR spectral assignments in nucleic acids. J Am Chem Soc 124(7):1160–1161

Price JL, Powers DL, Powers ET, Kelly JW (2011) Glycosylation of the enhanced aromatic sequon is similarly stabilizing in three distinct reverse turn contexts. Proc Natl Acad Sci U S A 108(34):14127–14132., S14127/14121-S14127/14127. https://doi.org/10.1073/pnas.1105880108

Sinclair AM, Elliott S (2005) Glycoengineering: the effect of glycosylation on the properties of therapeutic proteins. J Pharm Sci 94(8):1626–1635. https://doi.org/10.1002/jps.20319

Teeter MM, Mazer JA, L'Italien JJ (1981) Primary structure of the hydrophobic plant protein crambin. Biochemistry 20(19):5437–5443. https://doi.org/10.1021/bi00522a013

Tirado-Rives J, Jorgensen WL (1990) Molecular dynamics of proteins with the OPLS potential functions. Simulation of the third domain of silver pheasant ovomucoid in water. J Am Chem Soc 112(7):2773–2781. https://doi.org/10.1021/ja00163a046

Unverzagt C, Kajihara Y (2013) Chemical assembly of N-glycoproteins: a refined toolbox to address a ubiquitous posttranslational modification. Chem Soc Rev 42(10):4408–4420

Varki A (2017) Biological roles of glycans. Glycobiology 27(1):3–49. https://doi.org/10.1093/glycob/cww086

Wuthrich K (1986) NMR of proteins and nucleic acids. Wiley, New York

Yabuki T, Kigawa T, Dohmae N, Takio K, Terada T, Ito Y, Laue ED, Cooper JA, Kainosho M, Yokoyama S (1998) Dual amino acid-selective and site-directed stable isotope labeling of the human c-Ha-Ras protein by cell-free synthesis. J Biomol NMR 11(3):295–306

Yamashita K, Kamerling JP, Kobata A (1983) Structural studies of the sugar chains of hen ovomucoid. Evidence indicating that they are formed mainly by the alternate biosynthetic pathway of asparagine-linked sugar chains. J Biol Chem 258(5):3099–3106

Chapter 2
Synthesis of Glycosylated Metal Complexes for Probing Carbohydrate-Carbohydrate Interactions

Teruaki Hasegawa

Abstract Densely packed carbohydrate clusters on cell surfaces play essential roles in varieties of bioprocesses. Little information has been, however, accumulated so far concerning their structural/functional details. In this chapter, we discuss artificial systems to investigate carbohydrate-carbohydrate interactions within/between the carbohydrate cluster(s). Among such artificial systems, much attention will be especially placed on glycosylated tris-bipyridine ferrous complexes for monitoring not only carbohydrate-carbohydrate interactions within the glycocluster but also their resultant conformational changes.

Keywords Carbohydrate-carbohydrate interactions · Tris-bipyridine ferrous complex · Dynamic combinatorial library · Circular dichroism

2.1 Carbohydrate-Carbohydrate Interactions

Glycosphingolipids (GSL) are naturally occurring amphiphiles having hydrophilic carbohydrate heads and hydrophobic ceramide (Cer) tails, and they form one of major components of cell membranes (Fig. 2.1) (Prinetti et al. 2009). In these decades, many organic chemists and biochemists have launched their increasing research effort to structural/functional analyses on GSL to find their critical roles in varieties of bioprocesses (e.g., fertilizations, differentiations, and cell-cell adhesions). It has been widely believed that their carbohydrate heads function as specific ligands for carbohydrate binding proteins, such as lectins, and these carbohydrate-protein interactions are main driving forces to mediate these bioprocesses. Although this is still common understanding in biochemistry, increasing interest has been placed on carbohydrate-carbohydrate interactions (CCI). In these newly recognized binding events, the carbohydrate heads of GSL interact with each other in specific

T. Hasegawa (✉)
Faculty of Life Sciences, Toyo University, Itakura-machi, Ora-gun, Gumma, Japan

Bio-Nano Electronics Research Centre, Toyo University, Kawagoe, Saitama, Japan
e-mail: t-hasegawa@toyo.jp

Y. Yamaguchi, K. Kato (eds.), *Glycobiophysics*, Advances in Experimental Medicine and Biology 1104, https://doi.org/10.1007/978-981-13-2158-0_2

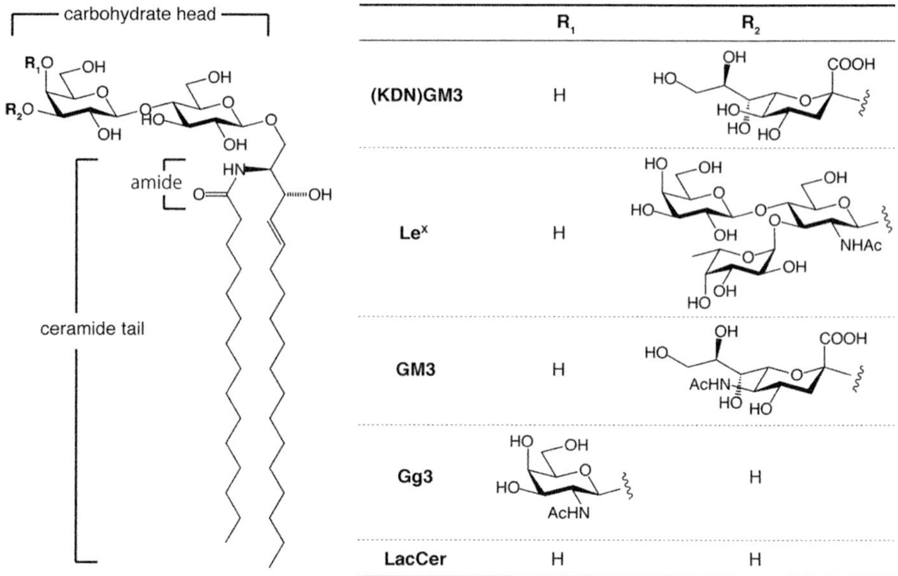

Fig. 2.1 Structures of some GSL in cell membrane to induce CCI

and cation-dependent manners. In detail, GSL in cell membranes laterally aggregate to form GSL-enriched microdomains presenting densely packed carbohydrate clusters (glycoclusters) on their surfaces (Fig. 2.2a, left). In addition to hydrogen bondings between amides in the ceramide tails, side-by-side interactions between the carbohydrate heads are main deriving forces to construct the microdomains. The interactions between the carbohydrate heads can be defined as *cis*-CCI. Herein, please be notified that *cis*-CCI does not mean parallel packings of two carbohydrates. On the contrary, varieties of carbohydrate packings (parallel, antiparallel, cross-shape, etc.) are conceivable in *cis*-CCI, depending on conformations of GSL in the cell membranes (Fig. 2.2a, right).

Investigation on *cis*-CCI is of great importance since it is a key factor to rationalize selective aggregations of certain GSL in cell membranes and the resultant microdomains with unique GSL compositions. In addition to *cis*-CCI at the same cell surface, intercellular CCI (*trans*-CCI) are also important research targets. For example, the GSL-enriched microdomains formed through the lateral aggregation interact with other GSL-enriched microdomains from adjacent cells in face-to-face manners through *trans*-CCI between the glycoclusters (Fig. 2.2b, left). Note again that *trans*-CCI also does not mean antiparallel packings of two carbohydrates. On the contrary, various carbohydrate packings are also conceivable in *trans*-CCI (Fig. 2.2b, right). It is now widely recognized that, although *trans*-CCI are relatively weak, they function as the first adhesive forces to initiate cell-cell adhesions those are then reinforced by strong carbohydrate-protein and protein-protein interactions.

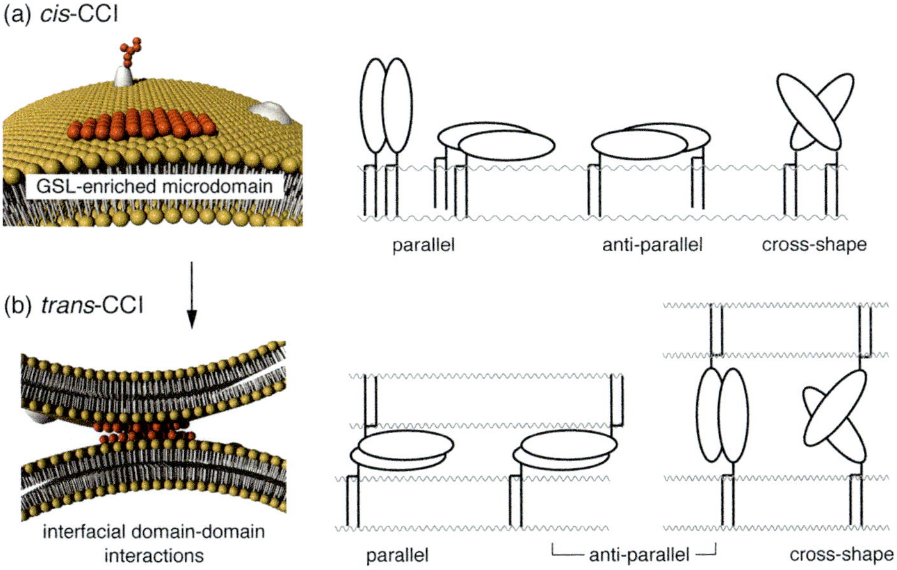

Fig. 2.2 Lateral aggregation of GSL in cell membranes to form the GSL-enriched microdomains (**a**) and the interfacial adhesions between such GSL-enriched microdomains (**b**)

The investigation on *cis*/*trans*-CCI has just started, but some excellent works demonstrated critical roles of *trans*-CCI in various bioprocesses in multicellular organisms. For example, binding of rainbow trout sperm to its egg is initiated by *trans*-CCI between (KDN)G_M3 on the sperm and Gg3 on the egg (See Fig. 2.1 for their structural details) (Yu et al. 2002). In addition, Le^X in mouse embryo interacts with each other in intercellular fashions, and these Le^X-Le^X interactions are responsible for compaction of the embryos at their morula stage (Eggens et al. 1989). These two examples demonstrate critical roles of CCI in fundamental bioprocesses (fertilizations and differentiations) in multicellular organisms. In this respect, investigation on CCI has high academic values since it gives essential information for deep understanding of these bioprocesses.

Investigation on CCI also has great impact on medicinal/pharmaceutical industry since CCI also participate in many diseases triggered by unfavorable cell-cell adhesions. For example, *trans*-CCI between GM3 and Gg3 as well as GM3 and LacCer are both responsible for adhesion between tumor cells and endothelial cells that is the very first step for the cancer metastases (Handa and Hakomori 2012). This fact suggests that investigation on CCI should offer essential information to design new drugs for preventing these diseases. In spite of such importance, little progress has been achieved in CCI chemistry, suffering from heterogenic and fluidic nature of cell membrane, highly diverse carbohydrate structures of GSL, and dynamic fluctuation of GSL levels in cell membranes. Weakness of CCI between two isolated carbohydrates also makes the investigations quite difficult. Suffering from all these difficulties, only few combinations of GSL (e.g., Le^X-Le^X, GM3-Gg3, and GM3-

LacCer) have been discovered so far to induce specific CCI. Considering the highly diverse carbohydrate structures of GSL, it is, however, quite reasonable to believe that many CCI still remain undiscovered in nature. In addition, although it is widely recognized that some cations are essential for CCI, detailed mechanism of these cations to induce CCI is still unclear. Finally, it should be emphasized herein that entirely no information has been obtained concerning the carbohydrate packings to induce CCI.

2.2 Artificial Systems for Probing CCI

Artificial systems are quite useful to avoid the aforementioned difficulties and to access molecular-level information on CCI. So far various artificial systems have been developed for probing CCI, and these systems can be divided into two categories, that is, intermolecular and intramolecular systems.

2.2.1 Intermolecular Systems

The intermolecular systems are relatively large glycoclusters based on inorganic/organic molecular templates (nanoparticles (de la Fuente et al. 2005; Reynolds et al. 2006; Zhao et al. 2012; Lorenz et al. 2012; Lai et al. 2016), micelles (Jayaraman et al. 2013), liposomes (Jayaraman et al. 2013; Pincet et al. 2001), linear polymers (Matsuura and Kobayashi 2004; Matsuura et al. 2004; Chen et al. 2010), dendrimers (Seah et al. 2009), etc.), and associations of these glycoclusters are monitored to evaluate CCI. Since their associations are usually accompanied by large entropic loss, huge numbers of carbohydrates and the multivalent CCI are essential to induce their associations. These intermolecular artificial systems well mimic the cell-cell adhesions in nature and give decisive information concerning carbohydrate and ion specificities to induce CCI. No information can be, however, obtained for carbohydrate-ion stoichiometries and carbohydrate packings to induce CCI.

2.2.1.1 Glycosylated Nanoparticles

The most well-known intermolecular system is gold nanoparticles (AuNP) presenting terminal trisaccharides of Le^X ($Le^X{}_3$-AuNP) (de la Fuente et al. 2005). Their transmission electron microscopic (TEM) observations in the presence of various cations showed selective aggregation of $Le^X{}_3$-AuNP in the presence of Ca^{2+}. The Ca^{2+}-induced aggregation of $Le^X{}_3$-AuNP was also confirmed through atomic force microscopic (AFM) observations and isothermal calorimetric (ITC) measurements. On the other hand, under the same condition, no aggregation was observed for AuNP presenting multiple Lac units (Lac-AuNP), showing stronger Le^X-Le^X

interaction than Lac-Lac one. It should be noted that, although Lac-Lac interaction is weaker than Le^X-Le^X one, the former is strong enough to induced aggregation of Lac-AuNP under certain conditions. In fact, D. A. Russell et al. carried out similar TEM observations using AuNP having multiple glucosides (Glc-AuNP) as a negative reference to find that Lac-AuNP also selectively aggregated on the additions of Ca^{2+} (Reynolds et al. 2006). Combining these two results, it is clear that Lac-Lac interaction is weaker than Le^X-Le^X one, but it is stronger than Glc-Glc one (Glc-Glc < < Lac-Lac < Le^X-Le^X). Again, these AuNP-based intermolecular systems offered decisive information concerning the relative strengths of the CCI and their cation-specificity. No information was, however, supplied concerning the carbohydrate-cation stoichiometries and the carbohydrate packings.

Silica nanoparticles (SiNP) carrying multiple carbohydrates were also developed for probing CCI. For example, SiNP having multiple 3′-sulfated-galactosides (3′SGal-SiNP) was prepared, and their bindings onto cell surfaces were directly visualized. Briefly, in this work, fluorescein isothiocyanate (FITC)-labeled SiNP were prepared and then treated with azide-terminated trimethoxysilane derivatives to afford azide-functionalized fluorescent SiNP. The subsequent Cu^+-mediated alkyne-azide cycloaddition (CuAAC) with alkyne-terminated 3′SGal gave 3SGal-SiNP (Zhao et al. 2012). Selective associations of 3′SGal-SiNP onto oligodendrocytes were successfully visualized through fluorescence microscopic observations. Recently, SiNP carrying fluorescent dyes and Gg3 trisaccharides (Gg3-SiNP) was also prepared, and their specific bindings onto GM3-immobilized Au plates were monitored through surface plasmon resonance (SPR) measurements (Lai et al. 2016). Bindings of Gg3-SiNP onto B16F10 melanoma cells were also visualized in this work through fluorescence microscopic observations.

2.2.1.2 Glycosylated Micelles, Liposomes, Linear Polymers, and Dendrimers

In addition to these inorganic nanoparticles (AuNP and SiNP), organic templates were also utilized, and micelles (Jayaraman et al. 2013), liposomes (Pincet et al. 2001), linear polymers (Matsuura and Kobayashi 2004; Matsuura et al. 2004; Chen et al. 2010), and dendrimers (Chen et al. 2010) having multiple carbohydrates have been so far prepared as the intermolecular systems. For example, interactions between linear polymers having Gg3 trisaccharides (Gg3-polymer) and GM3 monolayers at air-water interface were monitored through π-A isotherms (Matsuura and Kobayashi 2004). In the same paper, SPR measurements were also conducted for probing CCI between the Gg3-polymers and GM3-immobilized Au surfaces. The π-A isotherm-based strategy was also accepted to detect CCI between dendrimers having Lac appendages and GM3 monolayers at air-water interface (Seah et al. 2009).

2.2.2 *Intramolecular Systems*

Most artificial systems so far reported are categorized as the intermolecular systems, and they are designed for probing CCI through the intermolecular associations. Their huge numbers of carbohydrates are, however, major impediments to access molecular-level information on CCI (carbohydrate-cation stoichiometries, carbohydrate packings, etc.). On the contrary, in the intramolecular systems, few carbohydrates are linked together with appropriate spacers to minimize entropic loss on their intramolecular associations, and their CCI-induced conformational changes are monitored to evaluate CCI. Numbers of carbohydrates to induce CCI can be greatly reduced in the intramolecular systems, and therefore, the molecular-level information can be readily obtained. For example, Schmidt et al. prepared two Le^{X}_{3} units linked together through a methylene glycol spacer (-OCH_2O-) to monitor their cation-induced conformational changes through 2D NMR measurements (Gege et al. 2002). Through these experiments, they found binding of this Le^{X}_{3} dimer with Ca^{2+} and the resultant cross-shaped Le^{X}_{3} packing.

This artificial system reported by Schmidt et al. is the first example of the intramolecular system showing the great advantage of the intramolecular systems for obtaining the molecular-level information (e.g., the Ca^{2+}-induced cross-shaped packing of Le^{X}_{3}). This Schmidt's system, however, requires specialized 2D NMR knowledge/technique for the conformational analysis. After this Schmidt's work, other research groups also developed intramolecular systems for probing CCI. In these systems, the spacers tethering the carbohydrates are circular dichroic (CD) active, and their CCI-induced conformational changes are readily monitored through their CD spectral measurements. With the best of my knowledge, two types of such intramolecular systems can be found in literature, that is, glycosylated peptides and glycosylated metal complexes. Before we start our discussion on the glycosylated metal complexes as a main topic of this chapter, we firstly focus our attention on the glycosylated peptides in the next section.

2.2.2.1 Glycosylated Peptides

The first glycosylated peptide for probing CCI was developed by the author with T. Sasaki (Hasegawa and Sasaki T 2003). In this work, an α-helical peptide with 19 residues containing 7 *O*-lactosylated Ser at (i) and (i + 3) or (i) and (i + 4) positions was developed to present a heptavalent Lac cluster on one side of its α-helical structure. Since lateral Lac-Lac interactions affect stability of the α-helix, intramolecular CCI can be readily probed through CD spectral analysis. In fact, Ca^{2+}-induced conformational change of this glycopeptide was detected, showing Ca^{2+}-induced Lac-Lac interaction. Similar peptide-based intramolecular system was also reported by T. Gallagher et al. to investigate intramolecular Le^{X}_{3}-Le^{X}_{3} interactions (Altamore et al. 2011). This research group also developed cyclopentyl/pyrrolidine foldamers with helical conformation presenting a trivalent Gal cluster on one side of its helical structure (Simpson et al. 2006). Intramolecular Gal-Gal interaction in this system was again investigated through CD spectral analysis.

2.2.3 *Glycosylated Metal Complexes*

In the peptide-based systems, their peptide main chains act not only as templates to construct well-defined glycoclusters but also as chromophores to monitor the CCI-mediated conformational changes. Especially, their helical architectures and helix-coil transitions are essential for them for probing CCI. In addition to the peptide-based systems, glycosylated metal complexes are also useful as intramolecular systems for probing CCI. In the latter systems, their unique structural/spectroscopic properties are also advantageous for probing CCI (See below for details).

2.2.3.1 Glycosylated μ-Oxo-bis-[5,15-eso-diphenylporphyrinatoiron(III)]

First example of the glycosylated metal complexes for probing CCI is a μ-oxo-bis-[5,15-meso-diphenylporphyrinatoiron(III)] having a tetravalent Lac cluster (Hasegawa et al. 2005). In this example, a porphyrinatoiron(III) having two Lac units (PorLac) was prepared through cyclization of *p*-formylphenyl-hepta-*O*-acetyl-β-lactoside and dipyrromethane, complexation of the resultant porphyrin derivative with $FeCl_2$, and the subsequent deacetylation with ammonia in a $MeOH/H_2O$ mixture (Fig. 2.3a). PorLac was then dimerized in basic aqueous solution to afford μ-oxo-bis-[5,15-meso-diphenylporphyrinatoiron(III)] having tetravalent Lac clusters (PorLac-dimer), in which two PorLac units were linked together with a Fe-O-Fe bond. In addition to its easy preparation through the self-assembly, its unique double-decker-type 3D structure and free rotations of each PorLac units make PorLac-dimer advantageous as a platform for probing Lac-Lac interactions between the upper and the lower PorLac units. In fact, PorLac-dimer showed unique UV-vis spectral changes on additions of Ca^{2+} (Fig. 2.3b). In detail, PorLac-dimer in aqueous *N*-methyl-2-pyrrolidone (NMP, pH 9.6, 50 mM Tris-HCl, [NMP] = 50% v/v) showed a newly appeared peak at around 525 nm on additions of $CaCl_2$. On the contrary, little colorimetric change was observed on the additions of other alkali/alkaline-earth metal chlorides (NaCl, KCl, nor $MgCl_2$), showing its Ca^{2+}-specific colorimetric change. In addition to referential experiments using a non-glycosylated analogue, CD spectral analyses and TEM observations on PorLac-dimer revealed that the Ca^{2+}-induced colorimetric change of PorLac-dimer arises from 1:2 binding of Ca^{2+} and Lac units to give parallel Lac-Ca^{2+}-Lac packings. These parallel Lac-Ca^{2+}-Lac packings fixed the conformation of PorLac-dimer, and the resultant "H-shaped" PorLac-dimers with large hydrophobic clefts were interdigitated through the hydrophobic interactions to induce the UV-vis spectral change (Fig. 2.3c).

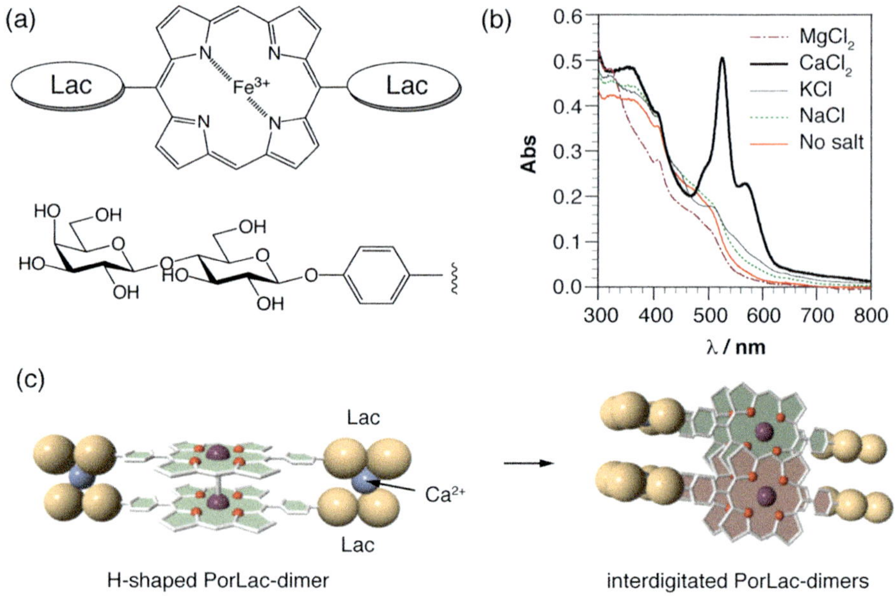

Fig. 2.3 Structure of PorLac (**a**), UV-vis spectral responses toward alkali/alkaline-earth metal chlorides (**b**), and proposed mechanism for the Ca^{2+}-induced UV-vis spectral change (**c**)

2.2.3.2 Glycosylated Ferrocenes

Second example of the glycosylated metal complexes for probing CCI is bis-lactosylated ferrocene (FcLac), in which a Lac unit is attached onto each of two cyclopentadienyl (Cp) rings of Fc core (Fig. 2.4a) (Otsuka et al. 2009). FcLac was prepared through Cu^+-mediated azide-alkyne cycloaddition (CuAAC) between 2-propargyl-β-lactosides (Lac-yn) and 1,1′-diazidemethylferrocene (Fc-N_3). In detail, commercially available lactose monohydrate was converted into Lac-yn through *O*-acetylation with acetic anhydride in pyridine, glycosylation with propargyl alcohol using BF_3OEt_2, and the subsequent deacetylation using aqueous ammonia. On the contrary, Fc-N_3 was prepared from 1,1′-bis-hydroxymethyl-ferrocene by being treated with NaN_3 in acetic acid. The subsequent CuAAC between Lac-yn and Fc-N_3 successfully afforded FcLac in a moderate yield.

In this work, intramolecular Lac-Lac interactions can be monitored through CD spectral analyses. In the absence of the alkali/alkaline-earth metal chloride, FcLac showed a negative CD peak at 220 nm, indicating its chirally twisted 1,3-triazole linkers and/or Cp rings. On the other hand, no peak was observed at around 300 nm, suggesting no adhesive force between the Lac units and the resultant free rotation of each mono-lactosylated Cp ring with the Cp-Fe-Cp axis (Fig. 2.4b). On the contrary, in the presence of NaCl and $CaCl_2$, FcLac showed a broad CD peak at around 300 nm, indicating a chirally twisted conformation of FcLac. Referential experiments using Fc having two maltoside appendages (FcMal) did not show any CD

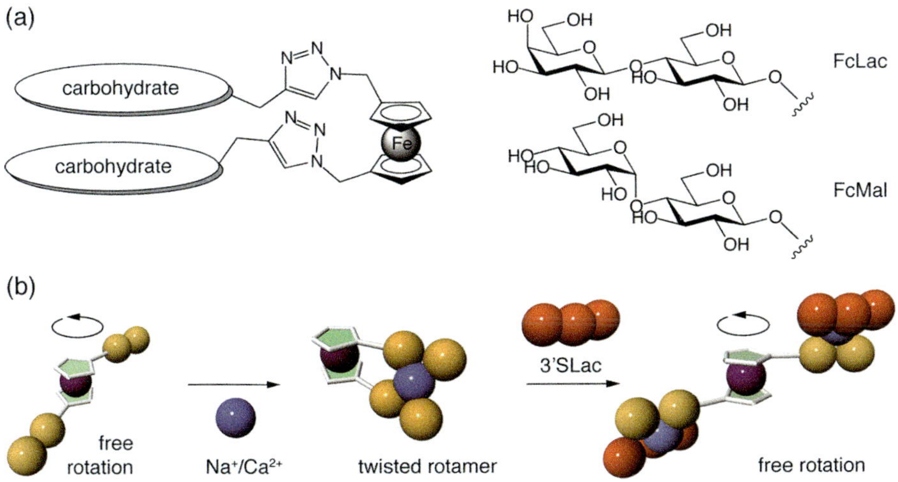

Fig. 2.4 Structure of FcLac/FcMal (**a**) and their conformational changes induced by the cations and 3'SLac (**b**)

spectral change on the additions of these salts, showing essential roles of Lac units in the salt-induced CD spectral changes. Interestingly, the salt-induced Lac-Lac interactions and the resultant conformational change were canceled by 3′-sialyllactose (3'SLac: carbohydrate head of GM_3). Note that GM3-LacCer interaction is the initial driving force to mediate the cancer metastases and it is stronger than LacCer-LacCer interaction. It is reasonable to assume that the intramolecular Lac-Lac interaction was replaced by a couple of intermolecular Lac-3'SLac interaction.

2.3 Glycosylated Tris-bipyridine Ferrous Complexes

2.3.1 Synthesis

The most recently reported example of the glycosylated metal complex for probing CCI is hexavalent glycoclusters having tris-bipyridine ferrous complex cores (Fig. 2.5a) (Nonaka et al. 2016). In this work, 2,2′-bipyridines (bpy) carrying two Lac or Mal units (bpyLac and bpyMal, respectively) were synthesized through condensation reactions between 1-*p*-aminophenyl-*per-O*-acety-glycosides (*p*AP-glycosides) and 2,2-bipyridine-5,5-dicarbonyl chloride (bpyCOCl) followed by deacetylation. Self-assemblies of these glycobpys onto Fe^{2+} gave the corresponding tris-bpy ferrous complexes presenting hexavalent Lac and Mal clusters ($[Fe(bpyLac)_3]^{2+}$ and $[Fe(bpyMal)_3]^{2+}$, respectively). In detail, lactose monohydrate and maltose were converted into the corresponding *p*AP-glycosides through *O*-acetylation with acetic anhydride in pyridine, bromination of their anomeric positions with HBr, S_N2 reactions with sodium *p*-nitrophenoxide, and the

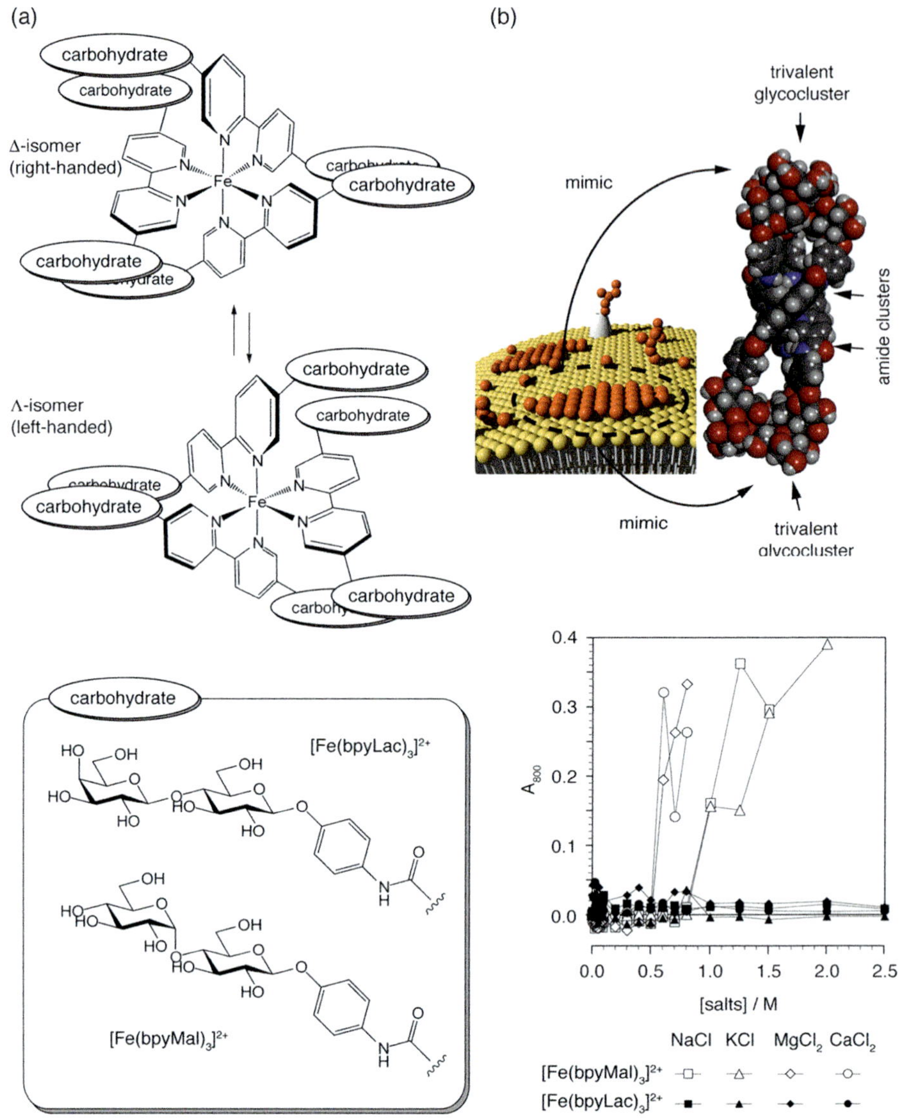

Fig. 2.5 Structures of the Δ- and Λ-isomeric metalloglycoclusters (**a**), their 3D structure obtained through the MD calculation (**b**), and plots of their A_{800} values against the coexisting salt concentrations (**c**)

subsequent hydrogenation with H_2 on Pd/C. On the other hand, for the synthesis of bpyCOCl, 5,5′-dimethyl-2,2′-bipyridine was oxidized with $KMnO_4$ in boiling water to give 2,2′-bipyridne-5,5′-dicarboxylic acid that was then converted into bpyCOCl with $SOCl_2$. The subsequent couplings of bpyCOCl and the *p*AP-glycosides were successfully achieved in THF containing triethylamine followed by deacetylation with aqueous ammonia to afford bpyLac and bpyMal. Since bpyLac and bpyMal were hardly soluble in aqueous media, white precipitates formed as the deacetylation proceeded, and therefore, pure bpyLac and bpyMal were readily retrieved from the reaction mixtures through filtration.

It is widely recognized that bpy and their derivatives are excellent bidentate ligands for Fe^{2+} and they can be readily converted into the corresponding tris-bipyridine ferrous complexes even with highly diluted Fe^{2+}. The water-insoluble natures of these glycobpys, however, strongly hindered their smooth complexations with Fe^{2+}. In fact, one-third equivalent of Fe^{2+} in aqueous glycobpy suspension was not enough to prepare the metalloglycoclusters. In these cases, the resultant mixtures were still colorless, and white precipitates still existed at the bottom of the flask, indicating their unsuccessful complexations. On the other hand, if we used excess amounts (3 eq.) of Fe^{2+} and the resultant mixtures were sonicated/heated repeatedly, the mixtures became transparent with reddish purple color. Through these procedures, $[Fe(bpyLac)_3]^{2+}$ and $[Fe(bpyMal)_3]^{2+}$ were successfully obtained from bpyLac and bpyMal, respectively. UV-vis spectra of these metalloglycoclusters showed metal-to-ligand charge transfer (MLCT) bands at around 544 nm, clearly confirming the successful conversion of the glycobpys into the corresponding metalloglycoclusters with $[Fe(bpy)_3]^{2+}$ cores.

2.3.2 Advantages of the Metalloglycoclusters for Probing CCIs

Great advantages of these metalloglycoclusters as the intramolecular systems arise from unique properties of their $[Fe(bpy)_3]^{2+}$ cores as shown below. (1) The tris-complexes are more stable than the corresponding bis- and mono-complexes, and therefore, the tris-complexes selectively/quantitatively form even in the presence of excess Fe^{2+}. (2) The $[Fe(bpy)_3]^{2+}$ cores have screw propeller-like 3D structures, in which three bpy units coordinate onto Fe^{2+} in right- (Δ) or left-handed (Λ) manners. (3) The Fe-N coordination bonds between Fe^{2+} and the bpy units are reversible at the ambient temperature, and therefore, the two stereoisomers (Δ and Λ) are under equilibrium. (4) The glycobpys are chiral, and therefore, the resultant tris-complexes are mixtures of diastereomeric stereoisomers (Δ and Λ) with different spatial carbohydrate packings and different stabilities. (5) Each of the diastereomeric stereoisomers shows a characteristic CD spectrum, and therefore, their Δ-Λ ratios can be readily monitored through CD spectral measurements. These all features make the metalloglycoclusters unique artificial systems for probing CCI. If one stereoisomer has a more stable carbohydrate packing than the other for intramolecular CCI, the salt-induced CCI should act as adhesive forces to stabilize this stereoisomer. The enriched stereoisomer can be readily monitored through CD spectral measurements.

Conformations of the metalloglycoclusters also emphasized their great advantages as the intramolecular systems to assess *cis*-CCI and the resultant carbohydrate packings within the glycoclusters. Molecular dynamics (MD) calculation on the metalloglycoclusters revealed their unique 3D structures having two trivalent glycoclusters at northern and southern poles of spherical $[Fe(bpy)_3]^{2+}$ cores (Fig. 2.5b). Each of these trivalent glycoclusters can be acted as minimum mimics of the natural glycoclusters on cell surfaces. Furthermore, since the trivalent glycoclusters and the $[Fe(bpy)_3]^{2+}$ core are close to each other, carbohydrate packings of the trivalent glycoclusters greatly affect Δ-Λ ratios of the $[Fe(bpy)_3]^{2+}$ cores. In other words, if *cis*-CCI induce particular carbohydrate packings of the glycoclusters, such CCI-induced conformational changes of the glycoclusters can be sensitively read out through CD spectral measurements. For example, in the absence of the salt, both $[Fe(bpyLac)_3]^{2+}$ and $[Fe(bpyMal)_3]^{2+}$ showed CD spectra having two characteristic positive-to-negative (+/−) and negative-to-positive (−/+) CD patterns on going from longer to shorter wavelengths at around 544 (MLCT band) and 315 nm (π-π* transition band), respectively. These CD spectral features are characteristic for Δ-$[Fe(bpy)_3]^{2+}$, indicating that the corresponding Δ-isomers predominantly exist in the mixtures. Intensities of these CD peaks were, however, much smaller than those of enantiopure Δ-$[Fe(bpy)_3]^{2+}$, implying their low Δ-excesses. These data suggest their flexible carbohydrate packings in pure water. On the contrary, addition of alkali/alkaline-earth metal salts induced unique CD spectral changes of the metalloglycoclusters, suggesting favorable carbohydrate packings of the glycoclusters in the presence of the salts. But, before starting discussion on the salt-induced conformational changes of the metalloglycoclusters, our attention will be focused onto salt-induced aggregations of the metalloglycoclusters.

2.3.3 *Salt-Induced Aggregation*

The metalloglycoclusters are highly soluble in pure water. Considering hardly water-soluble nature of their precursors (bpyLac and bpyMal), the high water solubilities of the metalloglycoclusters clearly arise from +2 charges in their $[Fe(bpy)_3]^{2+}$ cores. The aqueous solution of $[Fe(bpyMal)_3]^{2+}$, however, became turbid on additions of salts. For example, when NaCl was added (>ca. 1.0 M), $[Fe(bpyMal)_3]^{2+}$ immediately aggregated to give precipitates. Other alkali/alkaline-earth metal chlorides also induced similar aggregations of $[Fe(bpyMal)_3]^{2+}$. To assess the aggregations in quantitative manners, absorbance at 800 nm (A_{800}) was utilized as an index for the aggregation. For example, in the case of $[Fe(bpyMal)_3]^{2+}$, its A_{800} value was almost zero in the absence of the salts, showing no aggregation of $[Fe(bpyMal)_3]^{2+}$. In the presence of NaCl, however, its A_{800} was greatly increased when [NaCl] was higher than 1.0 M. Critical concentration (CC) of NaCl to induce the aggregations can be, therefore, estimated to 1.0 M (Fig. 2.5c). Likewise, the CC of $MgCl_2$, KCl, and $CaCl_2$ were estimated to 0.6, 1.0, and 0.6 M, respectively.

The salt-induced aggregation of $[Fe(bpyMal)_3]^{2+}$ is quite contradictive to the excellent water solubility of $[Fe(bpyLac)_3]^{2+}$ even under high concentrations of these salts; that is, $[Fe(bpyLac)_3]^{2+}$ showed no aggregation in aqueous solutions with much higher salt concentrations ([salts] = 2.0 M), showing ionic strength of the media as a negligible factor for the aggregation. We carried out some referential experiments to reveal mechanism of the salt-induced aggregation of $[Fe(bpyMal)_3]^{2+}$ to find that NaBr did not induce the aggregation. This datum shows that not Na^+ but Cl^- is responsible for the aggregation. The Cl^--induced aggregation of $[Fe(bpyMal)_3]^{2+}$ was also supported by the CC values; that is, the CC of the alkaline-earth metal chlorides (ca. 0.6 M for $MgCl_2$ and $CaCl_2$) were almost half to those of the alkali metal chlorides (1.0 M for NaCl and KCl). In addition, MD calculation on a molecular system composed of $[Fe(bpyMal)_3]^{2+}$ and Cl^- gave a clear picture of their binding, in which one Cl^- is captured within the amide cluster through electrostatic interactions between the former and the partially positive (δ+) hydrogen atoms (-N*H*CO-) of the latter. The $[Fe(bpy)_3]^{2+}$ core also electrostatically stabilizes the Cl^- captured within the amide cluster.

As already mentioned, the highly water-soluble nature of $[Fe(bpyMal)_3]^{2+}$ arises from its +2 charges. It is, therefore, reasonable to assume that the binding of Cl^- onto $[Fe(bpyMal)_3]^{2+}$ partially neutralizes the +2 charges to lower its water solubility. This proposed mechanism quite reasonably explains the Cl^--induced aggregation of $[Fe(bpyMal)_3]^{2+}$; it is, however, still difficult to rationalize the highly water-soluble nature of $[Fe(bpyLac)_3]^{2+}$ even under the high salt concentrations. Informative data to rationalize the selective $[Fe(bpyMal)_3]^{2+}$ aggregation was obtained through CD spectral analyses as discussed below.

2.3.4 *Salt-Induced CD Spectral Change*

The CD spectral analyses on the metalloglycoclusters in aqueous solutions containing the alkali/alkaline-earth metal chlorides (~0.4 M) showed great impacts of these salts on their CD spectra. In these experiments, the salt concentrations were lower than their CC, and therefore, the aqueous solutions were still transparent without any aggregation.

2.3.4.1 Salt-Induced CD Spectral Change of $[Fe(bpyMal)_3]^{2+}$

In the case of $[Fe(bpyMal)_3]^{2+}$, it showed magnified CD spectra in the presence of the alkali/alkaline-earth metal chlorides; that is, its CD spectra kept the positive-to-negative and negative-to-positive CD signals at almost the same wavelength (544 and 315 nm, respectively) although their CD intensities were highly magnified depending on concentrations of the coexisting salts (Fig. 2.6a). Plots of CD intensities at 302 nm (CD_{302}) against the salt concentrations showed Langmuir-type profiles (Fig. 2.6b), indicating 1:1 bindings between $[Fe(bpyMal)_3]^{2+}$ and the ions. Furthermore, there was little difference among these Langmuir-type profiles, indicating that not the alkali/alkaline-earth metal cation (Na^+, Mg^{2+}, K^+, or Ca^{2+}) but the anion (Cl^-) binds to $[Fe(bpyMal)_3]^{2+}$ to induce the CD spectral changes. Referential experiments using $FeCl_2$ supported the anion-induced CD spectral changes; that is,

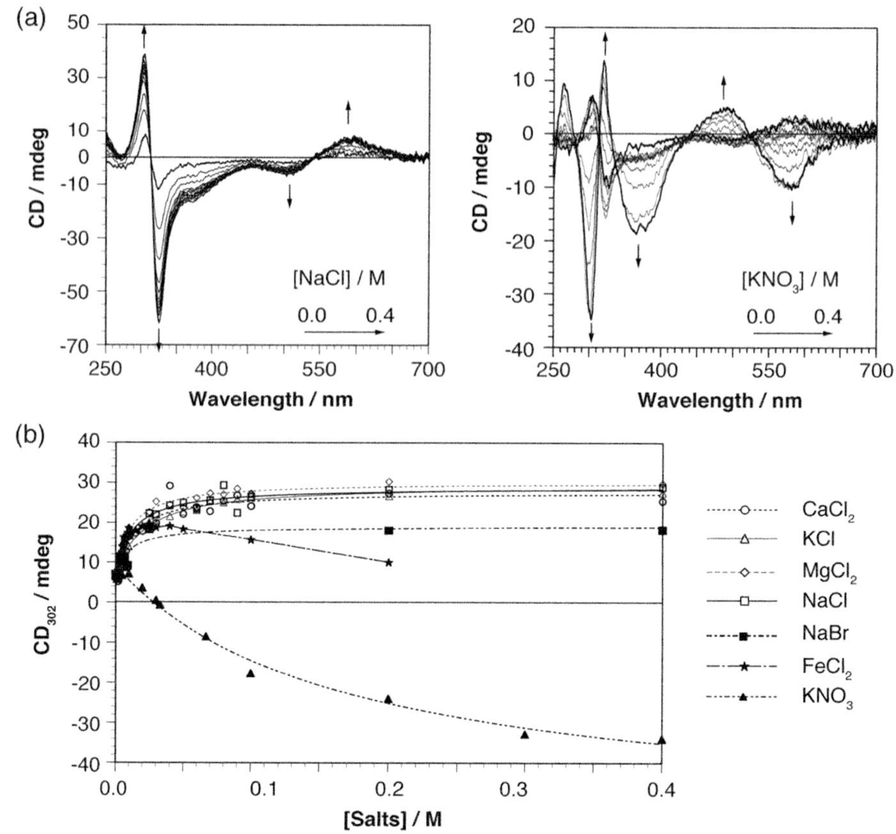

Fig. 2.6 CD spectral changes of $[Fe(bpyMal)_3]^{2+}$ on the additions of NaCl and KNO_3 (**a**) and plots of CD_{302} values of $[Fe(bpyMal)_3]^{2+}$ against the concentrations of the coexisting salts (**b**)

CD_{302} of $[Fe(bpyMal)_3]^{2+}$ was also magnified with increasing concentrations of $FeCl_2$, when $[FeCl_2]$ is lower than 40 mM. Decreased CD signals under high Fe^{2+} concentrations (>40 mM) are assignable to partial dissociation of $[Fe(bpyMal)_3]^{2+}$. Referential experiments using NaBr also supported the Cl^--induced CD spectral changes; that is, NaBr also magnified the CD signals although this NaBr-induced CD changes was much smaller than that induce by NaCl.

Of great interest, $[Fe(bpyMal)_3]^{2+}$ showed a totally different CD spectral response toward NO_3^-. For example, in the presence of KNO^3 (>0.05 M), $[Fe(bpyMal)_3]^{2+}$ showed CD spectra that were almost mirror images to that in pure water. In spite of a newly appeared negative peak at around 370 nm, this NO_3^--induced CD spectral change indicates that Λ-$[Fe(bpyMal)_3]^{2+}$ was greatly stabilized/enriched in the presence of NO_3^-.

These experimental data indicate the followings. (1) The CD spectral changes arise from the binding of $[Fe(bpyMal)_3]^{2+}$ with the anions (Cl^-, Br^- or NO_3^-). (2) In the binding events, only one anion binds to one $[Fe(bpyMal)_3]^{2+}$ with 1:1 stoichiometries. (3) The bindings of halides (Cl^- and Br^-) and nitrate (NO_3^-) enrich Δ- or Λ-$[Fe(bpyMal)_3]^{2+}$, respectively. In other words, $[Fe(bpyMal)_3]^{2+}$ dynamically changes their spatial carbohydrate packings in response to the anions. The last finding is of quite importance since the GSL-enriched microdomains on cell surfaces also have amide clusters (See Fig. 2.1). This structural similarity suggests anion response of the cell surface glycoclusters to change their carbohydrate packings.

2.3.4.2 Salt-Induced CD Spectral Change of $[Fe(bpyLac)_3]^{2+}$

In the case of $[Fe(bpyLac)_3]^{2+}$, it showed unique two-phase CD spectral changes that were totally different from the Langmuir-type binding profiles observed for $[Fe(bpyMal)_3]^{2+}$. For example, in the case of $CaCl_2$, CD signals of $[Fe(bpyLac)_3]^{2+}$ were slightly weakened with increasing $CaCl_2$ concentration until $[CaCl_2]$ reached to ca. 25 mM (first phase, Fig. 2.7). On the contrary, when $[CaCl_2]$ was higher than 25 mM, its CD signals turned to be magnified with increasing $CaCl_2$ concentration (second phase). The other alkali/alkaline-earth metal chlorides (NaCl, $MgCl_2$, and KCl) also induced similar CD spectral changes. These two-phase CD spectral changes indicate that Λ- and Δ-$[Fe(bpyLac)_3]^{2+}$ were stabilized under low (~25 mM) and high (25 mM ~) salt concentrations, respectively.

Referential experiments were carried out to reveal mechanisms of these two-phase CD spectral changes. When $FeCl_2$ was used in the experiments, CD signals of $[Fe(bpyLac)_3]^{2+}$ were simply reduced with increasing $FeCl_2$ concentration. These data suggest that Cl^- and the alkali/alkaline-earth metal cations (Na^+, Mg^{2+}, K^+, or Ca^{2+}) are responsible for the first and the second phase CD spectral changes, respectively. Of great interest, the Cl^- binding in the first phase is a prerequisite for the second phase CD spectral changes; that is, NaBr and KNO_3 induced the first phase CD spectral changes only, and they were not followed by the second phase ones. These data indicate the followings. (1) Various anions (Cl-, Br-, and NO_3^-) bind to $[Fe(bpyLac)_3]^{2+}$ to induce the first phase CD spectral change. (2) Among these anions, the Cl^- bindings in the first phase are essential for the subsequent (second phase) CD spectral change induced by the cations. It is reasonable to assume that the binding of Cl^- onto the amides induces preorganization of the Lac clusters into those suitable for the cation bindings. An additional experiment was also carried out using free lactose as an additive to find that coexisting free Lac (0.25 M) inhibited the NaCl-induced two-phase CD spectral change of $[Fe(bpyLac)_3]^{2+}$ to some extent, proving that the Lac units are responsible for the cation bindings (Nonaka et al. 2016). Considering that the second binding event was not observed for $[Fe(bpyMal)_3]^{2+}$, these data indicate that the βGal nonreducing terminals of $[Fe(bpyLac)_3]^{2+}$ are responsible for the cation bindings.

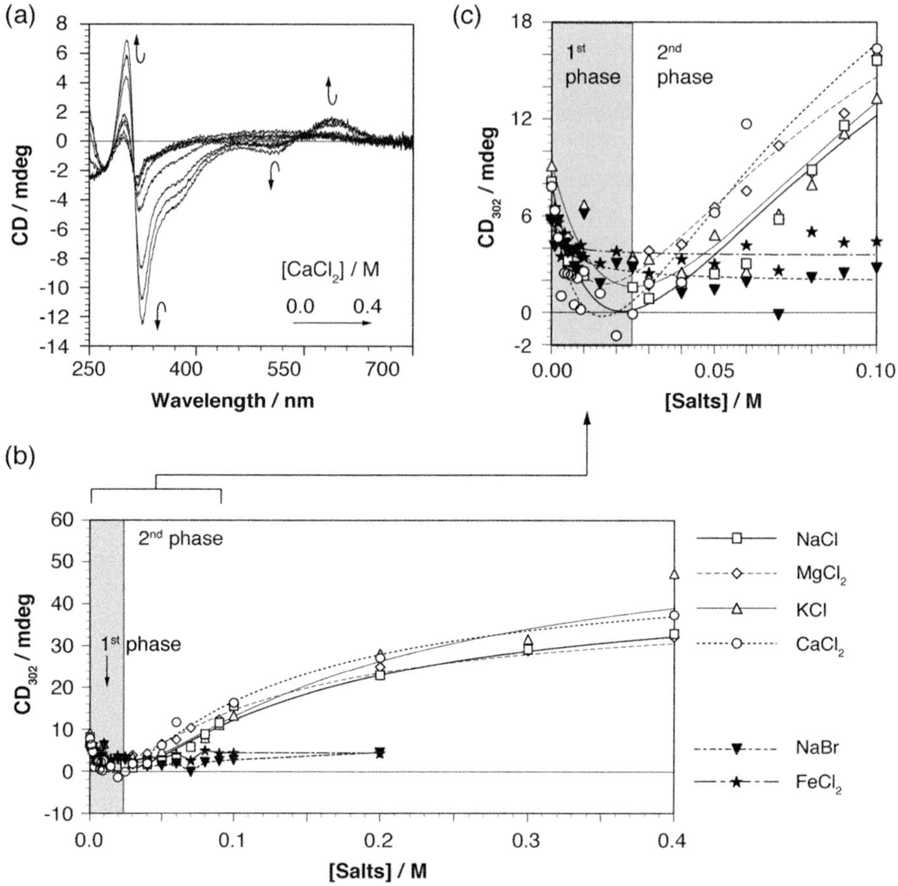

Fig. 2.7 CD spectral changes of $[Fe(bpyLac)_3]^{2+}$ on the additions of $CaCl_2$ (**a**) and plots of their CD_{302} values against the concentrations of the coexisting salts (**b** and **c**)

2.3.5 Detailed Mechanism of the Binding Between the Metalloglycoclusters and the Ions

The most conceivable mechanism to reasonably explain the salt responses is illustrated in Fig 2.8. In the first binding events, the Cl^- bindings to $[Fe(bpyMal)_3]^{2+}$ and $[Fe(bpyLac)_3]^{2+}$ induce their Langmuir-type and the first phase CD spectral changes, respectively. In the case of $[Fe(bpyMal)_3]^{2+}$, the subsequent binding with the cations does not occur, and therefore, the resultant Cl^- adduct ($[Fe(bpyMal)_3Cl]^+$) aggregates suffering from its lowered water solubility. On the other hand, in the case of $[Fe(bpyLac)_3]^{2+}$, its Cl^- adduct ($[Fe(bpyLac)_3Cl]^+$) immediately binds to the cations (M^{n+}; n = 1 or 2), and the resultant complexes ($[Fe(bpyLac)_3MCl]^{m+}$; m = 2 or 3) do not form any aggregation because of their retained/enhanced water solubility originating from the +2 or + 3 charges.

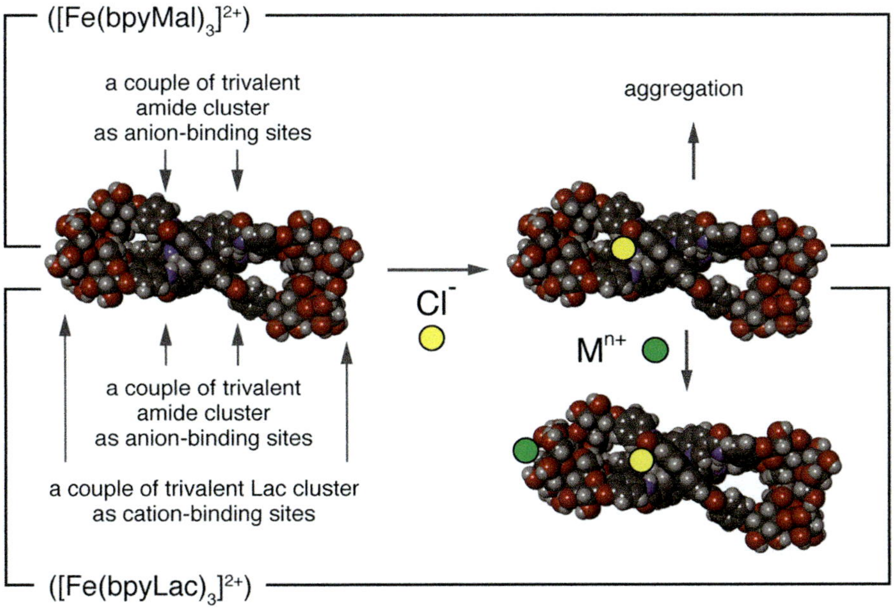

Fig. 2.8 Detailed mechanisms to rationalize both the anion binding of $[Fe(bpyMal)_3]^{2+}$ to induce the CD spectral changes with Langmuir-type profiles and the resultant aggregation (top) and anion/cation binding of $[Fe(bpyLac)_3]^{2+}$ to induce the two-phase CD spectral changes (bottom)

2.4 Summary

In this chapter, we focus our attention on the artificial systems, especially the glycosylated metal complexes, for probing CCI. As already mentioned, the *cis*/*trans*-CCI are essential not only for the GSL localizations in the cell membranes but also for the cell-cell adhesions. It is, however, quite difficult to probe CCI, and only few molecular-level information have been accumulated so far in CCI chemistry. Especially, no information has been obtained concerning *cis*-CCI and CCI-mediated carbohydrate packings within the glycoclusters on cell surfaces. In this respect, the glycosylated metal complexes having the $[Fe(bpy)_3]^{2+}$ cores are quite advantageous to monitor *cis*-CCI within the glycocluster and the CCI-mediated conformational change of the glycoclusters. Continuous studies using the glycosylated metal complexes and other well-designed artificial models are highly required for progress in this research field.

References

Altamore TM, Fernández-García C, Gordon AH, Hübscher T, Promsawan N, Ryadnov MG, Doig AJ, Woolfson DN, Gallagher T (2011) Randomcoil: α-helix equilibria as a reporter for the LewisX–LewisX interaction. Angew Chem Int Ed 50:11167–11171

Chen Q, Cui Y, Zhang T, Cao J, Han B (2010) Fluorescent conjugated polyfluorene with pendant lactopyranosyl ligands for studies of Ca^{2+} mediated carbohydrate-carbohydrate interaction. Biomacromolecules 11:13–19

de la Fuente JM, Eaton P, Barrientos AG, Menéndez M, Penadés S (2005) Thermodynamic evidence for Ca^{2+} mediated self-aggregation of Lewis X gold glyconanoparticles. A model for cell adhesion via carbohydrate-carbohydrate interaction. J Am Chem Soc 127:6192–6197

Eggens I, Fenderson B, Toyokuni T, Dean B, Stroud M, Hakomori S (1989) Specific Interaction between Le and Le determinants. J Biol Chem 264:9476–9484

Gege C, Geyer A, Schmidt RR (2002) Carbohydrate-carbohydrate recognition between Lewis X blood group antigens, mediated by calcium ions. Eur J Org Chem 2475:2485

Handa K, Hakomori S (2012) Carbohydrate to carbohydrate interaction in development process and cancer progression. Glycoconj J 29:627–637

Hasegawa T, Sasaki T (2003) Glyco-helix: parallel lactose-lactose interactions stabilize an α-helical structure of multi-glycosylated peptide. Chem Commun 8:978–979

Hasegawa T, Numata M, Asai M, Takeuchi M, Shinkai S (2005) Colorimetric calcium-response of β -lactosylated μ-oxo-bis-[5,15-mesodiphenylporphyrinatoiron(III)]. Tetrahedron 61:7783–7788

Jayaraman N, Maiti K, Naresh K (2013) Multivalent glycoliposomes and micelles to study carbohydrate–protein and carbohydrate–carbohydrate interactions. Chem Soc Rev 42:4640–4656

Lai C, Hütter J, Hsu C, Tanaka H, Varela-Aramburu S, De Cola L, Lepenies B, Seeberger PH (2016) Analysis of carbohydrate−carbohydrate interactions using sugar-functionalized silicon nanoparticles for cell imaging. Nano Lett 16:807–811

Lorenz B, de Cienfuegos LÁ, Oelkers M, Kriemen E, Brand C, Stephan M, Sunnick E, Yüksel D, Kalsani V, Kumar K, Werz DB, Janshoff A (2012) Model system for cell adhesion mediated by weak carbohydrate−carbohydrate interactions. J Am Chem Soc 134:3326–3329

Matsuura K, Kobayashi K (2004) Analysis of GM3-Gg3 interaction using clustered gycoconjugate models constructed from glycolipid monolayers and artificial glycoconjugate polymers. Glycoconj J 21:139–148

Matsuura K, Oda R, Kitakouji H, Kiso M, Kitajima K, Kobayashi K (2004) Surface plasmon resonance study of carbohydrate-carbohydrate interaction between various gangliosides and Gg3-carrying polystyrene. Biomacromolecules 5:937–941

Nonaka Y, Uruno R, Dai F, Matsuoka R, Nakamura M, Iwamura M, Iwabuchi H, Okada T, Chigira N, Amano Y, Hasegawa T (2016) Hexavalent glycoclusters having tris-bipyridine ferrous complex cores as minimum combinatorial libraries for probing carbohydrate-carbohydrate interactions. Tetrahedron 72:5456–5464

Otsuka A, Sakurai K, Hasegawa T (2009) Ferrocenes with two carbohydrate appendages at the upper and lower rings are useful for investigating carbohydrate–carbohydrate interactions. Chem Commun 36:5442–5444

Pincet F, Bouar TL, Zhang Y, Esnault J, Mallet J, Perez E, Sinay P (2001) Ultraweak sugar-sugar interactions for transient cell adhesion. Biophys J 80:1354–1358

Prinetti A, Loberto N, Chigorno V, Sonnino S (2009) Glycosphingolipid behaviour in complex membranes. Biochim Biophys Acta 1788:184–193

Reynolds AJ, Haines AH, Russell DA (2006) Gold glyconanoparticles for mimics and measurement of metal ion-mediated carbohydratecarbohydrate interactions. Langmuir 22:1156–1163

Seah N, Santacroce PV, Basu A (2009) Probing the lactose·GM3 carbohydrate-carbohydrate interaction with glycodendrimers. Org Lett 11:559–562

Simpson GL, Gordon AH, Lindsay DM, Promsawan N, Crump MP, Mulholland K, Hayter BR, Gallagher T (2006) Glycosylated foldamers to probe the carbohydrate–carbohydrate interaction. J Am Chem Soc 128:10638–10639

Yu S, Kojima N, Hakomori S, Kudo S, Inoue S, Inoue Y (2002) Binding of rainbow trout sperm to egg is mediated by strong carbohydrate-to−+carbohydrate interaction between (KDN)GM3 (deaminated neuraminyl ganglioside) and Gg3-like epitope. PNAS 99:2854–2859

Zhao J, Liu Y, Park H, Boggs JM, Basu A (2012) Carbohydrate-coated fluorescent silica nanoparticles as probes for the galactose/3-sulfogalactose carbohydrate–carbohydrate interaction using model systems and cellular binding studies. Bioconjug Chem 12:1166–1173

Chapter 3
Unraveling of Lipid Raft Organization in Cell Plasma Membranes by Single-Molecule Imaging of Ganglioside Probes

Kenichi G. N. Suzuki, Hiromune Ando, Naoko Komura, Takahiro Fujiwara, Makoto Kiso, and Akihiro Kusumi

Abstract Gangliosides are involved in a variety of physiological roles and particularly in the formation and function of lipid rafts in cell membranes. However, the dynamic behaviors of gangliosides have not been investigated in living cells owing to the lack of fluorescent probes that behave like their parental molecules. This has recently been resolved by developing new fluorescent ganglioside analogues that act similarly to their parental molecules, synthesized by only chemical methods. We performed single fluorescent-molecule imaging and revealed that ganglioside probes dynamically enter and exit rafts containing CD59, a glycosylphosphatidylinositol (GPI)-anchored protein, both before and after stimulation. The residency time of our ganglioside probes in CD59 oligomers was 48 ms after stimulation. The residency times in CD59 homodimer and monomer rafts were 40 and 12 ms, respectively. These results reveal the first direct evidence that GPI-anchored receptors and gangliosides interact in a cholesterol-dependent manner. Furthermore, they demonstrate that gangliosides continually move in and out of rafts that contain CD59 in an extremely dynamic manner and at a much higher frequency than expected. In this chapter, we review methods for the development and single-molecule imaging of new fluorescent ganglioside analogues and discuss how raft domains are formed, both before and after receptor engagement.

K. G. N. Suzuki (✉) · H. Ando · N. Komura
Center for Highly Advanced Integration of Nano and Life Sciences (G-CHAIN), Gifu University, Gifu, Japan

Institute for Integrated Cell-Material Sciences (WPI-iCeMS), Kyoto University, Kyoto, Japan
e-mail: kgsuzuki@gifu-u.ac.jp

T. Fujiwara · M. Kiso
Institute for Integrated Cell-Material Sciences (WPI-iCeMS), Kyoto University, Kyoto, Japan

A. Kusumi
Institute for Integrated Cell-Material Sciences (WPI-iCeMS), Kyoto University, Kyoto, Japan

Membrane Cooperativity Unit, Okinawa Institute of Science and Technology (OIST), Onna-son, Okinawa, Japan

Y. Yamaguchi, K. Kato (eds.), *Glycobiophysics*, Advances in Experimental Medicine and Biology 1104, https://doi.org/10.1007/978-981-13-2158-0_3

Keywords Fluorescent ganglioside probes · Raft domains · Single-molecule fluorescence tracking · GPI-anchored protein · Cell plasma membrane

3.1 Introduction

Gangliosides are subtypes of glycosphingolipids that contain one or more sialic acid (N-acetylneuraminic acid) groups. Gangliosides that contain one, two, three, or four sialic acid moieties are termed GM, GD, GT, and GQ, respectively. They are also defined using numbers to indicate their migratory position after thin-layer chromatography (e.g., GM1, GM3, GT1b). More than 40 types of gangliosides have been identified, with the location and number of sialic acids groups present being different for each. Gangliosides are known to be involved in a variety of important physiological functions in cell plasma membranes (PMs). Representative functional examples include raft domain formation through interaction with cholesterol (Fujita et al. 2007; Lingwood et al. 2011; Dalton et al. 2017; Galimzyanov et al. 2017) and the invasion of microbial toxins, viruses, and bacteria into cells (Smith et al. 2004; Fleming et al. 2014). Gangliosides have also been shown to regulate receptor activation (Yoon et al. 2006; Ohkawa et al. 2010; Coskun et al. 2011; Berntsson et al. 2013; Dam et al. 2017). More specifically, previous studies have indicated that GM3 inhibits activation of epidermal growth factor receptors (EGFRs) (Coskun et al. 2011). It has also been reported that activation of nerve growth factor (NGF) receptors (Trk) is enhanced by association with GM1 (Fukuda et al. 2015; Aureli et al. 2016). The ganglioside GD3 was found to augment adhesion signaling by recruiting integrins to lipid rafts (Ohkawa et al. 2010), and GQ1b was observed to increase tyrosine phosphorylation of N-methyl-d aspartate (NMDA) subunit B, which may subsequently enhance NMDA receptor synaptic activation (Jung et al. 2010).

However, despite the key involvement of gangliosides in such important cellular processes, our knowledge about their locations and distributions, associations with other molecular species, and dynamic behaviors in cell PMs has remained limited. Gangliosides have not been directly observed in PMs of living cells, and previous studies have been performed using fixed PMs or by examining how the expression levels of gangliosides affect receptor activities and other cellular processes. This is primarily because of the lack of fluorescent ganglioside probes that behave like endogenous ones (Eggeling et al. 2009; Chinnapen et al. 2012; Sezgin et al. 2012). GM1 analogues conjugated with fluorescent labels involving the fatty acid or glycan have been synthesized (Schwarzmann et al. 2005; Marushchak et al. 2007; Eggeling et al. 2009; Polyakova et al. 2009; Chinnapen et al. 2012). However, these analogues did not behave in a similar manner as the parental molecules in terms of partitioning into raft-like domains (liquid-ordered domains) in artificial liquid order-disorder phase-separated lipid membranes, such as giant unilamellar vesicles or in giant plasma membrane vesicles (GPMVs) at lower temperatures (Sezgin et al. 2012), and in terms of their affinities for GM1 binding to cholera toxin subunit B (CTB) (Chinnapen et al. 2012). In addition, CTB itself has been used to examine the location of GM1. However, CTB is pentavalent and can cross-link gangliosides, changing their lateral distribution in the PMs of cells (Hammond et al. 2005; Lingwood et al. 2008; Kaiser et al. 2009; Stefl et al. 2012). Even if PMs are fixed using 4% paraformaldehyde and 0.5% glutaraldehyde, a significant fraction of lipid molecules remain mobile (Kusumi and Suzuki 2005; Tanaka et al. 2010) and can cross-link ganglioside molecules. Similarly, it is difficult to perform immunohistochemical staining of gangliosides, as labeling using primary and secondary antibodies also results in cross-linking, even after cell fixation. This results in the generation of bright, fluorescent patches of gangliosides (Kusumi and Suzuki 2005; Tanaka et al. 2010). This could potentially be avoided by using the fab domains of primary and secondary antibodies, but this approach has not been pursued. Owing to all these various factors, it has been extremely difficult to investigate the location and distribution of gangliosides in cell PMs.

To resolve this issue, we employed entirely chemical synthesis methods (Ando et al. 2003, 2005; Imamura et al. 2009; Iwayama et al. 2009; Tamai et al. 2011; Fujikawa et al. 2011; Nakashima et al. 2012) to develop fluorescent ganglioside analogues that behave in a similar manner as their parental gangliosides (Komura et al. 2016). In this chapter, we overview the studies on the dynamic behaviors of gangliosides in cell PMs. In addition, we introduce studies on our new fluorescent probes by single-molecule imaging, which examined how raft domains are formed before and after receptor stimulation.

3.2 Single-Molecule Imaging

Single-molecule imaging was performed with an objective-type total internal reflection microscope (Tokunaga et al. 1997) that was constructed on an inverted microscope (Olympus IX-70) (Fig. 3.1). 488 nm and 594 nm laser beams, attenuated with neutral density filters, were expanded by two lenses (L1 and L2 in Fig. 3.1), focused at the back focal plane of the objective lens with an L3 lens (Fig. 3.1), and then steered onto the edge of a high numerical aperture objective lens (UAPON 100XOTIRF; numerical aperture = 1.49). The laser beam was totally internally reflected at the coverslip-medium interface (incident angle, ~66 degree), and an evanescent field was formed on the surface of the coverslip (Axelrod et al. 1984) (Fig. 3.1). The stray excitation light was blocked with an interference band-pass filter (BF in Fig. 3.1) placed after the dichroic mirror (DM). The two-color fluorescence emission signal was split by DM and projected onto two-stage microchannel plate intensifier (C8600–03; Hamamatsu photonics) (II), and the intensified images were detected by two sCMOS cameras (ORCA flash 4.0, Hamamatsu photonics). Single-molecule imaging at high temporal resolution was performed by using a specially designed CMOS sensor-based camera (Photron).

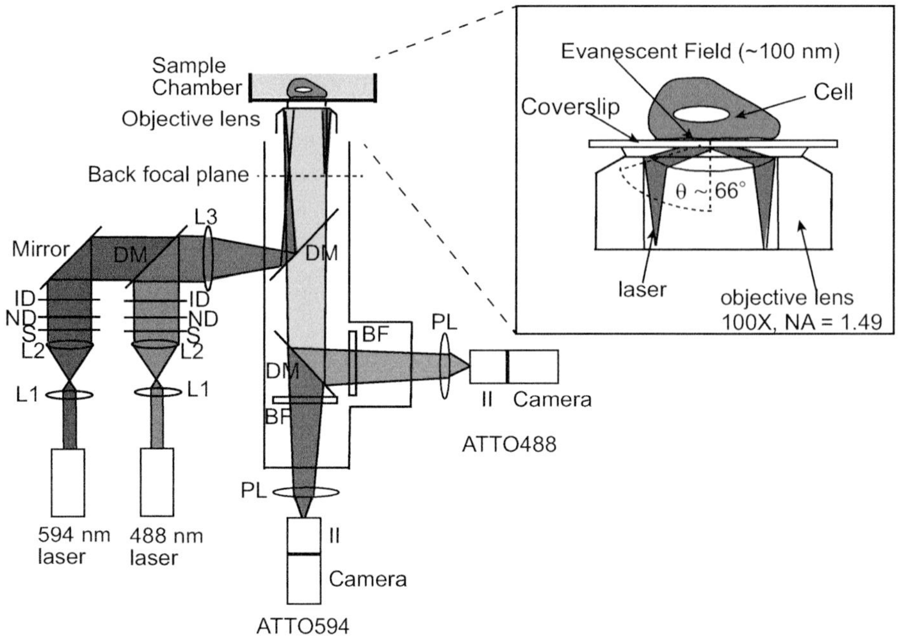

Fig. 3.1 Schematic drawing of the total internal reflection fluorescence microscope to perform simultaneous observation of two different molecules such as ATTO488 and ATTO594 at the level of single molecules. (Left) Optical ray diagram *L1* and *L2* beam expander, *L3* focusing lens, *PL* projection lens, *S* electric shutter, *ND* neutral density filter, *ID* iris diaphragm, *DM* dichroic mirror, *BF* band-pass filter, *II* image intensifier. (Right) Schematic drawing around the sample chamber and the objective lens. The laser was totally internally reflected at the coverslip-medium interface, and an evanescent field was formed

3.3 Chemical Synthesis of Ganglioside Probes

The method to synthesize fluorescent GM3 and GM1 analogues (Fig. 3.2a and b) was developed on the basis of our previously established method (Ando et al. 2003; Imamura et al. 2009; Tamai et al. 2011). The developed method to synthesize fluorescent GM1 and GM3 was then applied for the synthesis of fluorescent GM2 and GD1b probes (Fig. 3.3a and b). A fluorescent reporter molecule (five different, distinguishable dyes were used) was conjugated at the C9 position of the sialic acid moiety (termed S9-type) of both GM3 and GM1, at the C6 position of the galactose (referred to as G6-type) of GM3, and at the C6 position of the terminal galactose (one of the two galactoses in the sugar chain (called termG6-type) of GM1) (Fig. 3.2a, b).

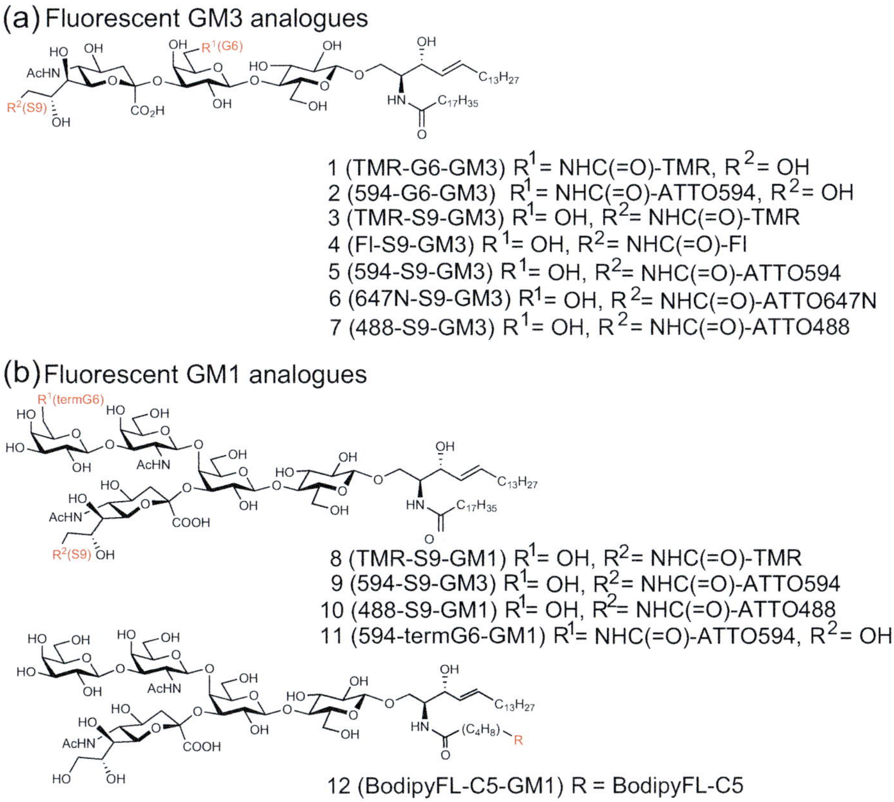

Fig. 3.2 Chemical structures of the fluorescent GM1 and GM3 probes we developed. (Komura et al. 2016)

3.4 Characterization of New Ganglioside Probes

The fluorescent ganglioside probes were compared with native molecules to assess partitioning into raft domains. Raft markers, such as sphingomyelin and glycosylphosphatidylinositol (GPI)-anchored proteins, are preferentially partitioned into the detergent-resistant membrane (DRM) fraction. These are typically insoluble in cold, nonionic detergents. Raft markers are also partitioned into liquid-ordered phase-like (Lo-like) domains in GPMVs, which contain the full complement of proteins and lipids found in native PMs, except for the actin-based membrane skeleton (Baumgart et al. 2007; Levental et al. 2011).

(a) Fluorescent GM2 analogues

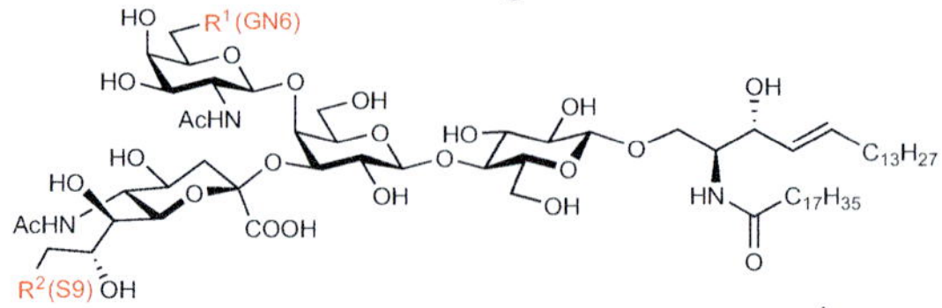

13 (594-S9-GM2) R^1= OH, R^2= NHC(=O)-ATTO594
14 (594-GN6-GM2) R^1 = NHC(=O)-ATTO594, R^2= OH

(b) Fluorescent GD1b analog

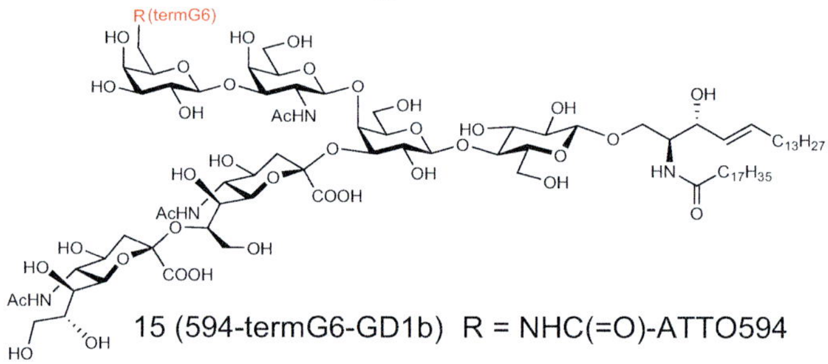

15 (594-termG6-GD1b) R = NHC(=O)-ATTO594

Fig. 3.3 Chemical structures of the fluorescent GM2 and GD1b probes we developed. (Komura et al. 2016)

3.4.1 Detergent Insolubility of Ganglioside Probes

Cold detergent insolubility of fluorescent ganglioside probes developed by us, in addition to commercially available analogues, was examined in human T24 cells. The cell PMs in which fluorescent ganglioside analogues and other fluorescent lipids were incorporated were treated with cold Triton X-100 and observed by epifluorescence microscopy (Kenworthy et al. 2004). S9 (GM3, GM2, and GM1), TermG6-GM1, GN6-GM2, and TermG6-GD1b analogues mainly partitioned into the DRMs (Fig. 3.4a and c). TMR-G6-GM3 and commercially available BodipyFL-C5-GM1 (Taylor et al. 2005; Lenne et al. 2006; Jones et al. 2010), in which a fatty acid chain is conjugated with BodipyFL and is widely used as a raft marker, solubilized almost completely in cold 1% Triton X-100, as did non-raft markers (BodipyFL-C5-PC, 594-DOPE [dioleoylphosphatidylcholine], and 647 N-DOPE) (Komura et al. 2016). In TMR-G6-GM3, the moderately hydrophobic dye TMR is attached to position C6 of galactose, while in BodipyFL-C5-PC, the hydrophobic dye BodipyFL is connected to the sn-2-acyl chain via a C5 alkyl chain of PC. For both of these markers, the hydrophobic dyes used may perturb their incorporation into DRMs. Analysis of DRM partitioning indicated that modification with hydrophilic fluorophores at the S9 position of GM3, including ATTO594, ATTO488, and Fl, retained a preference for DRM partitioning. These results are also consistent with those of a previous report that hydrophobic dyes introduce artifacts into

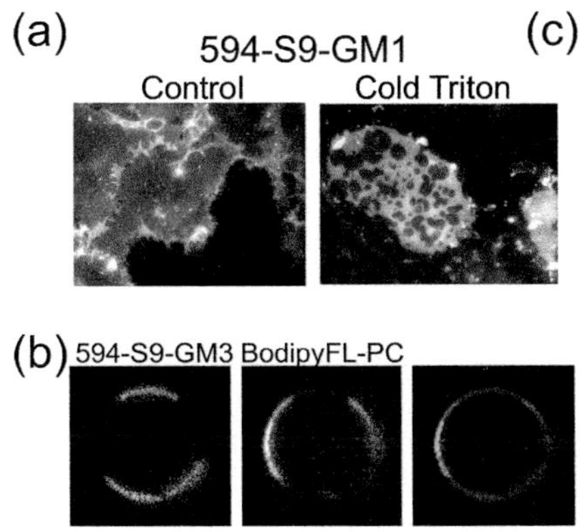

(c)

Fluorescent ganglioside	DRM vs. Intact Ratio (mean ± SE)	Lo/Ld Ratio in GPMV (mean ± SE)	CD59-patch Colocalization
Fl-S9-GM3 (**4**)	1.1 ± 0.05	8.8 ± 1.5	+
TMR-S9-GM3 (**3**)	1.0 ± 0.10	1.5 ± 0.08	+
594-S9-GM3 (**5**)	1.0 ± 0.05	4.0 ± 0.14	+
488-S9-GM3 (**7**)	1.1 ± 0.06	4.6 ± 0.58	+
647N-S9-GM3 (**6**)	0.066 ± 0.008	0.26 ± 0.02	-
TMR-G6-GM3 (**1**)	0.029 ± 0.006	1.7 ± 0.14	-
594-G6-GM3 (**2**)	0.49 ± 0.04	1.7 ± 0.09	±
Cy3-CTB	1.2 ± 0.08	6.7 ± 0.49	+
TMR-S9-GM1 (**8**)	1.0 ± 0.05	1.9 ± 0.25	+
594-S9-GM1 (**9**)	1.1 ± 0.08	3.3 ± 0.11	+
488-S9-GM1 (**10**)	1.1 ± 0.06	3.0 ± 0.23	+
594-termG6-GM1(**11**)	1.0 ± 0.05	4.7 ± 0.35	+
594-S9-GM2 (**13**)	1.0 ± 0.05	4.1 ± 0.48	+
594-GN6-GM2 (**14**)	1.1 ± 0.07	4.4 ± 0.43	+
594-termG6-GD1b (**15**)	1.0 ± 0.04	4.2 ± 0.40	+
BodipyFL-GM1 (**12**)	0.042 ± 0.009	0.62 ± 0.05	-

Fig. 3.4 Partitioning of fluorescent ganglioside probes. (**a**) Epi-fluorescence images of 594-S9-GM1 before and after treatment of T24 cells with cold-Triton X-100. (**b**) Fluorescence images of 594-S9-GM3 and BodipyFL-PC, indicating that 594-S9-GM3 was partitioned into the Lo-like phase while BodipyFL-PC was partitioned into the Ld-like phase in GPMVs. (**c**) Partitioning of fluorescent ganglioside probes into the DRM fraction (second column, far left), Lo-like phase (second column, far right) and CD59 patches (far right column). The numbers in parentheses in the far left column correspond to those in Figs. 3.1 and 3.2

single-molecule tracking experiments owing to non-specific binding (Zanetti-Domingues et al. 2013).

3.4.2 Analysis of Partitioning into Lo-Like and Ld-Like Domain in GPMVs

Partitioning of the fluorescent ganglioside probes into Lo-like and Ld-like domains in GPMVs was consistent with analyses of partitioning in DRMs (Fig. 3.4b and c). These results showed that ganglioside analogues modified with hydrophilic fluorophores at the terminal position of the glycan group interacted with raft domains, just like the native gangliosides.

A discrepancy between the partitioning patterns of several probes into either the DRM or Lo phase was also observed. For example, TMR-G6-GM3 was almost completely solubilized in cold Triton X-100 but had an average Lo/Ld ratio of 1.7. This may be due to the fact that these assays were performed with different cell types (T24 cells for DRM and RBL-2H3 cells for Lo/Ld) or because Triton X-100 promotes domain formation (Heerklotz 2002). Therefore, TMR-G6-GM3 may be segregated from these induced domains.

3.4.3 Recognition of Sugar-Chain of Gangliosides by the Binding Proteins

The dissociation constants (K_d) between ganglioside analogues and sugar-binding proteins were measured by surface plasmon resonance (SPR) analysis. The binding affinities of CTB to fluorescent GM1s, and those of wheat germ agglutinin (WGA) to fluorescent GM3s incorporated into 1-palmitoyl-2-oleoyl-sn-glycerophosphorylcholine (POPC) vesicles (10 Mol%), were determined by SPR equilibrium analysis. GM1 and GM3 conjugated with ATTO594 at the S9 positions exhibited comparable affinity constants to native molecules for both of the sugar-binding proteins CTB and WGA. These results suggested that conjugation of ATTO594 at the S9 position of GM1 and GM3 hampered neither binding of CTB to GM1, nor binding of WGA to GM3 (Komura et al. 2016). On the other hand, it was reported that the affinity of CTB for a fluorescent GM1 analogue conjugated with Alexa568 at the glycerol side chain of the sialic acid (with the removal of two hydroxyl groups and one carbon atom) was reduced by a factor of approximately 10 (Chinnapen et al. 2012). This difference may be due to the location of the dye molecule within the GM1 molecule and/or the particular dye used.

In previous studies to investigate the dynamic behaviors and functions of gangliosides in cell PMs, CTB labeled with fluorescence (e.g., Cy3-CTB) has been widely used as a GM1-specific marker. However, GM1 molecules are cross-linked by CTB, which modifies the distribution of GM1 and their behavior in PMs. Single-molecule tracking showed that Cy3-CTB complexed with GM1 migrated at smaller diffusion coefficients (0.01–0.06 $\mu m^2/s$) than any of our fluorescent ganglioside probes (0.31–0.36 $\mu m^2/s$) or phospholipids (0.20–0.22 $\mu m^2/s$). This was due to the clustering of GM1 molecules after binding to the five CTB carbohydrate-binding pockets. These results explicitly indicate that CTB alters the dynamic behavior of GM1. In this context, fluorescent GM1 probes, in which a fluorescent dye was directly attached to the lipid via the fatty acid moiety, have been found to be unsuitable as raft markers (Schwarzmann et al. 2005; Marushchak et al. 2007; Eggeling et al. 2009; Polyakova et al. 2009).

We also found that the most frequently used marker, BodipyFL-C5-GM1 (Taylor et al. 2005; Lenne et al. 2006; Jones et al. 2010; Svistounov et al. 2012), is partitioned into neither the DRM nor Lo-like phase in GPMVs, although its diffusion coefficients were similar to those of our fluorescent gangliosides and phospholipids. Incorporation of a dye molecule into the lipid tail moiety may weaken lipid interactions with other molecules residing in the DRMs and Lo-like phase. As described above, a fluorescent GM1 analogue conjugated with Alexa568 at the glycerol side chain of the sialic acid (with the removal of two hydroxyl groups and one carbon atom) showed much lower affinity (~tenfold less) for CTB than the native molecules (Chinnapen et al. 2012). This indicates the importance of retaining the glycerol chain of sialic acid when gangliosides are modified. Likewise, another head group-labeled GM1 probe, in which the C5 amino position of sialic acid was conjugated to ATTO647N, was mostly partitioned into the Ld-like phase in GPMVs (Sezgin et al. 2012).

3.5 Behavior of Ganglioside Analogues in Cell PMs

3.5.1 Partitioning of Ganglioside Probes into GPI-Anchored Protein Clusters

Since ganglioside-binding proteins such as CTB and WGA induce clusters of gangliosides, the distribution and dynamic behaviors of gangliosides in the living cell PMs cannot be investigated using these proteins. Therefore, we used our new fluorescent ganglioside probes to study the partitioning of these analogues in lipid rafts. GPI-anchored proteins are representative raft markers and coalesced into large patches after cells were incubated with primary and secondary antibodies (Harder et al. 1998). We used a complement control factor, CD59, as a GPI-anchored protein marker of lipid rafts.

Our confocal fluorescence microscopy observations revealed that our new ganglioside probes partitioned into both the DRM and Lo-like domains and co-localized with large CD59 patches. Note that such co-localization has been observed in previous studies only when ganglioside-binding proteins such as CTB, WGA, and antibodies were employed to induce clusters. These results showed that the affinity of GM3 analogues for rafts was highest for 594-S9-GM3, 488-S9-GM3, and Fl-S9-GM3, and the affinities were reduced from TMR-S9-GM3 to TMR-G6-GM3. For GM1 analogues, 594-S9-GM1, 594-termG6-GM1, and 488-S9-GM1 were all equivalent and had higher affinity than TMR-S9-GM1. Meanwhile, 647 N-S9-GM3, TMR-G6-GM3, and BodipyFL-GM1 partitioned into the non-raft fraction, probably due to the detrimental effects of hydrophobic dyes on lipid interactions (Zanetti-Domingues et al. 2013) and should not be used as lipid raft markers. Based on these results, we synthesized and evaluated the GM2 and GD1b fluorescent probes, 594-S9-GM2, 594-GN6-GM2, and 594-termG6-GD1b. We found that these analogues also partitioned into large, induced lipid rafts.

3.5.2 Single-Molecule Imaging of Ganglioside Probes Inside and Outside CD59 Patches

Confocal fluorescence microscopy observation revealed that the ganglioside probes we developed were co-localized with rafts of CD59 patches in PMs. However, it is still unknown whether these probes dynamically entered and exited the raft domains. Our single-molecule tracking of 594-S9-GM1 and GM3 showed that both probes diffused in and outside the raft domains of CD59 patches. Each individual molecule of ganglioside probe was maintained within the boundaries of CD59 patches and the bulk regions, rather than passing through. On the other hand, a non-raft phospholipid probe, 594-DOPE, was located outside the CD59 patches for most of the time and repelled at the boundaries of the patches. These results indicated that our ganglioside probes were concentrated in the CD59 patches by a factor of 4–10; meanwhile, 594-DOPE was almost completely excluded. These results concur with the confocal fluorescence microscopy observations and show that the large CD59 patches form raft domains by temporarily recruiting other raft molecules such as gangliosides.

The 594-S9-GM3 and GM1 molecules in the large CD59 patches often underwent temporal confinement with lifetimes of ~100 ms, a phenomenon termed "Temporal Arrest of LateraL diffusion" or "TALL." Meanwhile, almost all the 594-S9-GM3 and GM1 molecules diffused continuously outside the CD59 patches when observed at 33-, 8-, and 0.5-ms resolutions (Komura et al. 2016). After actin depolymerization, the time fractions of TALL for 594-S9-GM3 and GM1 inside the CD59 patches were significantly reduced, suggesting that the induced large CD59 patches might interact with the actin-based membrane skeleton. Alternatively, they may induce actin depolymerization, and TALL occurs through indirect interactions between gangliosides and actin filaments. This may occur due to interaction between gangliosides and CD59 clusters that are immobilized on actin filaments by an unknown transmembrane protein (Suzuki 2012).

A recent study reported that the formation of monomeric GFP-GPI (mGFP-GPI) clusters 3 μm in diameter by using anti-GFP antibody-coated micropattern did not result in concentration of CD59, another GPI-anchored protein (Sevcsik et al. 2015). The density of mGFP-GPI in the micropattern changed between 500~1000 molecules/μm^2 and occupied 0.5~11% of the area of the spots. In our earlier study (Komura et al. 2016), CD59 was cross-linked by primary and secondary antibodies, which generate relatively small clusters less than 3 μm in diameter. These occupied some 7.7~20% of the area of the clusters if the average distance between CD59 molecules was 9.6 nm (Werner et al. 1972) or 6.0 nm (Hewat and Blaas 1996) and the radius of CD59 protein moiety was 1.5 nm (Sevcsik et al. 2015). Higher densities of CD59 were able to enrich the ganglioside probes we developed (Komura et al. 2016). This is also consistent with the large drop in the diffusion coefficients of the ganglioside probes when the nearest neighbor distance of steric obstacles was less than 10 nm (Sevcsik et al. 2015).

3.5.3 GPI-Anchored Protein Clusters Recruit Ganglioside Probes to Stabilize the Cluster Rafts

We investigated raft formation under more physiological conditions. Upon ligation, CD59 forms stable tetramers that recruit and activate intracellular signaling molecules such as phospholipase Cγ (PLCγ), Src family kinase (SFKs), and trimeric G proteins (Suzuki et al. 2007a, b, 2012, 2013). The recruitment of these signaling molecules eventually triggers intracellular IP_3-Ca^{2+} responses. CD59 tetramers form under certain physiological conditions, and we refer to them as "CD59-cluster signaling rafts." To investigate whether CD59 tetramers recruit other raft lipids and form stable rafts, we carried out two-color single-molecule imaging of fluorescent ganglioside probes (GM1 and GM3) and of the CD59 clusters that are formed upon ligation.

Both 594-S9-GM1 and GM3 probes were recruited to the CD59 cluster-signaling rafts and remained there for approximately 100 ms, on average (Fig. 3.4c). Meanwhile, a non-raft, unsaturated phospholipid probe, 594-DOPE, co-localized with CD59-cluster rafts for much shorter periods. After cholesterol depletion, or when the non-raft chimeric mutant protein CD59-TM was employed, the lifetimes of the co-localization with ganglioside probes were reduced to about a half. These results showed that the driving force of the co-localization of CD59 with gangliosides was not an interaction between the headgroup regions of CD59 and gangliosides but, rather, was raft-lipid interactions and the alkyl chains. This notion is also supported by a recent study involving the interaction of sphingomyelin and saturated phosphatidylcholine probes with CD59 clusters (Kinoshita et al. 2017).

3.5.4 *Stabilization of GPI-Anchored Protein Homodimers by Raft-Lipid Interactions*

CD59 forms transient homodimers with a lifetime of approximately 160 ms, on average (Suzuki et al. 2012; Suzuki et al. 2013). The lifetime of CD59 homodimers was decreased by approximately one half after cholesterol depletion or after replacing the GPI anchoring chain with a non-raft-associating transmembrane domain. These results suggested that CD59 homodimers are stabilized by raft-lipid interactions.

Single-molecule imaging of 594-S9-GM3 and GM1 indicated that both ganglioside probes are transiently recruited to CD59 homodimers for about 80 ms and to CD59 monomers for only 50 ms (Fig. 3.5a–c). Short-lived association between gangliosides and both CD59 monomers and homodimers may be due to interactions between glycans and/or the protein moieties of CD59 and/or lipid components. It is intriguing that the interaction periods of CD59-cluster signaling rafts and ganglioside (approximately 100 ms) are longer than those of CD59 homodimers (80 ms) and much greater than those of CD59 monomers (50 ms). These results are in good agreement with our previous results that CD59-cluster signaling rafts and CD59 homodimers are stabilized by raft-lipid interactions and that raft-lipid interactions may be mediated or strengthened by ganglioside recruitment.

It has been obscured whether clusters of raft-associated molecules can recruit other raft components, although when additional raft elements were cross-linked, both the initial and the subsequent raft molecules co-localized with each other

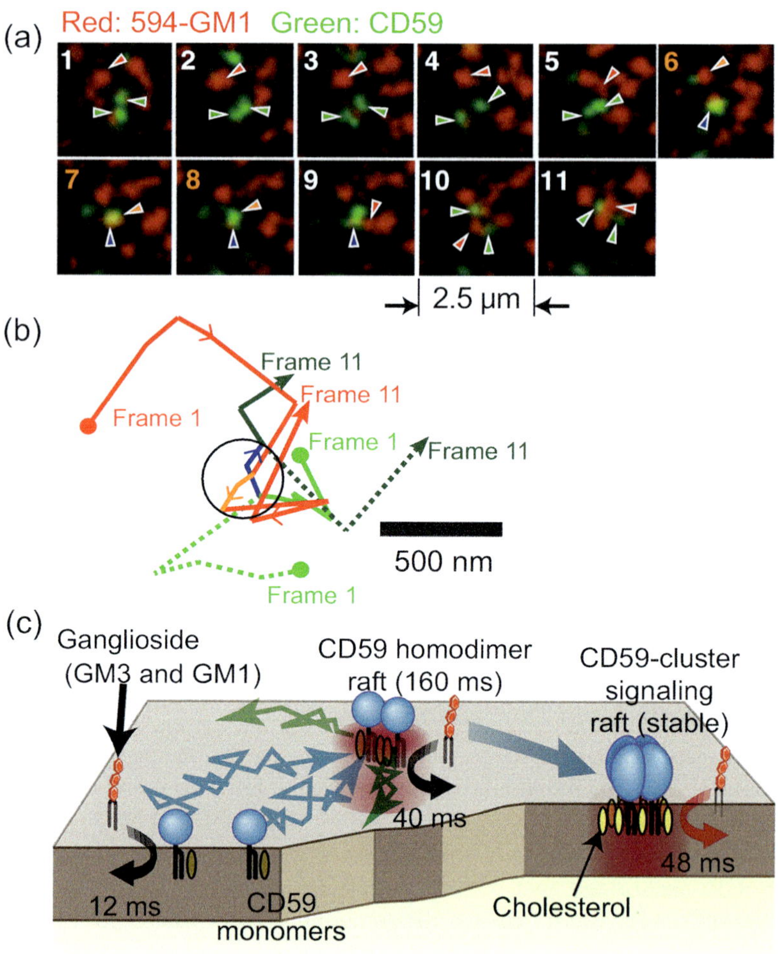

Fig. 3.5 Single-molecule imaging of 594-S9-GM1 and CD59. (**a**) Typical image sequence of a 594-S9-GM1 (red spots) and two CD59 molecules (green spots) in CHO-K1 cell PMs. CD59 formed homodimers between frames 6 and 9, and a 594-GM1 molecule was co-localized with the CD59 homodimer between frames 6 and 8. (**b**) Trajectories of the molecules shown in (**a**). (**c**) Schematic figure showing the transient interaction of gangliosides (GM1 and GM3) with CD59 monomers, transient homodimer rafts, and CD59 cluster-signaling rafts lasting for 12, 40, and 48 ms, respectively

(Harder et al. 1998; Kusumi and Suzuki 2005; Mayor et al. 1994; Tanaka et al. 2010). It has been unknown whether these secondary non-cross-linked raft elements can be recruited to clusters of raft-associated molecules. Simultaneous two-color single-molecule observations of CD59 and non-cross-linked ganglioside probes have shown that non-cross-linked gangliosides are transiently recruited to CD59 monomers, CD59 homodimers, CD59-cluster signaling rafts, and large CD59 patches. These results are in good agreement with those of another study showing that mGFP-GPI dimers induced by a monoclonal antibody transiently recruited other non-cross-linked GPI-anchored proteins (Huang et al. 2015).

3.5.5 Evaluation of the Temporal Confinement of Ganglioside Probes

It has been reported that ganglioside probes are temporarily confined in small domains in cell PMs for 10~20 ms, whereas control non-raft phospholipid probes scarcely exhibited TALL (Eggeling et al. 2009; Honigmann et al. 2014; Lenne et al. 2006; Sahl et al. 2010; Wenger et al. 2007). In these previous studies, the time fraction of temporal confinement of raft lipid probes was reduced after cholesterol depletion. However, as mentioned above, the ganglioside probes used in these earlier works partitioned into non-raft fractions, such as the non-DRMs and Ld phase (Sezgin et al. 2012; Komura et al. 2016). Therefore, we investigated whether our new ganglioside probes partitioned into the raft fraction or were confined to small domains in the cell PMs. To accomplish this, we carried out single-molecule imaging of 594-S9-GM1, 594-S9-GM3, and 594-DOPE at 0.5 ms intervals at 23°C in PtK2, CHO-K1, T24, and NRK cell types. The TALL of a variety of gangliosides was measured by previously developed methods (Sahl et al. 2010) that were further refined through our studies (Suzuki et al. 2012).

The time fractions in which diffusion was confined in small and immobile membrane domains for longer than 5 ms were rarely detected for all the gangliosides and non-raft phospholipids in four types of cells (T24, CHO-K1, NRK and COS7). The proportions of time spent in the 100-nm-diameter domains were estimated to be 1.7~3.1% for 594-S9-GM1 and GM3 and 4.6~5.2% for 594-DOPE in each of the four cell types. These results explicitly demonstrate that our new GM1 and GM3 probes are excellent raft markers and scarcely spent time trapped in small immobile membrane domains for longer than 5 ms. This result was also supported by a recent study that new sphingomyelin (SM) and distearoyl phosphatidylcholine (DSPC) probes, which are extremely good raft markers, also hardly exhibited TALL in small and immobile membrane domains (100 nm in diameter) when evaluated by high-speed observations at 0.5 ms resolution (Kinoshita et al. 2017), which was performed by using a specially designed CMOS sensor-based camera (Photron). However, these results are apparently inconsistent with some previous findings. (Eggeling et al. 2009; Honigmann et al. 2014; Lenne et al. 2006; Sahl et al. 2010; Wenger et al. 2007). For example, Sahl et al. (2010) reported that their SM probe (ATTO647N-SM) had a TALL with a lifetime of 17 ms and the time fraction the probe spent in small, immobile domains was approximately 70% (with an apparent trapped domain 12 nm in diameter) in the basal PMs of PtK2 cells. A non-raft unsaturated phospholipid probe (ATTO647N-DOPE) only exhibited TALL with a lifetime of 3 ms, and the time fraction spent in small, immobile domains was 30% (with an apparent trapped domain 22 nm in diameter). We employed the same method as Sahl et al. (2010) to detect TALL using various means but were only able to determine a short period of TALL (Komura et al. 2016; Kinoshita et al. 2017).

We also incorporated the lipid probes into cell PMs without bovine serum albumin (BSA) and observed single molecules in cell PMs with low confluency (less than 50%). This was only possible when the fluorophore of the probe was hydrophilic and negatively charged because ganglioside probes containing hydrophobic dyes strongly bind to glass surfaces, irrespective of the use or not of BSA (our unpublished result). In addition, the confluency of PtK2 cells used by Sahl et al. (2010) was much higher (cells were seeded at 80% confluency, and lipid probes were observed at 100% cell confluency). Since PtK2 cells form tight junctions at high confluences (Ando-Akatsuka et al. 1996), lipid probes added from cell apical sides accumulate in the apical membrane and are rarely observed in the basal membranes (our unpublished data). Since Sahl et al. (2010) observed diffusion of their ganglioside probes at the basal membranes, the lipid composition and membrane structures they observed may be different from those examined in our study. This may be one of the reasons our results are at odds with Sahl et al. (2010).

A large fraction of the trajectories of ganglioside probes was categorized as simple Brownian diffusion when observed at 0.5-ms resolution (Komura et al. 2016). However, when we observed diffusion of single molecules of lipids and transmembrane proteins at higher resolution such as 25 μs/frame, the trajectories were categorized to a different type of motion. When observed at 25 μs/frame, the molecules underwent temporary confinement in a boundary region of the membrane with an average diameter of 30–230 nm. This confinement was interrupted by infrequent "hops" (1–50 ms) to adjacent yet still temporary regions that we have termed compartments (Kusumi et al. 2005; Suzuki et al. 2005). We have termed this behavior "hop diffusion" as the membrane molecules undergo macroscopic diffusion by "hopping" to adjacent compartments, although they freely diffuse within a compartment. In future work, we will investigate the interactions between gangliosides, rafts, and membrane structures in more detail by observations at greater time resolution.

3.6 Conclusion

Ganglioside probes that partition into raft domains (DRM, Lo-like domains, and CD59 patches) and bind to proteins with similar affinity as the parental molecules were developed by an entirely chemical-based synthesis method after extensive efforts. The conjugation of hydrophilic fluorescent dyes at the appropriate positions within gangliosides is critical for retaining their properties. Hydrophobic dyes disturb the partitioning of the ganglioside probes into raft domains and induce nonspecific binding to the glass surface, as reported previously (Zanetti-Domingues et al. 2013). Single-molecule imaging revealed that the fluorescent ganglioside probes we developed were temporarily recruited to CD59 monomers, short-lived homodimeric CD59 cluster-signaling rafts, and large antibody-induced CD59 patches. This is the first direct evidence that gangliosides dynamically associate with a GPI-anchored protein in a cholesterol-dependent manner without cross-linking of gangliosides. Such findings would not have been possible without naturally behaving fluorescent ganglioside analogues and single-molecule imaging.

Gangliosides play important roles in various physiological functions, including the regulation of receptor activity. However, the specific mechanisms in these roles are still unknown in living cell PMs because gangliosides have never been directly observed in living cell PMs owing to a lack of functional analogues. The new fluorescent ganglioside analogues we developed will facilitate studies to elucidate such mechanisms.

Acknowledgment This work was supported in part by Grants-in-Aids for Innovative Areas grants to K.G.M.S. [2311001 and 16H01358] and H.A. [24110505 and 26110704], various scientific research grants from the Japan Society for the Promotion of Science (Kiban B awarded to K.G.N.S. [24370055 and 15H04351], H.A. [15H04495], and M.K. [22380067]; Wakate A awarded to H.A. [23688014]; and Kiban S awarded to A.K. [16H06386]), a grant from the Core Research for Evolutional Science and Technology (CREST) project "Creation of Fundamental Technologies for Understanding and Control of Biosystem Dynamics" funded by the Japan Science and Technology Agency (JST) awarded to A.K., a grant from the Naito Foundation and Daiichi-Sankyo Foundation awarded to K.G.N.S., and a grant from the Mizutani Foundation awarded to K.G.N.S. and H.A. The WPI-iCeMS of Kyoto University is supported by the World Premier International Research Center Initiative (WPI) of MEXT.

References

Ando H, Koike Y, Ishida H et al (2003) Extending the possibility of an *N*-Troc-protected sialic acid donor toward variant sialo-glycoside synthesis. Tetrahedron Lett 44(36):6883–6886

Ando H, Koike Y, Koizumi S et al (2005) 1,5-Lactamized sialyl acceptors for various disialoside syntheses: novel method for the synthesis of glycan portions of Hp-s6 and HLG-2 gangliosides. Angew Chem Int Ed 44(41):6759–6763

Ando-Akatsuka Y, Saitou M, Hirase T et al (1996) Interspecies diversity of the occludin sequence: cDNA cloning of human, mouse, dog, and rat-kangaroo homologues. J Cell Biol 133(1):43–47

Aureli M, Mauri L, Ciampa MG et al (2016) GM1 ganglioside: past studies and future potential. Mol Neurobiol 53(3):1824–1842

Axelrod D, Burghardt TP, Thompson NL (1984) Total internal reflection fluorescence. Annu Rev Biophys Bioeng 13:247–268

Baumgart T, Hammond AT, Sengupta P et al (2007) Large-scale fluid/fluid phase separation of proteins and lipids in giant plasma membrane vesicles. Proc Natl Acad Sci U S A 104(9):3165–3170

Berntsson RP, Peng L, Dong M et al (2013) Structure of dual receptor binding to botulinum neurotoxin B. Nat Commun 4:2058

Chinnapen DJ, Hsieh WT, te Welscher YM et al (2012) Lipid sorting by ceramide structure from plasma membrane to ER for the cholera toxin receptor ganglioside GM1. Dev Cell 23(3):573–586

Coskun U, Grzybek M, Drechsel D et al (2011) Regulation of human EGF receptor by lipids. Proc Natl Acad Sci U S A 108(22):9044–9048

Dalton G, An SW, Al-Juboori SI et al (2017) Soluble klotho binds monosialoganglioside to regulate membrane microdomains and growth factor signaling. Proc Natl Acad Sci U S A 114(4):752–757
Dam DH, Wang XQ, Sheu S et al (2017) Ganglioside GM3 mediates glucose-induced suppression of IGF-1 receptor-Rac1 activation to inhibit keratinocyte motility. J Invest Dermatol 137(2):440–448
Eggeling C, Ringemann C, Medda R et al (2009) Direct observation of the nanoscale dynamics of membrane lipids in a living cell. Nature 457(7233):1159–1162
Fleming FE, Bohm R, Dang VT et al (2014) Relative roles of GM1 ganglioside, N-acetylneuraminic acids, and α2β1 integrin in mediating rotavirus infection. J Virol 88(8):4558–4571
Fujikawa K, Nakashima S, Konishi M et al (2011) The first total synthesis of ganglioside GalNAc-GD1a, a target molecule for autoantibodies in Guillain-Barre syndrome. Chemistry 17(20):5641–5651
Fujita A, Cheng J, Hirakawa M et al (2007) Gangliosides GM1 and GM3 in the living cell membrane form clusters susceptible to cholesterol depletion and chilling. Mol Biol Cell 18(6):2112–2122
Fukuda Y, Fukui T, Hikichi C et al (2015) Neurotropin promotes NGF signaling through interaction of GM1 ganglioside with Trk neurotrophin receptor in PC12 cells. Brain Res 1596:13–21
Galimzyanov TR, Lyushnyak AS, Aleksandrova VV et al (2017) Line activity of ganglioside GM1 regulates the raft size distribution in a cholesterol-dependent manner. Langmuir 33(14):3517–3524
Hammond AT, Heberle FA, Baumgart T et al (2005) Crosslinking a lipid raft component triggers liquid ordered-liquid disordered phase separation in model plasma membranes. Proc Natl Acad Sci U S A 102(18):6320–6325
Harder T, Scheiffele P, Verkade P et al (1998) Lipid domain structure of the plasma membrane revealed by patching of membrane compartments. J Cell Biol 141(4):929–942
Heerklotz H (2002) Triton promotes domain formation in lipid raft mixtures. Biophys J 83(5):2693–2701
Hewat EA, Blaas D (1996) Structure of a neutralizing antibody bound bivalently to human rhinovirus 2. EMBO J 15(7):1515–1523
Honigmann A, Mueller V, Ta H et al (2014) Scanning STED-FCS reveals spatiotemporal heterogeneity of lipid interaction in the plasma membrane of living cells. Nat Commun 5:5412
Huang H, Simsek MF, Jin W et al (2015) Effect of receptor dimerization on membrane lipid raft structure continuously quantified on single cells by camera based fluorescence correlation spectroscopy. PLoS One 10:e0121777
Imamura A, Ando H, Ishida H et al (2009) Ganglioside GQ1b: efficient total synthesis and the expansion to synthetic derivatives to elucidate its biological roles. J Org Chem 74(8):3009–3023
Iwayama Y, Ando H, Ishida H et al (2009) A first total synthesis of ganglioside HLG-2. Chemistry 15(18):4637–4648
Jones R, Howes E, Dunne PD et al (2010) Tracking diffusion of GM1 gangliosides and zona pellucida binding molecules in sperm plasma membranes following cholesterol efflux. Dev Biol 339(2):398–406
Jung WR, Kim HG, Shin MK et al (2010) The effect of ganglioside GQ1b on the NMDA receptor signaling pathway in H19-7 cells and rat hippocampus. Neuroscience 165(1):159–167
Kaiser HJ, Lingwood D, Levental I et al (2009) Order of lipid phases in model and plasma membranes. Proc Natl Acad Sci U S A 106(39):16645–16650
Kenworthy AK, Nichols BJ, Remmert CL et al (2004) Dynamics of putative raft-associated proteins at the cell surface. J Cell Biol 165(5):735–746
Kinoshita M, Suzuki KGN, Matsumori N et al (2017) Raft-based sphingomyelin interactions revealed by new fluorescent sphingomyelin analogs. J Cell Biol 216(4):1183–1204
Komura N, Suzuki KGN, Ando H et al (2016) Raft-based interactions of gangliosides with a GPI-anchored receptor. Nat Chem Biol 12(6):402–410
Kusumi A, Suzuki K (2005) Toward understanding the dynamics of membrane-raft-based molecular interactions. Biochim Biophys Acta 1746(3):234–251

Kusumi A, Nakada C, Ritchie K et al (2005) Paradigm shift of the plasma membrane concept from the two-dimensional continuum fluid to the partitioned fluid: high-speed single-molecule tracking of membrane molecules. Annu Rev Biophys Biomol Struct 34:351–378

Lenne PF, Wawrezinieck L, Conchonaud F et al (2006) Dynamic molecular confinement in the plasma membrane by microdomains and the cytoskeleton meshwork. EMBO J 25(14):3245–3256

Levental I, Grzybek M, Simons K (2011) Raft domains of variable properties and compositions in plasma membrane vesicles. Proc Natl Acad Sci U S A 108(28):11411–11416

Lingwood D, Ries J, Schwille P et al (2008) Plasma membranes are poised for activation of raft phase coalescence at physiological temperature. Proc Natl Acad Sci U S A 105(29):10005–10010

Lingwood D, Binnington B, Rog T et al (2011) Cholesterol modulates glycolipid conformation and receptor activity. Nat Chem Biol 7(5):260–262

Marushchak D, Gretskaya N, Mikhalyov I et al (2007) Self-aggregation-an intrinsic property of G(M1) in lipid bilayers. Mol Membr Biol 24(2):102–112

Mayor S, Rothberg KG, Maxfield FR (1994) Sequestration of GPI-anchored proteins in caveolae triggered by cross-linking. Science 264(5167):1948–1951

Nakashima S, Ando H, Saito R et al (2012) Efficiently synthesizing lacto-ganglio-series gangliosides by using a glucosyl ceramide cassette approach: the total synthesis of ganglioside X2. Chem Asian J 7(5):1041–1051

Ohkawa Y, Miyazaki S, Hamamura K et al (2010) Ganglioside GD3 enhances adhesion signals and augments malignant properties of melanoma cells by recruiting integrins to glycolipid-enriched microdomains. J Biol Chem 285(35):27213–27223

Polyakova SM, Belov VN, Yan SF et al (2009) New GM1 ganglioside derivatives for selective single and double labelling of the natural glycosphingolipid skeleton. Eur J Org Chem 2009(30):5162–5177

Sahl SJ, Leutenegger M, Hilbert M et al (2010) Fast molecular tracking maps nanoscale dynamics of plasma membrane lipids. Proc Natl Acad Sci U S A 107(15):6829–6834

Schwarzmann G, Wendeler M, Sandhoff K (2005) Synthesis of novel NBD-GM1 and NBD-GM2 for the transfer activity of GM2-activator protein by a FRET-based assay system. Glycobiology 15(12):1302–1311

Sevcsik E, Brameshuber M, Folser M et al (2015) GPI-anchored proteins do not reside in ordered domains in the live cell plasma membrane. Nat Commun 6:6969

Sezgin E, Levental I, Grzybek M et al (2012) Partitioning, diffusion, and ligand binding of raft lipid analogs in model and cellular plasma membranes. Biochim Biophys Acta 1818(7):1777–1783

Smith DC, Lord JM, Roberts LM et al (2004) Glycosphingolipids as toxin receptors. Semin Cell Dev Biol 15(4):397–408

Stefl M, Sachl R, Humpolickova J et al (2012) Dynamics and size of cross-linking-induced lipid nanodomains in model membranes. Biophys J 102(9):2104–2113

Suzuki KGN (2012) Lipid rafts generate digital-like signal transduction in cell plasma membranes. Biotechnol J 7(6):753–761

Suzuki K, Ritchie K, Kajikawa E et al (2005) Rapid hop diffusion of a G-protein-coupled receptor in the plasma membrane as revealed by single-molecule techniques. Biophys J 88(5):3659–3680

Suzuki KGN, Fujiwara TK, Edidin M et al (2007a) Dynamic recruitment of phospholipase C gamma at transiently immobilized GPI-anchored receptor clusters induces IP3-Ca2+ signaling: single-molecule tracking study 2. J Cell Biol 177(4):731–742

Suzuki KGN, Fujiwara TK, Sanematsu F et al (2007b) GPI-anchored receptor clusters transiently recruit Lyn and G alpha for temporary cluster immobilization and Lyn activation: single-molecule tracking study 1. J Cell Biol 177(4):717–730

Suzuki KGN, Kasai RS, Hirosawa KM et al (2012) Transient GPI-anchored protein homodimers are units for raft organization and function. Nat Chem Biol 8(9):774–783

Suzuki KGN, Kasai RS, Fujiwara TK et al (2013) Single-molecule imaging of receptor-receptor interactions. Methods Cell Biol 117:373–390

Svistounov D, Warren A, McNerney GP et al (2012) The relationship between fenestrations, sieve plates and rafts in liver sinusoidal endothelial cells. PLoS One 7:e46134
Tamai H, Ando H, Tanaka HN et al (2011) The total synthesis of the neurogenic ganglioside LLG-3 isolated from the starfish Linckia laevigata. Angew Chem Int Ed Engl 50(10):2330–2333
Tanaka KAK, Suzuki KGN, Shirai YM et al (2010) Membrane molecules mobile even after chemical fixation. Nat Methods 7(11):865–866
Taylor DR, Watt NT, Perera WS et al (2005) Assigning functions to distinct regions of the N-terminus of the prion protein that are involved in its copper-stimulated, clathrin-dependent endocytosis. J Cell Sci 118(21):5141–5153
Tokunaga M, Kitamura K, Saito K et al (1997) Single molecule imaging of fluorophores and enzymatic reactions achieved by objective-type total internal reflection fluorescence microscopy. Biochim Biophys Res Commun 235:47–53
Wenger J, Conchonaud F, Dintinger J et al (2007) Diffusion analysis within single nanometric apertures reveals the ultrafine cell membrane organization. Biophys J 92(3):913–919
Werner TC, Bunting JR, Cathou RE et al (1972) The shape of immunoglobulin G molecules in solution. Proc Natl Acad Sci U S A 69(4):795–799
Yoon SJ, Nakayama K, Hikita T et al (2006) Epidermal growth factor tyrosine kinase is modulated by GM3 interaction with N-linked GlcNAc termini of the receptor. Proc Natl Acad Sci U S A 103(50):18987–18991
Zanetti-Domingues LC, Tynan CJ, Rolfe DJ et al (2013) Hydrophobic fluorescent probes introduce artifacts into single molecule tracking experiments due to non-specific binding. PLoS One 8:e74200

Chapter 4
MALDI Mass Spectrometry Imaging of N-Linked Glycans in Tissues

Richard R. Drake, Connor A. West, Anand S. Mehta, and Peggi M. Angel

Abstract Matrix-assisted laser desorption ionization time-of-flight mass spectrometry (MALDI-TOF MS) has been used for two decades to profile the glycan constituents of biological samples. An adaptation of the method to tissues, MALDI mass spectrometry imaging (MALDI-MSI), allows high-throughput spatial profiling of hundreds to thousands of molecules within a single thin tissue section. The ability to profile N-glycans within tissues using MALDI-MSI is a recently developed method that allows identification and localization of 40 or more N-glycans. The key component is to apply a molecular coating of peptide-N-glycosidase to tissues, an enzyme that releases N-glycans from their protein carrier. In this chapter, the methods and approaches to robustly and reproducibly generate two-dimensional N-glycan tissue maps by MALDI-MSI workflows are summarized. Current strengths and limitations of the approach are discussed, as well as potential future applications of the method.

Keywords Formalin-fixed paraffin-embedded tissue · Fucosylation · Glycomics · Glycoprotein · MALDI imaging mass spectrometry · N-linked glycosylation · Sialylation

Abbreviations

FFPE	Formalin-fixed paraffin-embedded
FT-ICR	Fourier transform ion cyclotron resonance
Fuc	Fucose
GlcNAc	N-acetylglucosamine
GnT-III	N-acetylglucosaminyltransferase 3
GnT-IV	N-acetylglucosaminyltransferase 4
GnT-V	N-acetylglucosaminyltransferase 5

R. R. Drake (✉) · C. A. West · A. S. Mehta · P. M. Angel
Department of Cell and Molecular Pharmacology and Experimental Therapeutics, Medical University of South Carolina, Charleston, SC, USA
e-mail: draker@musc.edu

Y. Yamaguchi, K. Kato (eds.), *Glycobiophysics*, Advances in Experimental Medicine and Biology 1104, https://doi.org/10.1007/978-981-13-2158-0_4

Hex	Hexose (e.g., mannose, glucose, galactose)
HexNAc	N-acetylhexosamine (e.g., N-acetylglucosamine or N-acetylgalactosamine)
LacNAc	Lactosamine (galactose and N-acetylglucosamine disaccharide)
MALDI-TOF MS	Matrix-assisted laser desorption ionization time-of-flight mass spectrometry
Man	Mannose
MS	Mass Spectrometry
MSI	Mass spectrometry imaging
NeuAc	N-acetylneuraminic acid (sialic acid)
NeuGc	N-glycolylneuraminic acid
PNGaseF	Peptide-N-glycosidase F
TFA	trifluoroacetic acid
TMA	Tissue microarray

4.1 Introduction

MALDI-MSI methods have been routinely utilized for spatial profiling and identification of proteins, lipids, and small metabolites directly in tissue (Angel and Caprioli 2013; Baker et al. 2017; Berin and Boughton 2017; Cornett et al. 2007). In 2013, our group was the first to report the use of MALDI-MSI to profile N-linked glycans in frozen tissues (Powers et al. 2013), followed by application of the method to formalin-fixed paraffin-embedded (FFPE) tissue blocks and tissue microarrays (TMA) (Powers et al. 2014; Drake et al. 2017). The key to the approach is the spraying of a molecular layer of recombinant peptide N-glycosidase F enzyme (PNGaseF) on the tissue of interest. After digestion and spraying of chemical matrix, the released N-glycans are detected by MALDI-FT-ICR MS or MALDI-TOF MS (Drake et al. 2017). Different imaging software options are used to visualize the two-dimensional distributions of individual N-glycans to specific histopathology regions and features within normal and diseased tissues. In addition to studies from our laboratories in cancer tissues (Powers et al. 2014, 2015; Drake et al. 2017) and cardiac tissues (Angel et al. 2017a), other groups have successfully used the approach to profile N-glycans in cancer (Everest-Dass et al. 2016) and noncancer tissues (Briggs et al. 2016; Gustafsson et al. 2015).

As the method has matured, several advantages of the approach are emerging, as well as continued challenges. The ability to use FFPE tissues and TMA obtained directly from pathology services and/or archives in biorepositories allows for analysis of unlimited sample options, especially for most types of cancers. The requirement for PNGaseF to release the N-glycans results in signal detection that is dependent on enzyme activity, making selection of N-glycans in the spectra straightforward. No purification and enrichment of the glycans are required prior to analysis. For most tissue samples analyzed thus far, 30 or more glycans are detected using standard MALDI-TOF instruments (Heijs et al. 2016; Holst et al. 2016) and 45 or more per sample when using a MALDI-FT-ICR instrument (Drake et al. 2017). This in turn facilitates the generation of different glycan panels associated with specific histopathology features and tissue subregions useful for differentiation of disease from normal regions. There are still limitations associated with MALDI analysis of N-glycans, such as loss of sialic acid residues and differentiation of isomeric structures. Some progress has been made in addressing these limitations as applied to MALDI-MSI of tissues, and these will be discussed in this chapter along with example data. A summary of the optimized techniques used for MALDI-MSI of N-glycans will be provided with example data, along with protocol and troubleshooting notes for new users. Lastly, a summary of new instrumentation options and potential for developing new clinical diagnostics using the method will be described.

4.2 Characterization of Tissue N-Glycans by MALDI-FT-ICR MS and MALDI-TOF MS

The MALDI-MSI workflows build on decades of research using MALDI-TOF MS as a characterization tool for N-glycans (Harvey 1999, 2015; Nishikaze 2017). An experimental flowchart of the basic steps in the MALDI-MSI analysis of N-glycans is shown in Fig. 4.1, and specific experimental details can be found in published studies (Angel et al. 2017b; Briggs et al. 2017; Holst et al. 2016). Relative to electrospray ionization (ESI) MS methods, the advantages of using MALDI-TOF for N-glycan analysis include ease of use and data interpretation, rapidity and throughput, low cost per assay, and accessibility to instrumentation (Nishikaze 2017). For all MALDI analyses, one of the key considerations is chemical matrix type, which influences ionization properties of the N-glycans. For N-glycans, 2,5-dihydroxybenzoic acid (DHB) and α-cyano-4-hydroxycinnamic acid (CHCA) are the most commonly used in positive ion mode, with CHCA being used primarily by our group (Powers et al. 2014).

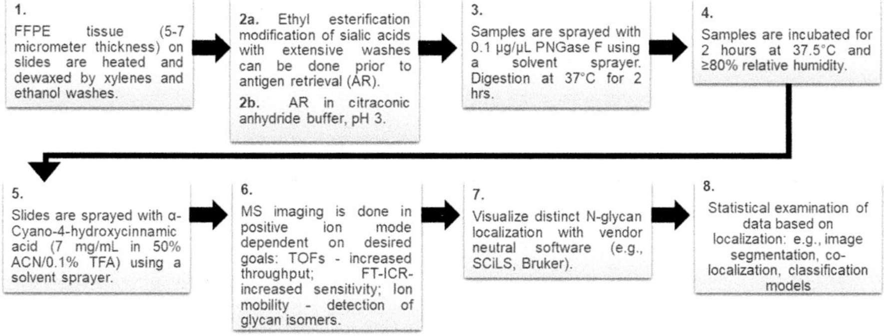

Fig. 4.1 A summary flowchart for the preparation and PNGaseF digestion of FFPE tissues

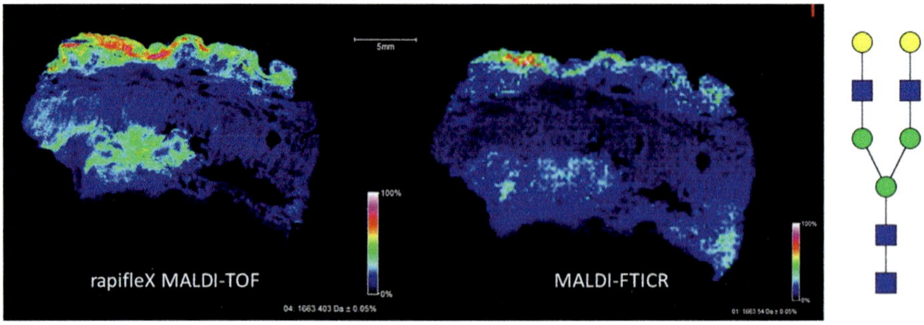

Fig. 4.2 High-speed MALDI-TOF (rapifleX, Bruker) versus MALDI-FT-ICR (solariX Bruker) IMS for N-linked glycans for serial esophageal cancer tissue. RapifleX IMS data (left panel) was acquired at a 50 μm spatial resolution in 2.5 h and SolariX IMS data (right panel) at a 200 μm spatial resolution in 4.1 h. For glycans, red triangle, fucose; blue square, N-acetylglucosamine; green circles, mannose; yellow circles, galactose; and purple diamonds, sialic acid

Instrumentation choice is also another key factor, as there are two main groups of instruments for MALDI-IMS: time of flight (TOF) and Fourier transform ion cyclotron resonance (FT-ICR). MALDI-TOF instruments provide high throughput with lower mass resolution, while MALDI-FT-ICR instruments provide a higher mass resolution with less throughput capabilities. The mass resolution capabilities of MALDI-FT-ICR instruments are critical to distinguish the many isotopic peaks of the N-glycans, with on-tissue resolving power of 85,000–160,000, with high sensitivity detection of below 200 femtomoles on-tissue. Recently, advances in instrumentation have resulted in commercially available high-speed MALDI-TOF instrumentation for high-throughput image acquisition from manufacturers like Bruker Daltonics, Shimadzu, and SimulTOF. These high-speed MALDI-TOF can perform MS tissue imaging with a 50x increase in speed over the MALDI-FT-ICR MS without oversampling (Ogrinc Potočnik et al. 2015). As an example, a comparison of one of these new high-speed instruments, the rapifleX MALDI-TOF and MALDI-FT-ICR, is shown in Fig. 4.2 for N-glycan imaging analysis of the same esophageal cancer tissue for a complex bi-antennary N-glycan. High-speed MS imaging should allow for faster 3D spatial rendering of a tissue, as well as potential clinical laboratory applications.

4.2.1 Protocol Notes

- Antigen retrieval of FFPE samples prior to PNGaseF was adapted directly from routine tissue slide protocols used in immunohistochemistry analyses. We have found that a low-cost vegetable steamer available at many retail stores works well for the heating steps during antigen retrieval (Angel et al. 2017b).
- For PNGaseF enzyme, be sure that there are minimal additives in the buffer (e.g., stabilizers, detergent) and low salt to minimize potential ion suppression effects.

– A key to MALDI-MSI is the spraying of a thin layer of PNGaseF for digestion and a thin layer of matrix. The automated solvent sprayers typified by those available from HTX Imaging and SunChrom provide reproducibility from sample to sample, as well as optimization of signal detection parameters.

4.3 Structural Classes of Tissue N-Glycans Detected by MALDI-MSI

There are many diverse types and structural classes of N-glycans that are readily detected from tissues by MALDI-MSI. Evaluation of structures is done based on motif and context of the N-linked glycan biosynthesis and processing pathways. A summary of the major components of these pathways is shown in Fig. 4.3. The diversity of the many possible N-glycan structures that are attached to glycoproteins is the result of a series of sequential glycosidase and glycosyltransferase reactions (reviewed in Rini et al. 2009; Stanley et al. 2009). After transfer of a lipid-linked precursor oligosaccharide co-translationally to a glycoprotein carrier, the glycan is trimmed sequentially in the ER and cis-Golgi by glucosidases and mannosidases to Man5GlcNAc2, which serves as the precursor to complex, branched, and hybrid glycan structures (see Fig. 4.3). Glycans that are not processed or incompletely

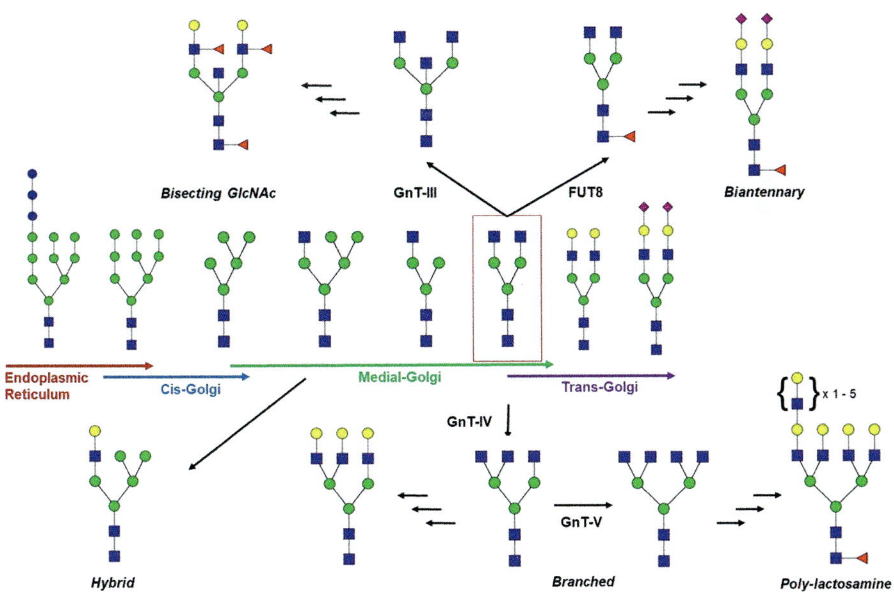

Fig. 4.3 Summary of biosynthesis of N-linked glycans. For glycans, red triangle, fucose; blue square, N-acetylglucosamine; green circles, mannose; yellow circles, galactose; purple diamonds, sialic acid

processed by the mannosidases are termed high-mannose glycans, containing 5–9 mannose residues (Man5–9GlcNAc2). To generate complex glycans, Man5GlcNAc2 is processed by a series of N-acetylglucosaminyltransferases to generate an initial bi-antennary complex glycan. Fucosyltransferase 8 (FUT8) catalyzes the addition of α1,6-linked fucose residues to the first GlcNAc residue bonded with asparagine on the glycoprotein (termed core fucosylation). Triantennary complex glycans are generated from the activity of N-acetylglucosaminyltransferase IV (GnT-IV), which adds a β1,4 GlcNAc branch, and tetra-antennary complex glycans are generated in turn from the activity of N-acetylglucosaminyltransferase V (GnT-V), which adds a β1,6 GlcNAc branch. N-acetylglucosaminyltransferase III (GnT-III) is an additional enzyme that transfers a bisecting GlcNAc residue onto the complex glycan (Fig. 4.3). In the trans-Golgi, further glycans are added to bi-, tri-, or tetra-antennary structures by the activity of different galactosyltransferases, fucosyltransferases (Schneider et al. 2017), and sialyltransferases (Pearce and Läubli 2016). Another common type of complex structure that is detected in many tissues is polylactosamine (LacNAc) glycans (Bern et al. 2013; Kinoshita et al. 2014) (Fig. 4.3). In unpublished studies, our group has detected by MALDI-FT-ICR structures containing as many as 5 LacNAc disaccharides on a tetra-antennary structure.

Improving and verifying detection of these many types of glycan structures has required much optimization. Thus far, structures containing mannose, galactose, glucose, N-acetylglucosamine, fucose, and sialic acids have been readily detected in mammalian tissues. In general, the base structures in regard to mannose, GlcNAc, and galactose content are easiest to detect in tissues. These same structures with a core fucose are also readily detectable, which was confirmed by on-tissue CID using the MALDI-FT-ICR MS (Powers et al. 2014). Other possible modifications that include sulfation, phosphorylation, and N-acetylgalactosamine can be detected but still require further optimization strategies to facilitate their routine detection. Xylose and other sugar modifications associated with N-glycans present in plants and other nonmammalian organisms remain an under-evaluated area of study in regard to MALDI-MSI approaches. Detection of branched N-glycans with multiple fucose and sialic acid constituents presents their own challenges in regard to stability, identification, and isomeric structures. In the next sections, observations and optimization strategies for different classes of N-glycans are presented.

4.3.1 Protocol Notes

- Using the preparation protocols noted for both frozen and FFPE tissues followed by MALDI-TOF MS, the glycans are detected as sodiated species, so an additional 22 m.u. (i.e., M + H/M + Na) for MALDI is included and reported for each total glycan mass. Additional sodium ions were detected with each sialic acid present. This shift in mass associated with sodium atoms can be used to distinguish single sialylated glycans (291 m.u. plus Na) from those with two fucose residues (292 m.u., no extra Na).

- The MALDI-FT-ICR configuration is most advantageous for identifying as many N-glycans present in a given tissue sample. The high mass accuracy, sensitivity, and isotope patterns that can be detected for each N-glycan are critical features provided for each N-glycan imaging analysis. This resolution and sensitivity comes at the cost of analysis time and data file sizes.
- Another advantage of using the MALDI-FT-ICR instrument is the ability to perform collision-induced dissociation (CID) on glycans directly from tissue. Ions of interest can be specifically isolated in a quadrupole followed by subsequent CID. The CID energies used primarily release constituent glycans intact, and not with cross-ring fragmentation (Powers et al. 2014). Released glycan masses represent hexoses (162.053 m.u.), HexNAc (203.079 m.u.), sialic acid (291.095 m.u.), fucose (146.058 m.u.), fucose-HexNAc (349.137 m.u.), and Hex-HexNAc (365.132 m.u.).
- It is important to confirm the N-glycan structures determined by MALDI-MS imaging using other carbohydrate analysis tools like lectins or glycosidase digestions with HPLC (Powers et al. 2015). Correlative LC-MS identification methods for N-glycans from adjacent tissue slices are helpful in further defining structural features (Briggs et al. 2017; Holst et al. 2016).

4.4 N-Linked Glycans Associated with Cancer and Cell Proliferation

MALDI-MSI applied to FFPE tumor tissue slides is very effective for identifying N-glycans that are localized to regions of tumor (Drake et al. 2017). Example MS images of the most abundant type of N-glycans detected in prostate cancer tissues are shown in Fig. 4.4. The first two structures in this figure represent high-mannose glycans. While N-glycans with high-mannose structures (GlcNAc2-Man5–9) represent an early stage in normal biosynthetic processing, as in Fig. 4.2, these high-mannose glycans are frequently detected in high abundance in most all solid tumor tissues analyzed by MALDI-MS imaging (Drake et al. 2017; Powers et al. 2014; Everest-Dass et al. 2016; Holst et al. 2016; Heijs et al. 2016). For normal biosynthesis, the high-mannose structures are degraded by mannosidases to only three mannose residues and then further processed to complex N-glycans. In tumor cells, it is apparent that high-mannose structures are bypassing normal ER and Golgi glycan processing. In sera from prostate cancer patients, autoantibodies to high-mannose N-glycans have been reported (Wang 2012; Wang et al. 2013), and specific C-type lectin receptors, especially in liver and on macrophages, are known to bind to them. Also in tumor tissues, glycan structures with three or less mannose residues, with and without core fucose, are termed pauci-mannose glycans, which are also readily detected on secreted glycoproteins and on the cell surface (Loke et al. 2016; Nyalwidhe et al. 2013). Defining the functional roles of high-mannose and

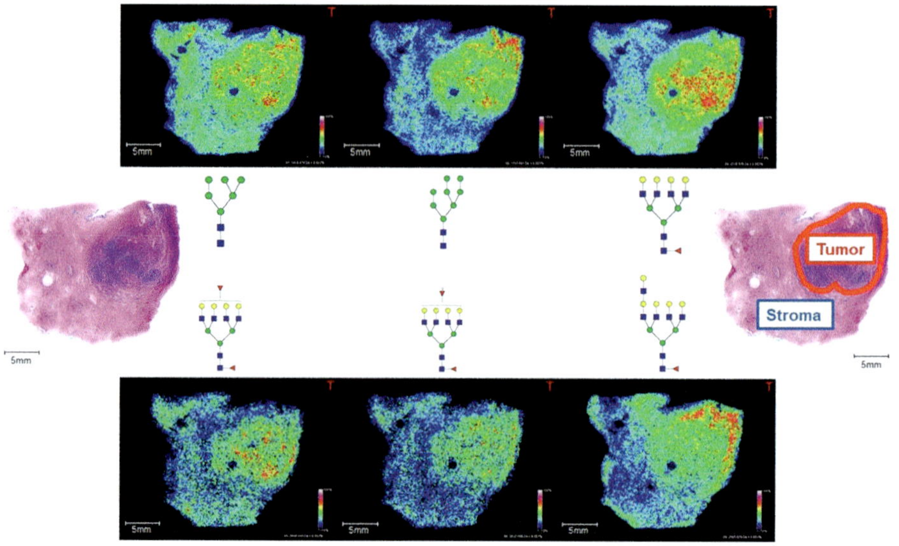

Fig. 4.4 Common tumor N-linked glycans. Shown are representative N-glycans associated with prostate tumors, including high-mannose (Man6 and Man8) and tetra-antennary N-glycans with zero, one, and two outer arm fucoses and/or a LacNAc

pauci-mannose N-glycans and their glycoprotein carriers is at the nexus of understanding basic tumor biology, glycoprotein processing, and tumor immunity.

The other main structural classes of tumor N-glycans detected are the representative tetra-antennary structures shown in Fig. 4.4. These tetrasaccharides are the result of GnT-V activity, and the increase in β1–6 GlcNAc branching of N-linked glycans is associated with the metastatic phenotype of multiple cancer types (Miwa et al. 2012; Pinho and Reis 2015; Schultz et al. 2012; Taniguchi and Kizuka 2015). Structurally, increased branching of glycans in cancers is typified by increased detection of sialyl Lewis X and sialyl Lewis A antigens, as well increases in polylactosamine modifications. These structures in turn are recognized by selectins and other carbohydrate lectins expressed on different tissues involved with immune cell binding and innate immunity and, in the case of extravasation, binding to cells in distant organ/tissue sites.

4.4.1 Protocol Notes

- It is critical to have a pathologist involved in studies to confirm the location and presence of tumor in a given sample. It is the ability to directly overlay the N-glycan distributions directly by MALDI that provides the location of glycans associated with tumor and non-tumor regions.

4.5 N-Linked Glycans Most Commonly Associated with Tissue Stroma Regions

The most abundant and easily detectable glycans by MALDI-MSI are those represented in the central portion of the biosynthetic pathway shown in Fig. 4.3, representing the spectrum from high-mannose to sialylated bi-antennary structures, with and without the core fucose. This statement is based on peak intensity comparisons across hundreds of tissues analyzed by MALDI-MSI. Core-fucosylated bi-antennary glycans are frequently the most abundant N-glycans detected in areas of tissue containing smooth muscle and collagen, while bi-antennary glycans lacking core fucose are found in the adjacent stroma areas (an example is shown in Fig. 4.5). In a given tissue, these glycans can be detected in tumor regions, but not with the frequency observed for the high-mannose and branched chain glycans. Because their molecular masses range from 1100 to 2100 daltons, these stroma glycans are easily detectable by MALDI-TOF, with Hex5HexNAc4, Hex4HexNAc4Fuc1, Hex5HexNAc4Fuc1, Hex5HexNAc4NeuAc1, and Hex5HexNAc4Fuc1NeuAc1 glycans being the most abundant detected from tissues.

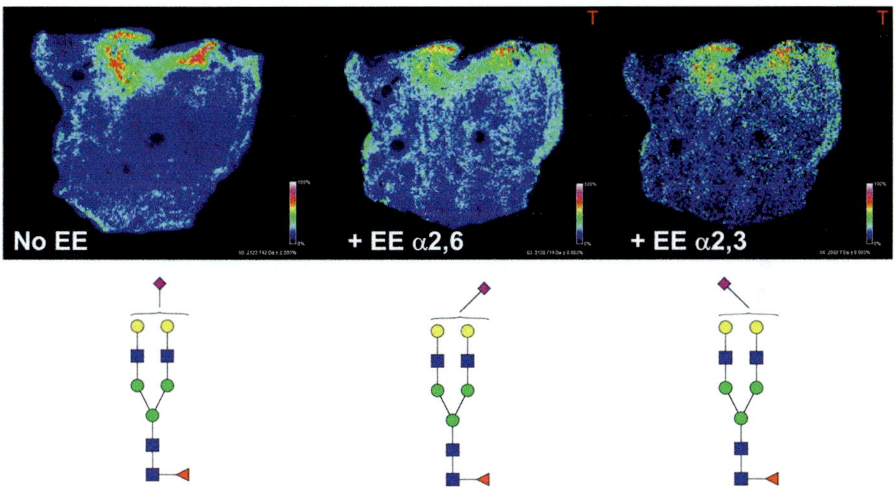

Fig. 4.5 Stabilization of sialic acids by in-tissue ethyl esterification. Ethyl esterification (EE) was adapted from a published protocol (Holst et al. 2016) for a prostate cancer FFPE tissue and analyzed by MALDI-FT-ICR IMS. Shown are the examples for a mono-sialylated core fucose bi-antennary glycan, without EE (left panel) or with EE for α2,6 (+28 amu) (middle panel) or α2,3 (+18 amu) (right panel) linkages

4.6 Stabilizing Terminal Sialic Acids for MALDI Analysis

N-glycans containing a terminal sialic residue in either α2,3 or α2,6 linkages are common tissue components and readily detectable by MALDI-MSI. A well-described complication in the detection of sialylated glycans by MALDI is their well-known lability due to in-source decay during the ionization process, especially in positive ion mode. The negative charge also leads to different Na salt adducts and lowered sensitivity (Reiding et al. 2014). When native glycans are analyzed by MALDI-TOF MS, modifications like permethylation or amidation that stabilize and help ionize sialic acid containing glycans are routinely used (Bern et al. 2013; Sekiya et al. 2005). For free glycans derived from fluids or cells, these common chemical modification methods require organic solvents and the need to separate reaction products, thus they are not feasible for on-tissue derivatization or maintenance of spatial distributions. An ethyl esterification approach that differentially modifies α2,3- or α2,6-linked sialic acids (Reiding et al. 2014) was adapted to MALDI-MSI for N-glycans in FFPE tissues, allowing distinct tissue distributions of sialylated glycans with both isomers to be determined (Holst et al. 2016). Relative to the unmodified parent sialylated glycans, an ethylation reaction results in an increase in 28 amu when an α-2,6 linkage is involved, and when an α-2,3 linkage is involved, a lactonization reaction will occur with a decrease of 18 amu. These mass shifts are easily detectable by MALDI-TOF MS and MALDI-FT-ICR MS.

Example MALDI-FT-ICR MSI N-glycan data for an ethyl esterified prostate cancer FFPE tissue using an adapted protocol from Holst et al. is shown in Figs. 4.5 and 4.6. In Fig. 4.5, images for a core-fucosylated bi-antennary glycan with one sialic acid (Hex5HexNAc4Fuc1NeuAc1) are shown without modification and with both α2,6 and α2,3 linkages. In particular for MALDI-MSI of tissues, detection of glycans with more than one sialic acid has proven difficult. This is effectively illustrated in the non-ethyl esterified image examples shown in Fig. 4.6 for three glycans: a Hex5HexNAc4NeuAc1, a Hex5HexNAc4NeuAc2, and a Hex6HexNAc5NeuAc3. The mono-sialylated glycan is readily detected by MALDI-FT-ICR MSI (top panel), as also evident in Fig. 4.5. However, without stabilization, it is clear that detection of the di-sialylated is decreased relative to the mono-sialylated glycan and the tri-sialylated glycan is undetectable. Ethyl esterification stabilization greatly facilitates detection of these multi-sialylated glycans (Fig. 4.6).

4.6.1 Protocol Notes

- The MALDI-FT-ICR instruments incorporate a cooling nitrogen gas stream within the source that can decrease sialic acid cleavage (O'Connor and Costello 2001), but it does not fully eliminate the problem (powers et al. 2013). Sialic acid derivatization and stabilization are still the most effective way to maximize detection of sialylated glycans by MALDI-IMS.

– These derivatization protocols represent an active area of research that is still evolving. The initial approach has proven to be robust and adaptable to different FFPE tissues. We have found that doing the chemical reactions in tissues prior to antigen retrieval and PNGaseF release allows extensive rinsing after reactions and more flexibility in the type of reactions that can be done while maintaining spatial localization as the N-glycans are still attached to the protein structure. Doing these modifications in fresh tissue is likely to result in other detection and side-reaction challenges and remains to be optimized.

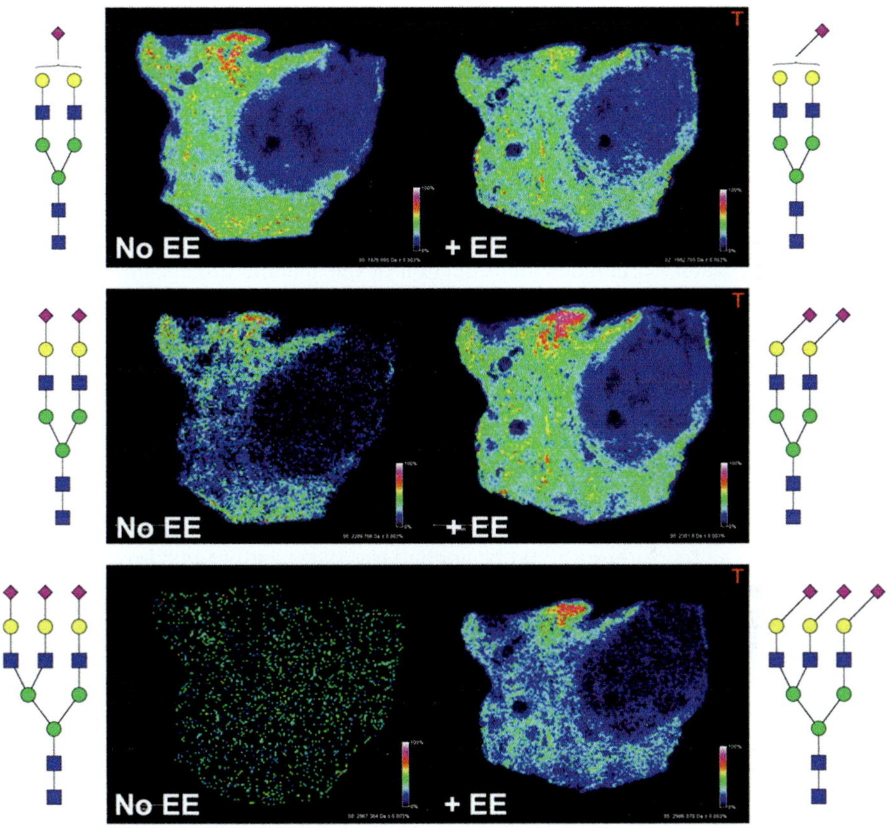

Fig. 4.6 Stabilization of multi-sialylated tissue glycans by ethyl esterification. MALDI-FT-ICR IMS data from the same ethyl esterified prostate tissue prepped as in Fig. 4.5 is shown for mono-, di-, and tri-sialylated bi-antennary glycans. Images in each panel on the left side represent non-ethyl esterified glycans, and the right side images are after EE. Also shown for the EE examples are the α-2,6 sialic acid linkage structures

4.7 Glycan Isomers and Fucosylation

The glycan structures shown in Fig. 4.3 for the different tri- and tetra-antennary fucosylated and sialylated glycans are compositional representations of what each glycan could be, and these do not account for the many different anomeric glycan linkages that are possible. Distinguishing these structural isomers has always been a challenge for mass spectrometry analysis of N-glycans and especially for MALDI-MS approaches. While accurate masses and glycan compositions can be easily determined, the location and anomeric linkages associated with fucose and sialic acid modifications are difficult to establish. The ethyl esterification examples highlight that multiple α2,6- or α2,3-linked sialic acids can be detected, but this still does not identify which branch of the glycan is occupied. For fucose modifications, this is particularly challenging as there are 13 known fucosyltransferase (FUT1–13) genes (Schneider et al. 2017), and fucose residues can be attached via α1,2 (FUT1,2), α1,3 (FUT3,4,6,7,9,10,11), or α1,4 (FUT3) linkages for outer arm modifications or α1,6 for core fucosylation linkages (FUT8). There are no differences in masses for these fucose linkages, and using MALDI-MS approaches, the presence of two or more fucose residues precludes identification of the specific linkages or which arm of the glycan the attachment occurs. There are two approaches that we expect can begin to address these fucosylation challenges in the context of MALDI-MSI workflows. One is the use of ion mobility MS instruments, which measures the mobility of gas-phase ions through an electric field in the presence of a buffer gas. The mobility of ions is based on their charge, shape, and size, and the method is increasingly being applied to the separation of glycan isomers (Fenn and McLean 2011; Gray et al. 2016), as well as imaging MS (Škrášková et al. 2016). The second approach is based on the ability to effectively spray enzymes on tissue (analogous to PNGaseF). Linkage-specific fucosidases could be applied to tissues, before PNGaseF digestion for comparative analysis with non-fucosidase treated tissues. This is feasible if the linkage specificity of the fucosidase is established, an area that still needs improvement and demonstration for the glycomic field in general.

4.7.1 Protocol Notes

- Using the MALDI-FT-ICR instrument, collision-induced dissociation (CID) of N-glycans under 2300 m.u. can be used to effectively distinguish core-fucosylated glycans (Powers et al. 2014) based on disaccharide fragmentation patterns. Demonstrating this for multi-fucosylated glycans has not been reported.

4.8 Current and Emerging Applications of N-Glycan Tissue Imaging

Our laboratories have cumulatively evaluated over 1000 different human tissue samples from multiple organ types using the basic PNGaseF digestion and MALDI-MS imaging workflows described herein. These methods have also been applied to animal models used in biomedical research, allowing studies on mechanisms and therapeutics that influence N-glycosylation changes over a wide range of applications including cancer treatments, transplant optimization, and cardiovascular diseases. Based on this experience, there are several research applications that we have identified, summarized in Fig. 4.7, grouped as different color codes. The major application of the technique is linked with the core premise of using an MS imaging approach for analysis of tissues, i.e., determine the location and number of N-glycans in a given tissue (Fig. 4.7, green panel). This data is linked with

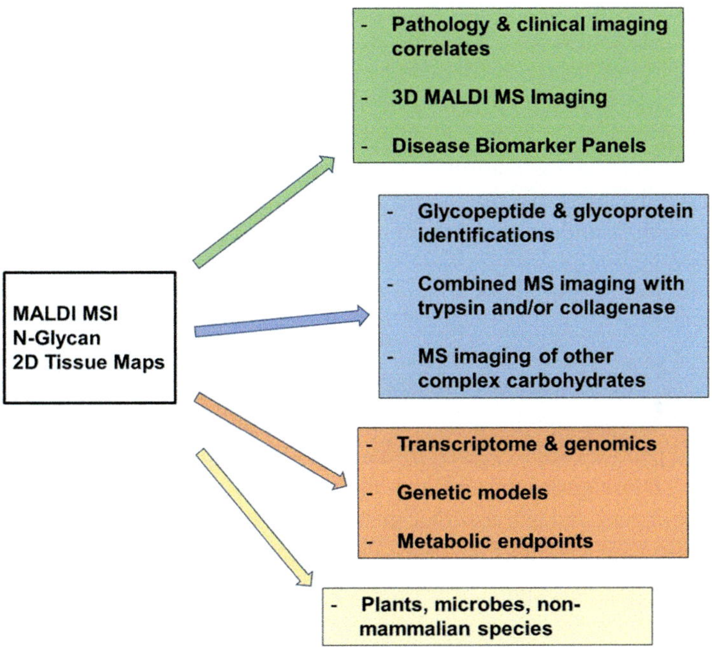

Fig. 4.7 Application options for the N-glycan MALDI-MSI method. Different current and future applications of the method are summarized in four color-coded groups: (1) Tissue localization and disease biomarkers (green panels), (2) glycoproteomic and glycomic analyses (blue panel), (3) Genomic and metabolomic correlates (orange panel), and (4) potential applications to plant, microbial, and other nonmammalian systems (pale yellow panel)

histopathology correlates of the tissue, which can be standard pathology stains like hematoxylin and eosin or immunohistochemical stains of specific protein targets. Also inherent in this determination is a list of N-glycans associated with different tissue regions that can be used as potential biomarker panels when multiple tissues are analyzed. This is particularly amenable to screening of many samples using tissue microarrays (Drake et al. 2017). The rapid analysis capabilities of the high-speed MALDI-TOF instrumentation in turn can be adapted to clinical chemistry laboratory assays as possible diagnostics. Emerging capabilities in this grouping include the ability to link molecular glycan data for specific tissue slices co-registered with clinical images obtained by magnetic resonance imaging. Determining 3D MS imaging of N-glycans in a given tissue remains to be reported, but it is certainly feasible and will be done in the near future.

Highlighted in the second panel of Fig. 4.7 (blue) are applications related to the analysis of N-linked glycoproteins and other types of complex carbohydrates. The tissue N-glycans described herein were all released by PNGaseF from a protein carrier. Therefore, the presence of N-glycans detected by MALDI-MSI in a particular tissue location, e.g., in a tumor region, also informs on tumor-specific glycoproteins. Identifying these glycoprotein carriers are also of great interest, and thus the glycan distribution tissue profiles can be used as localization maps. The most direct approach would be to target regions of interest for digestion with trypsin, followed by extraction and enrichment of glycopeptides from these regions. The biggest limitation of this approach is the minimal amount of glycopeptides that are extractable from small focal areas of a thin tissue slice. A current alternative approach is to use MS imaging of peptides in situ, combined with extracted peptide sequencing identification, to map protein expression to tissue localization. This distribution map of the peptides can be combined with the N-glycan map to correlate areas of overlap (Angel et al. 2017b). Recent studies (Angel et al. 2017b; Heijs et al. 2016) describe a sequential approach, one in which a FFPE tissue is first treated with PNGaseF to release N-glycans, which are then imaged, followed by trypsin digestion of the same tissue and peptide imaging. An adjacent tissue slice is treated similarly, but glycans and peptides are extracted for tandem MS analysis and peptide sequencing. A new MALDI imaging approach that uses collagenase digestion to map the distribution of collagens and extracellular matrix proteins (Angel et al. 2017c) could also be applicable with PNGaseF analyses. Alternatively, other carbohydrate digestion enzymes could be applied to tissues to map fine structures of heparan sulfates or chondroitin sulfate.

The third panel in Fig. 4.7 (orange) highlights the connection of N-glycans to their underlying genetic control and metabolic pathways. The different N-glycans shown in the biosynthetic schematic in Fig. 4.3 are the result of the sequential activity of glycosyltransferases and glycosidases. It is feasible that the activity of these enzymes is controlled at the transcriptome level or other genomic components, which in turn regulates the composition of the N-glycome in each region (Neelamegham and Mahal 2016). A previous study comparing the transcriptome of glycan biosynthetic genes in different mouse organs with the most abundant N-glycans detected in each tissue determined that the levels of many transcripts could be correlated with the levels of specific glycan structures (Nairn et al. 2008). Several studies have been published that have evaluated the transcriptome profiles of glycan biosynthesis genes to predict what glycan structures would be present (Kawano et al. 2005), including glycosylation reaction network analysis strategies (Liu and Neelamegham 2014; Neelamegham and Mahal 2016). The N-glycome is also linked to basic glucose and N-acetylglucosamine metabolism via the hexosamine biosynthetic pathway (Ryczko et al. 2016). There are now comprehensive metabolomic networks available for tissue metabolites following in vivo ^{13}C-glucose labeling (Bruntz et al. 2017), which can readily be linked to metabolic precursors of N-linked biosynthesis and the related hexosamine pathway. Additionally, there is an increasing number of genetic mouse models emerging from CRISPR/Cas9 technology, as well as related models with organ-specific inducible regulation. The ability to examine individual organs that have gene knockout or knock-ins, or multiple organs from full knockout mice, represents many new opportunities to evaluate the role of glycosylation in these models. In addition to glycosyltransferases and glycosidase genes, any gene target that is linked to glycan metabolism, glycoproteins, and extracellular matrix proteins can potentially affect the N-glycome in these models. It is feasible to now link multiple genetic models with comprehensive transcriptome and metabolome information to evaluate the effect of the N-glycome in target tissues.

The last panel in Fig. 4.7 represents the vast opportunity to apply this approach to nonmammalian samples. Descriptions of N-glycan imaging by MSI have yet to be reported for any other species beyond human and rodent tissues. Beyond evaluation of different plants, amphibians, or reptiles, there is an opportunity to evaluate the effect that microbial infections or microbiome constituents have on the glycomes of affected tissues. It is expected that application and adaptation of the glycan imaging workflow to these systems will provide important new information about the diversity, distribution, and localization of novel N-glycomes.

Acknowledgment This work was supported by the National Institutes of Health/National Cancer Institute R21 CA185799 to RRD and the National Institutions of Health/National Institute of General Medical Sciences P20GM103542 to PMA.

Literature Citations

Angel PM, Caprioli RM (2013) Matrix-assisted laser desorption ionization imaging mass spectrometry: in situ molecular mapping. Biochemistry 52(22):3818–3828

Angel PM, Baldwin HS, Gottlieb Sen D, Su YR, Mayer JE, Bichell D, Drake RR (2017a) Advances in MALDI imaging mass spectrometry of proteins in cardiac tissue, including the heart valve. Biochim Biophys Acta 1865(7):927–935

Angel PM, Mehta A, Norris-Caneda K, Drake RR (2017b) MALDI imaging mass spectrometry of N-glycans and tryptic peptides from the same formalin-fixed, paraffin-embedded tissue section. Methods Mol Biol In press

Angel PM, Comte-Walters S, Ball LE, Talbot K, Mehta A, Brockbank KGM, Drake RR (2017c) Mapping extracellular matrix proteins in formalin-fixed, paraffin-embedded tissues by MALDI imaging mass spectrometry. J Proteome Res 17(1):635–646
Baker TC, Han J, Borchers CH (2017) Recent advancements in matrix-assisted laser desorption/ionization mass spectrometry imaging. Curr Opin Biotechnol 43:62–69
Berin A, Boughton BA (2017) Spatial metabolite profiling by matrix-assisted laser desorption ionization mass spectrometry imaging. Adv Exp Med Biol 965:291–322
Bern M, Brito AE, Pang PC, Rekhi A, Dell A, Haslam SM (2013) Polylactosaminoglycan glycomics: enhancing the detection of high-molecular-weight N-glycans in matrix-assisted laser desorption ionization time-of-flight profiles by matched filtering. Mol Cell Proteomics 12(4):996–1004
Briggs MT, Kuliwaba JS, Muratovic D, Everest-Dass AV, Packer NH, Findlay DM, Hoffmann P (2016) MALDI mass spectrometry imaging of N-glycans on tibial cartilage and subchondral bone proteins in knee osteoarthritis. Proteomics 16(11–12):1736–1741
Briggs MT, Ho YY, Kaur G, Oehler MK, Everest-Dass AV, Packer NH, Hoffmann P (2017) N-Glycan matrix-assisted laser desorption/ionization mass spectrometry imaging protocol for formalin-fixed paraffin-embedded tissues. Rapid Commun Mass Spectrom 31(10):825–841
Bruntz RC, Lane AN, Higashi RM, Fan TW (2017) Exploring cancer metabolism using stable isotope-resolved metabolomics (SIRM). J Biol Chem 292(28):11,601–11,609
Cornett DS, Reyzer ML, Chaurand P, Caprioli RM (2007) MALDI imaging mass spectrometry: molecular snapshots of biochemical systems. Nat Methods 4(10):828–833
Drake RR, Powers TW, Jones EE, Bruner E, Mehta AS, Angel PM (2017) MALDI Mass Spectrometry Imaging of N-Linked Glycans in Cancer Tissues. Adv Cancer Res 134:85–116
Everest-Dass AV, Briggs MT, Kaur G, Oehler MK, Hoffmann P, Packer NH (2016) N-Glycan MALDI imaging mass spectrometry on formalin-fixed paraffin-embedded tissue enables the delineation of ovarian cancer tissues. Mol Cell Proteomics 15(9):3003–3016
Fenn LS, McLean JA (2011) Structural resolution of carbohydrate positional and structural isomers based on gas-phase ion mobility-mass spectrometry. Physical Chemistry Chemical Physics 13(6):2196–2205
Gray CJ, Thomas B, Upton R, Migas LG, Eyers CE, Barran PE et al (2016) Applications of ion mobility mass spectrometry for high throughput, high resolution glycan analysis. Biochim Biophys Acta 1860(8):1688–1709
Gustafsson OJ, Briggs MT, Condina MR, Winderbaum LJ, Pelzing M, McColl SR et al (2015) MALDI imaging mass spectrometry of N-linked glycans on formalin-fixed paraffin-embedded murine kidney. Anal Bioanal Chem 407:2127–2139
Harvey DJ (1999) Matrix-assisted laser desorption/ionization mass spectrometry of carbohydrates. Mass Spectrom Rev 18:349–450
Harvey DJ (2015) Analysis of carbohydrates and glycoconjugates by matrix-assisted laser desorption/ionization mass spectrometry: an update for 2009–2010. Mass Spectrom Rev 34(3):268–422
Heijs B, Holst S, Briaire-de Bruijn IH, van Pelt GW, de Ru AH, van Veelen PA et al (2016) Multimodal mass spectrometry imaging of N-Glycans and proteins from the same tissue section. Anal Chem 88:7745–7753
Holst S, Heijs B, de Haan N, van Zeijl RJ, Briaire-de Bruijn IH, van Pelt GW et al (2016) Linkage-specific in Situ Sialic Acid derivatization for N-Glycan mass spectrometry imaging of formalin-fixed paraffin-embedded tissues. Anal Chem 88(11):5904–5913
Kawano S, Hashimoto K, Miyama T, Goto S, Kanehisa M (2005) Prediction of glycan structures from gene expression data based on glycosyltransferase reactions. Bioinformatics 21:3976–3982
Kinoshita M, Mitsui Y, Kakoi N, Yamada K, Hayakawa T, Kakehi K (2014) Common glycoproteins expressing polylactosamine-type glycans on matched patient primary and metastatic melanoma cells show different glycan profiles. J Proteome Res 13(2):1021–1033

Liu G, Neelamegham S (2014) A computational framework for the automated construction of glycosylation reaction networks. PLoS One 9(6):e100939

Loke I, Kolarich D, Packer NH, Thaysen-Andersen M (2016) Emerging roles of protein mannosylation in inflammation and infection. Mol Asp Med 51:31–55

Miwa HE, Song Y, Alvarez R, Cummings RD, Stanley P (2012) The bisecting GlcNAc in cell growth control and tumor progression. Glycoconj J 29(8–9):609–618

Nairn AV, York WS, Harris K, Hall EM, Pierce JM, Moremen KW (2008) Regulation of glycan structures in animal tissues: transcript profiling of glycan-related genes. J Biol Chem 283(25):17,298–17,313

Neelamegham S, Mahal LK (2016) Multi-level regulation of cellular glycosylation: from genes to transcript to enzyme to structure. Curr Opin Struct Biol 40:145–152

Nishikaze T (2017) Sensitive and structure-informative N-Glycosylation analysis by MALDI-MS; ionization, fragmentation, and derivatization. Mass Spectrom 6(1):A0060

Nyalwidhe JO, Betesh LR, Powers TW, Jones EE, White KY, Burch TC et al (2013) Increased bisecting N-acetylglucosamine and decreased branched chain glycans of N-linked glycoproteins in expressed prostatic secretions associated with prostate cancer progression. Proteomics Clinical Applications 7:677–689

O'Connor PB, Costello CE (2001) A high pressure matrix-assisted laser desorption/ionization fourier transform mass spectrometry ion source for thermal stabilization of labile biomolecules. Rapid Commun Mass Spectrom 15:1862–1868

Ogrinc Potočnik N, Porta T, Becker M, Heeren RM, Ellis SR (2015) Use of advantageous, volatile matrices enabled by next-generation high-speed matrix-assisted laser desorption/ionization time-of-flight imaging employing a scanning laser beam. Rapid Commun Mass Spectrom 29(23):2195–2203

Pearce OM, Läubli H (2016) Sialic acids in cancer biology and immunity. Glycobiology 26(2):111–128

Pinho SS, Reis CA (2015) Glycosylation in cancer: mechanisms and clinical implications. Nat Rev Cancer 15(9):540–555

Powers TW, Jones EE, Betesh LR, Romano PR, Gao P, Copeland JA et al (2013) Matrix assisted laser desorption ionization imaging mass spectrometry workflow for spatial profiling analysis of N-linked glycan expression in tissues. Anal Chem 85:9799–9806

Powers TW, Neely BA, Shao Y, Tang H, Troyer DA, Mehta AS et al (2014) MALDI imaging mass spectrometry profiling of N-glycans in formalin-fixed paraffin embedded clinical tissue blocks and tissue microarrays. PLoS One 9(9):e106255

Powers TW, Holst S, Wuhrer M, Mehta AS, Drake RR (2015) Two-Dimensional N-Glycan distribution mapping of hepatocellular Carcinoma tissues by MALDI-Imaging mass spectrometry. Biomolecules 5(4):2554–2272

Reiding KR, Blank D, Kuijper DM, Deelder AM, Wuhrer M (2014) High-throughput profiling of protein N-glycosylation by MALDI-TOF-MS employing linkage-specific sialic acid esterification. Analytical Chemistry 86:5784–5793

Rini J, Esko J, Varki A (2009) Chapter 5: Glycosyltransferases and Glycan-processing Enzymes. In: Varki A, Cummings RD, Esko JD, Freeze HH, Stanley P, Bertozzi CR, Hart GW, Etzler ME (eds) Essentials of Glycobiology, 2nd edn. Cold Spring Harbor Laboratory Press, Cold Spring Harbor

Ryczko MC, Pawling J, Chen R, Abdel Rahman AM, Yau K, Copeland JK, Zhang C, Surendra A, Guttman DS, Figeys D, Dennis JW (2016) Metabolic reprogramming by Hexosamine biosynthetic and Golgi N-Glycan branching pathways. Sci Rep 6:23043

Schneider M, Al-Shareffi E, Haltiwanger RS (2017) Biological functions of fucose in mammals. Glycobiology 27(7):601–618

Schultz MJ, Swindall AF, Bellis SL (2012) Regulation of the metastatic cell phenotype by sialylated glycans. Cancer Metastasis Rev 31(3–4):501–518

Sekiya S, Wada Y, Tanaka K (2005) Derivatization for stabilizing sialic acids in MALDI-MS. Anal Chem 77(15):4962–4968

Škrášková K, Claude E, Jones EA, Towers M, Ellis SR, Heeren RM (2016) Enhanced capabilities for imaging gangliosides in murine brain with matrix-assisted laser desorption/ionization and desorption electrospray ionization mass spectrometry coupled to ion mobility separation. Methods 104:69–78

Stanley P, Schachter H, Taniguchi N (2009) Chapter 8: N-Glycans. In: Varki A, Cummings RD, Esko JD, Freeze HH, Stanley P, Bertozzi CR, Hart GW, Etzler ME (eds) Essentials of Glycobiology, 2nd edn. Cold Spring Harbor Laboratory Press, Cold Spring Harbor

Taniguchi N, Kizuka Y (2015) Glycans and cancer: role of N-glycans in cancer biomarker, progression and metastasis, and therapeutics. Adv Cancer Res 126:11–51

Wang D (2012) N-glycan cryptic antigens as active immunological targets in prostate cancer patients. J Proteomics Bioinformatics 5:90–95

Wang D, Dafik L, Nolley R, Huang W, Wolfinger RD, Wang LX et al (2013) Anti-oligomannose antibodies as potential serum biomarkers of aggressive prostate cancer. Drug Dev Res 74:65–80

Chapter 5
Isomeric Separation and Characterisation of Glycoconjugates

Kathirvel Alagesan, Arun Everest-Dass, and Daniel Kolarich

Abstract Individual monosaccharides can be linked in a variety of different combinations to form complex glycoconjugates. In contrast to DNA and proteins, glycoconjugate synthesis does not follow any template but is the consequence of the concerted action of various enzymes such as transferases and glycosidases. Thus, tools for glycoconjugate sequencing need to differentiate individual monosaccharide identity, linkage and anomericity to investigate and understand glycoconjugate function. In this chapter we provide a concise overview on the most commonly used and robust tools to separate and characterise glycoconjugate isomers.

Keywords Glycomics · Glycoproteomics · *N*-glycan · *O*-glycan · Porous graphitized carbon · PGC · HILIC · HPLC

Abbreviation

CCS	Collision cross section
CE	Capillary electrophoresis
DNA	Deoxyribonucleic acid
DP	Degree of polymerisation
ESI	Electrospray ionisation
Gal	Galactose
GlcNAc	*N*-Acetylglucosamine
GSL	Glycosphingolipid
HILIC	Hydrophilic interaction chromatography
HPLC	High-performance liquid chromatography

Kathirvel Alagesan and Arun Everest-Dass have contributed equally with all other contributors

K. Alagesan · A. Everest-Dass · D. Kolarich (✉)
Institute for Glycomics, Griffith University, Southport, QLD, Australia
e-mail: d.kolarich@griffith.edu.au

Y. Yamaguchi, K. Kato (eds.), *Glycobiophysics*, Advances in Experimental Medicine and Biology 1104, https://doi.org/10.1007/978-981-13-2158-0_5

IP	Ion pairing
mAb	Monoclonal antibody
MALDI	Matrix-assisted laser desorption/ionisation
MS	Mass spectrometry
NeuAc	*N*-Acetylneuraminic acid
NMR	Nuclear magnetic resonance
PGC	Porous graphitised carbon
RNA	Ribonucleic acid
RP	Reverse phase

5.1 Introduction

In comparison with genome and proteome sequencing, glycan sequencing is unique with intriguing challenges. First, the biosynthesis of glycoconjugates is a dynamic, extremely complex process that also varies among different cell types of a single organism. Second, glycosylation is a non-template-driven process involving a coordinated expression of a cascade of various glycosyltransferases, glycosidases, transporters and other enzymes active in the cytosol, endoplasmic reticulum and Golgi apparatus and thus, cannot really be predicted (Rini and Esko 2015). However, knowledge gathered over the past decades of glycobiology lead to a better understanding of the enzymes and their genes involved in glycan biosynthesis and degradation processes. Third, the mammalian glycan repertoire relies on a limited pool of monosaccharides, which are often stereoisomers and therefore cannot be distinguished by their mass alone (Cummings 2009; Seeberger 2015). This is in stark contrast to amino acids, where, except Leucine and Isoleucine, every amino acid is represented by a specific mass value. The fact that each monosaccharide contains multiple hydroxyl groups that can be involved in glyosidic linkages to form both linear and branched polymer structures with varying anomeric linkage (α and β) is another major difference to amino acid-based biopolymers. Unlike linear synthesised molecules such as DNA/RNA (4096 for a hexamer) or peptides (6.4^7 hexapeptides), an oligosaccharide containing six hexoses can give rise to as many as $\sim 10^{12}$ linear and branched isomers (Laine 1994). Also, each monosaccharide unit can carry additional modifications such as methylation, sulphation, acetylation or phosphorylation, to name a few. Last but not least, glycans are often found attached to other biomolecules such as proteins and lipids, further contributing to the complexity and functional diversity that needs to be analytically captured to study the role glycosylation plays for cell function in health and disease (Kolarich et al. 2012; Almeida and Kolarich 2016).

Glycans play essential roles in biological and disease-related events (Varki and Gagneux 2015; Varki 2017; Pang et al. 2011; Wassarman 2011; Haltiwanger 2002; Dwek 1995; Muramatsu 1993; Fukuda 1996; Furukawa et al. 2001; Fuster and Esko 2005; Ohtsubo and Marth 2006; Varki 1993; Varki and Lowe 2009; Saeland and van Kooyk 2011; Rudd et al. 2001), and to understand their biological functions, structure elucidation of either released oligosaccharides or their glycoconjugates (glycopeptides/glycoproteins/glycolipids, etc.) is a prerequisite. This has been a major driving force in the recent development of various sophisticated analytical tools for glycan sequencing. The functional diversity embedded in glycan structures is profoundly demonstrated in the case of the influenza virus. The human influenza virus preferentially binds to α2–6-linked sialic acid receptors, whereas the avian influenza virus prefers binding to α2–3-linked ones (de Graaf and Fouchier 2014). In the context of galectins, α2–6-linked sialic acid-carrying glycoconjugates have been shown to be involved in blocking galectin binding and enhancing tumour cell survival (Schultz et al. 2012; Zhuo and Bellis 2011). These examples are just demonstrative for the wide range of functions that glycoconjugates are involved in, which makes reliable, sensitive and selective analytical techniques to study these even more important.

5.2 A Summary on Glycoanalytical Techniques

From the early beginnings of glycobiology nuclear magnetic resonance (NMR) spectroscopy has been the gold standard to obtain detailed glycan structure information and has paved the way to our current understanding of glycan biosynthesis (Duus et al. 2000). Despite recent advances to improve sensitivity and enable analysis of more complex mixtures, NMR analyses remain time-consuming, still require comparably large amounts of sample and are essentially unsuitable for the analysis of complex samples or high-throughput screening of clinical samples. Over the past four decades, a variety of orthogonal analytical technologies have developed to be the backbone of today's glycomics: capillary electrophoresis (CE), high-performance liquid chromatography (HPLC), mass spectrometry (MS) or combinations/variations of CE or HPLC with MS are currently the methods of choice for most glycan profiling approaches. They are capable of separating complex glycan and glycoconjugate samples, analyse the individual molecules and also, in some cases, allow the differentiation between structure isomers. In this book chapter, we will provide a short overview of the most commonly used glycan sequencing techniques with an emphasis on techniques that allow differentiation of isomeric glycan features.

5.3 Liquid Chromatography Separation of Glycans

HPLC has been a widely used separation technique for glycan isomer characterisation. The multiplexing and coupling capabilities of HPLC to other analytical techniques either on the front end through other chromatographic methods or back end to mass spectrometry (MS) or fluorescent detectors have made it a highly attractive separation technology that is also widely used in the biopharma field. Although isomer characterisation by HPLC is considered challenging, various studies have shown that separation and characterisation of closely related structures can be achieved (Kolarich et al. 2015; Hua et al. 2013; Everest-Dass et al. 2013; Wuhrer et al. 2009). Classical reversed-phase (RP) HPLC separation is frequently employed in the analysis of glycopeptides and fluorescently derivatised glycans but is not suited to separate native, underivatised glycans because of their high polarity. Here, alternative techniques such as ion exchange (IE), porous graphitised carbon (PGC) and hydrophilic interaction liquid chromatography (HILIC) are the preferred methods employed by researchers around the world.

5.3.1 Reversed-Phase Chromatography (RP)

Reversed-phase chromatography is possibly one of the most widely used separation techniques and can be employed to separate small molecules as well as large proteins, depending on the conditions and stationary phase. As a consequence, a plethora of different RP stationary phases are available and are being developed, with varying molecular features and applicability for a wide variety of separations. Unless glycans are tagged with a hydrophobic (and usually fluorescent or UV-absorbing) molecule or naturally attached to a more hydrophobic moiety (e.g. glycopeptides or glycolipids), native glycans are usually too hydrophilic to be adequately retained by conventional RP stationary phases. Reductive amination of glycans with a hydrophobic, fluorescent tag has become a standard procedure used by many labs around the work to enable retention and thus, separation of glycans by RP. A large variety of tags with different molecular features have been described over the past decades, such as 2-aminopyridine (2-AP), 2-aminobenzamide (2-AB) and 2-aminobenzoic acid (2-AA), to name a few (Fig. 5.1) (Pabst et al. 2009; Marino et al. 2010). A comprehensive overview of the most commonly used tags can be found in the work by Pabst et al., who systematically examined and compared the sensitivity of these tags, in particular with respect to subsequent mass spectrometric detection (Fig. 5.1) (Pabst et al. 2009). While glycan retention on RP stationary phases is mainly mediated via the hydrophobicity of the tag, differentiation of structural isomers relies upon the individual structural features such as difference in position and/or linkage of a fucose residue (Fig. 5.2) (Bleckmann et al. 2011).

Ion pairing (IP) is another strategy that can be employed to retain and separate glycans on RP stationary phases. Charged additives in the eluting buffer act as ion pairing agents that increase the retention of oppositely charged analytes. This approach has successfully been employed in the separation of highly negatively charged glycans such as glycosaminoglycans, which were than detected by MS (Kuberan et al. 2002; Thanawiroon et al. 2004).

Another opportunity to increase glycan hydrophobicity is the permethylation of glycans, where each free hydroxyl group is replaced by a methyl group (Ciucanu and Kerek 1984; Ciucanu and Costello 2003). Permethylated *N*-glycans have also successfully been separated by RP chromatography using, e.g. chip-based C18 columns followed by online mass spectrometric (MS) detection, as permethylated glycans are not fluorescent (Bielik and Zaia 2011; Huang et al. 2017). Despite the fact

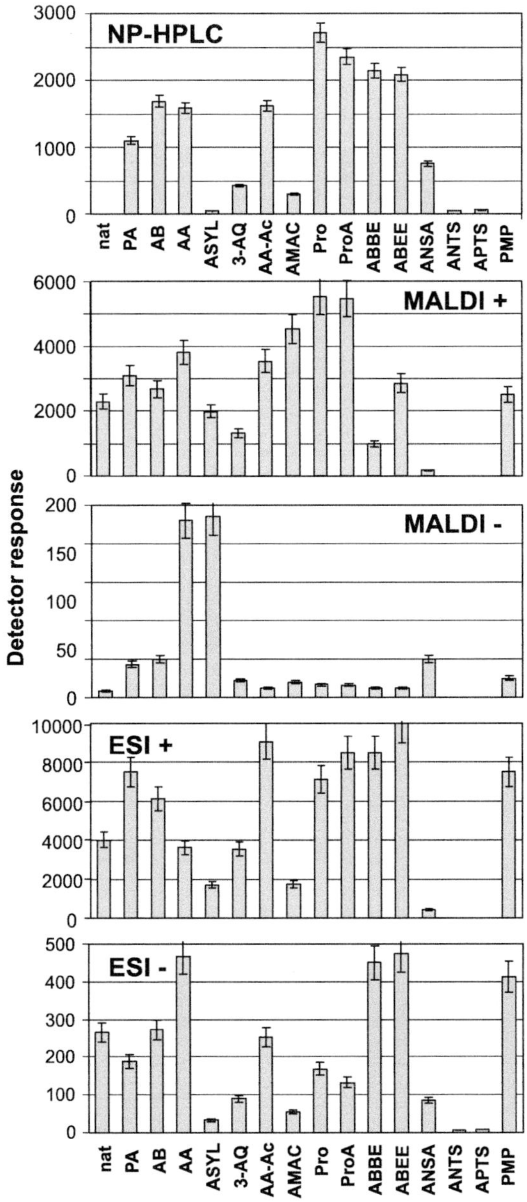

Fig. 5.1 Detection sensitivity of variously labelled oligosaccharides. The NP–HPLC panel shows the peak areas obtained in NP–HPLC at pH 4.4 (or 7.3 for PA-labelled glycans). The MALDI + and MALDI – panels depict the peak heights of the MALDI base peaks of the derivatives in positive and negative mode, respectively. The most prominent signals in positive mode MALDI were $[M + Na]^+$. In negative mode MALDI–MS, $[M–H]^-$ ions constituted the base peaks. The ESI + and ESI – panels give the peak heights obtained in positive and negative mode ESI–MS, respectively. $[M + 2H]^{2+}$ and $[M–2H]^{2-}$ ions dominated. Abbreviations: nat no labelling, PA 2-aminopyridine, AB 2-aminobenzamide, AA aminobenzoic acid (anthranilic acid), ASYL 4-amino-salicylic acid, 3-AQ 3-aminoquinoline, AA-Ac 3-(acetylamino)-6-aminoacridine, AMAC 2-aminoacridone, Pro procaine (4-aminobenzoic acid 2-diethylaminoethyl ester), ProA procainamide (N-(2-diethylamino)ethyl-4-aminobenzamide), ABEE 2-aminobenzoic acid ethyl ester, ABBE 2-aminobenzoic acid butyl ester, ANSA 5-amino-2-naphthalene sulfonic acid, ANTS 2-aminonaphthalene trisulfonate, APTS 8-aminopyrene-1,3,6-trisulfonic acid, APTS 8-aminopyrene-1,3,6-trisulfonic acid, PMP 1-phenyl-3-methyl-5-pyrazolone. (Reproduced from Pabst et al. (2009) with permission of the publisher)

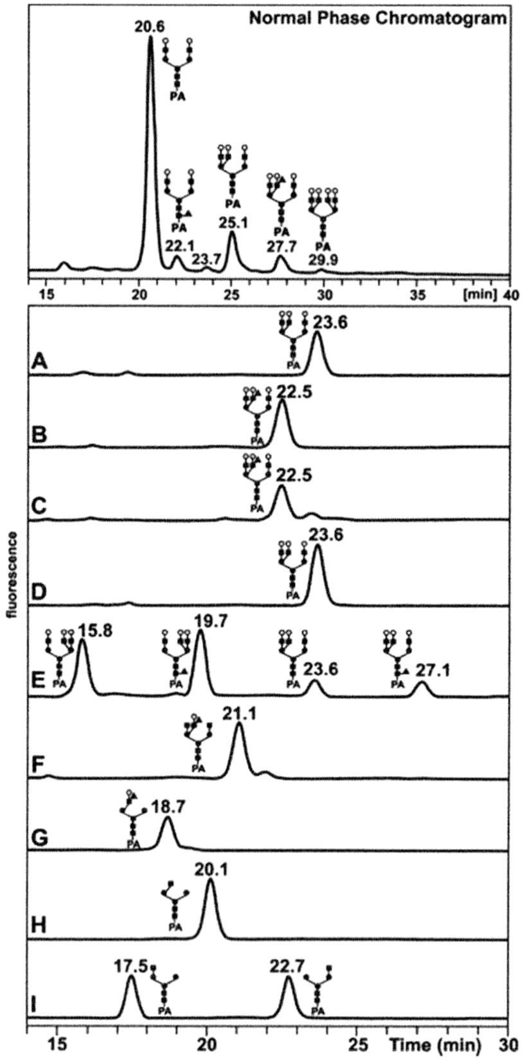

Fig. 5.2 Structural analysis of triantennary N-glycans from alpha1-proteinase inhibitor (A1PI). Top panel: Pyridylaminated asialo-N-glycans from plasma A1PI were at first separated according to size by normal-phase (NP-) HPLC. Cartoons represent the respective glycan structures as identified by MS and 2-D-HPLC. Panels A–I show RP chromatograms of the NP peaks of triantennary and fucosylated triantennary structures eluting at 25.1 and 27.7 min, respectively, and of standard PA-glycans. (**a**) Triantennary glycan (NP peak at 25.1 min) and (**b**) fucosylated triantennary glycan (NP peak at 27.7 min) on RP. (**c**) Digestion of fucosylated triantennary glycan with α1–6-fucosidase did not change the retention time. (**d**) Digestion of fucosylated triantennary glycan with α1–3/4-fucosidase shifted the retention time to that of the plain triantennary N-glycan. (**e**) PA-labelled triantennary N-glycans from human erythropoietin where the core α1–6 fucose had been partially removed by fucosidase. Two-branch isomers of triantennary structures with and without core α1–6 fucose exhibit different elution positions in accordance with Tomiya et al. (1988). This comparison

that permethylated glycans produce highly informative tandem mass spectra, this approach sacrifices separation resolution as the intrinsic hydrophobicity differences of permethylated glycans limit sufficient chromatographic separation based on glycan structure differences. Further issues can arise from incomplete permethylation, possible desialylation of acidic glycans and artefacts such as epimerisation of the reducing end (Ashline et al. 2007).

5.3.2 Hydrophilic Interaction Chromatography (HILIC)

In contrast to RP, HILIC uses the glycans' hydrophilic features for their separation. Glycan hydrophilicity depends on various structural characteristics such as size, charge, composition, structure and linkage. Depending on the individual HILIC stationary phases glycan retention and separation is mediated by hydrogen bonding, ionic interactions, and dipole–dipole interactions (Wuhrer et al. 2009; Alagesan et al. 2017). During the analyte loading, high organic solvent concentrations are used to promote these interactions between the analytes and the stationary phase. A gradient increasing the aqueous solvent is then used to elute the analytes depending on their hydrophilic interactions with the (mostly) amide-derivatised silica stationary phases.

This principle has successfully been employed as a major backbone technology in the early glycomics investigations and contributed significantly to our current knowledge on immunoglobulin glycosylation (Royle et al. 2008; Royle et al. 2003; Arnold et al. 2004; Arnold et al. 2005), major human plasma glycoproteins (Kolarich et al. 2006; Garner et al. 2001) and also has been employed in the high-throughput screening of serum/plasma or immunoglobulin G (IgG) (Adamczyk et al. 2014; Pucic et al. 2011; Saphire et al. 2003; Theodoratou et al. 2016). Despite its advances, this approach has its limitations in particular for the analysis of highly complex mixtures. Depending on the sample complexity, five or more different glycan molecules can co-elute and result in a single chromatographic peak (Pucic et al. 2011). These issues can at least partially be resolved by sequential application of exoglycosidase digestions and reanalysis of these samples or by coupling with orthogonal

Fig. 5.2 (continued) both indicated that the triantennary A1PI glycan was not core-fucosylated and also implied its branch structure. (**f**) Digestion of fucosylated triantennary glycan with β-galactosidase resulted in the removal of two galactose residues. (**g**) Further digestion of the above glycan with jack bean hexosaminidase removed two GlcNAc residues. (**h**) Simultaneous digestion of the glycan from panel G with β-galactosidase and almond fucosidase after heat inactivation of the previously used enzymes removed one galactose and one fucose residue. Product had the mass of the pentasaccharide core with one GlcNAc, but its retention time did not match that of the two standard PA-glycans shown in I. Therefore, the fucose must have been linked to the β1–4 GlcNAc. I: Reference N-glycans carrying β1–2-linked GlcNAc on the α1–3 (17.5 min) or 1–6 arm (22.7 min) of the core N-glycan. (Reproduced from Kolarich et al. (2006) with permission of the publisher) detection methodologies such as MS. This, however, significantly impacts high-throughput capabilities, can induce artefacts and misinterpretations when incomplete exoglycosidase digestion is occurring and also requires considerable amounts of sample material.

As the hydrophilic properties of glycans are the major driver for binding and separation to the stationary phase, HILIC can also be employed to separate underivatised glycans when coupled to online MS detection as shown by Wuhrer et al. (2004). This principle has also recently been employed by Mancera-Arteu et al. who used a zwitterionic-hydrophilic μZIC-HILIC separation coupled with ESI MS/MS detection in negative ion mode to characterise glycan isomers. This allowed them to separate and characterise sialic acid and fucose linkage-type *N*-glycan isomers (Mancera-Arteu et al. 2017).

5.3.3 Porous Graphitised Carbon Chromatography (PGC)

PGC represents an entirely different type of stationary phase chemistry that has unique properties for glycan separation and analysis. In 1979 Knox and Gilbert introduced a new form of chromatographic matrix called porous glassy carbon. It had limited applicability due its moderate chromatographic performance in GC and HPLC. Porous glassy carbon was further improved by Knox et al. to produce porous graphitised carbon (PGC) that displayed good stability and chromatographic performance, in particular for polar analytes (Knox et al. 1986).

The underlying interactions involved in PGC chromatographic behaviour are only vaguely understood, but it has been established that hydrophobic, ionic, polar and molecular features all influence analyte retention (West et al. 2010; Kolarich et al. 2015; Everest-Dass et al. 2013; Campbell et al. 2014; Stavenhagen et al. 2015; Pabst and Altmann 2008; Pabst et al. 2012; Miller et al. 2017; Adamczyk et al. 2018). As a consequence, PGC chromatography has shown superior resolution for native, non-labelled, glycans. Releasing the glycan from its conjugate will unavoidably result in the formation of α- and β-anomers at the glycan reducing end, unless stabilised by reduction of reductive amination labelling with fluorescent tags. Such α- and β-anomers are easily separated by PGC-LC into two distinct peaks. Thus glycan reduction has become a standard procedure for this approach to ensure that just a single peak is being detected for a specific glycan structure (Jensen et al. 2012). Since MS is the preferred detection approach for PGC separation due to its sensitivity and versatility, glycan reduction comes with an additional advantage as it introduces a specific mass tag on the reducing end that facilitates fragment peak assignment in tandem MS spectra.

PGC finds its applications in sample preparation and glycan analyses. Packer et al. showed that sequential elution of neutral and acidic glycans from solid-phase PGC is possible using specific additives such as 0.05%(v/v) trifluoroacetic acid to the mobile phase, making it a versatile and effective way to separate these two glycan classes (Packer et al. 1998) but also desalt the sample (Jensen et al. 2012). Pabst and Altmann further studied the influence of ionic strength, pH and temperature on the retention of

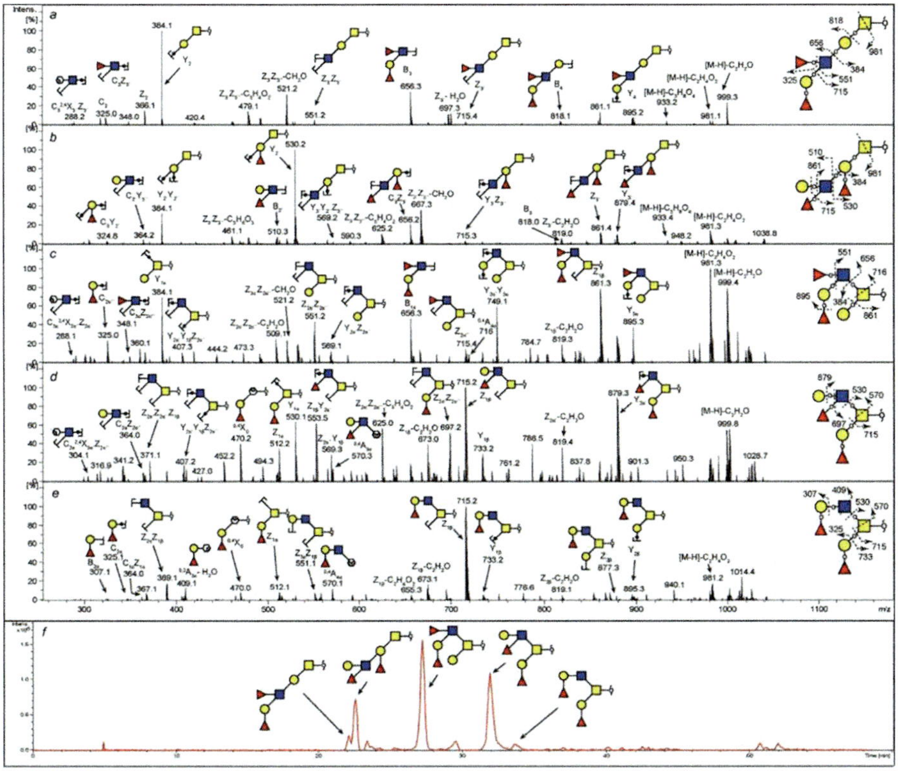

Fig. 5.3 Identification and characterisation of neutral O-glycan isomers by PGC-LC ESI MS/MS. Singly charged negative ion tandem spectra of m/z 1041.3^{1-} isomers eluting at 22 min (**a**), 26 min (**b**), 32 min (**c**) and 33.6 min (**d**). Extracted ion chromatogram of m/z 1041.3^{1-} isomers (**f**). Negative mode fragment spectra with PGC-LC separation aids in the identification and characterisation of closely related structures. (Reproduced from Everest-Dass et al. (2013) with permission of the publisher)

glycans by PGC (Pabst and Altmann 2008). When the ionic strength is reduced while the pH is maintained, acidic *N*-glycans are stronger retained by the PGC stationary phase. Increasing the pH also resulted in a stronger retention of acidic *N*-glycans on PGC while binding and elution of neutral glycans was unaffected. A temperature increase, however, resulted in a more effective retention of all *N*-glycans.

The PGC-LC ESI MS/MS approach has successfully been employed by several laboratories around the world to investigate either protein-specific or tissue−/body fluid-specific, global glycosylation patterns (Kolarich et al. 2015; Jensen et al. 2012; Wongtrakul-Kish et al. 2013). Since no glycan labelling is required, it is equally suitable to separate any glycan type. Thus, PGC-LC has been applied to analyse *N*- and *O*-glycans but also to characterise the glycan portion of glycolipids and glycosaminoglycan fragments, as also reviewed recently by Stavenhagen et al. in the context of clinical glycomics (Stavenhagen et al. 2015). Several studies on both *N*- and *O*-glycans have described the baseline resolution of many structural glycan isomers in PGC (Jensen et al. 2012; Pabst et al. 2007; Gao et al. 2015; Abrahams et al. 2018; Karlsson et al. 2004a, b) (Fig. 5.3). Numerous studies have used PGC to perform global *N*- and *O*-glycomics on cancer cell lines and tissues (Abrahams et al. 2015; Anugraham et al. 2015; Gustafsson et al. 2015; Lee et al. 2010, 2011). Using a capillary flow PGC-LC-ESI-MS with negative ion mode setup, Anugraham et al. were able to show that α2–6 NeuAc and bisecting *N*-acetyl-glucosamine-type *N*-glycans of ovarian cancer cell lines (SKOV3, IGROV1, A2780 and OVCAR3) were specific and abundant when compared to non-cancerous ovarian epithelial cells (HOSE6.3 and HOSE17.1) (Anugraham et al. 2014). In a follow-up study, they showed that glycan structural isomers that corresponded to LacdiNAc-type (GalNAcβ1-4GlcNAc) motifs were unique to serous ovarian cancers and could differentiate between serous ovarian and peritoneal cancer tissues (Anugraham et al. 2017). The Karlsson group have shown over numerous studies the advantage of using the PGC-LC-MS setup in the analysis of mucin type *O*-glycans in cancer and other diseases (Adamczyk et al. 2018; Flowers et al. 2017; Chaudhury et al. 2016). In a study by Flowers et al., *O*-glycans released from synovial glycoproteins during acute and chronic arthritic conditions were compared and immune-reactive glycans identified. They found that the sulphated core 1 *O*-glycan (Galβ1–3GalNAcol) had different isomeric profiles between the two conditions. Patients with acute reactive arthritis showed the presence of three isomers, while patients with chronic rheumatoid arthritis showed only a single 3-Gal-linked sulphate isomer (Flowers et al. 2013). They further developed a selected reaction monitoring (SRM) method to differentiate and relatively quantitate the core 1 *O*-glycan and the sulphated core 1 *O*-glycan Gal- and GalNAc-linked isomers, thereby demonstrating a statistically significant increase in sulphation of salivary MUC7 from rheumatoid arthritis patients.

PGC-nanoLC ESI MS/MS has also been the basis for the most sensitive glycomics approach to date. Hinneburg and co-workers were the first to obtain in-depth glycomic profiles for both *N*- and *O*-glycans from as low as 1000 cells obtained by laser microdissection from formalin-fixed, paraffin-embedded (FFPE) histopathological tissue sections obtained from hepatocellular carcinoma. This approach enabled them to investigate the cancer glycome, revealing that hepatocellular carcinoma exhibited a drastically changed glycan profile that was highly increased in particular pauci-mannosidic *N*-glycans and fucosylated core 2 type *O*-glycans (Hinneburg et al. 2017). We also used this technology to screen the *N*- and *O*-glycome of non-melanoma type skin cancers and normal human skin (Moginger et al. 2018). These basal and squamous cell carcinomas show an increase in oligomannose type *N*-glycans but exhibited overall less pronounced glycome differences compared to hepatocellular carcinoma. Interestingly, the peculiar ability of PGC to separate structure isomers allowed Möginger and co-workers to identify that core-fucosylation was about three times higher on diantennary *N*-glycans carrying at least one α2–3-linked sialic acid compared to *N*-glycans that just carried α2–6 NeuAc.

Next to global glycomics, PGC-LC is equally well suited to study protein-specific glycosylation. Both, purified glycoproteins as well as glycoproteins separated by 1D or 2D-electrophoresis can easily be analysed (Wilson et al. 2002), and since recent developments in PGC-nanoLC ESI MS/MS provides the necessary sensitivity (Hinneburg et al. 2017; Moginger et al. 2018), even protein bands containing less than 500 ng of glycoprotein can still be sufficient to determine global glycan profiles. This has successfully been employed to study the components of secretory IgA from human breast milk (Deshpande et al. 2010), human plasma proteins such as butyrylcholinesterase (Kolarich et al. 2008) or corticosteroid-binding globulin (Sumer-Bayraktar et al. 2011; Sumer-Bayraktar et al. 2012), human as well as recombinant immunoglobulins (Stadlmann et al. 2008) and human cancer-associated glycoproteins such as E-cadherin (Carvalho et al. 2016) or bacterial flagellins (Rath et al. 2018), to name a few examples.

In addition to the detailed analysis of protein-derived glycans, PGC-LC-MS analysis has also successfully been used to differentiate and characterise heparan sulphate (HS) oligosaccharides (Miller et al. 2017) and glycosphingolipid (GSL)-derived glycans (Anugraham et al. 2015; Karlsson et al. 2010). The recent study by Miller et al. exhibits the effective and sensitive separation of digested HS oligosaccharides by PGC. They showed the separation of complex mixture with various degrees of polymerisation (dp) ranging from dp4 to dp8. Baseline separation was observed for dp4 and dp6 isomers. The elution order was affected by sulphation levels, chain length and conformation. Their work demonstrates that heterogeneous mixtures of HS oligosaccharides can be efficiently separated and sensitively analysed using a PGC-LC-MS setup. Similarly, separation of isomeric or isobaric glycans is particularly important in the differentiation of GSL-derived glycans. Karlsson et al. exhibited this through the separation and characterisation of isomeric glycans derived from isoglobotetraosylceramide (iGb4; GalNAcβ-3Galα-3Galβ-4Glcβ-1Cer) and globotetraosylceramide (Gb4; GalNAcβ-3Galα-4Galβ-4Glcβ-1Cer). Anugraham et al. also revealed the separation of both acidic and neutral GSL-derived glycans in a PGC-LC-MS-based setup. This approach was notably applied to identify isomeric blood group antigens that showed distinct structure differences between serous ovarian and endometrioid peritoneal cancer tissues (Anugraham et al. 2015). Thus PGC-LC ESI MS/MS opens novel opportunities to study glycosyltransferase activities to obtain a better understanding of how the complex glycosylation machinery is producing these glycoconjugates.

Although mostly used for the analysis of non-labelled, non-derivatised glycans, recently the separation of permethylated glycans by PGC has also been reported (Huang et al. 2017; Costello et al. 2007). The packing of PGC into nanoscale chromatography chips for nanoLC-MS-based analysis of glycans (Hua et al. 2011) has shown a considerable sensitivity promise in the rapid analysis of glycans, albeit sacrificing isomer separation due to the significantly increased hydrophobicity of the permethylated glycan molecules. Each of the above-mentioned chromatography methods have their unique advantageous and disadvantageous in the separation and characterisation of glycans and maybe suitably applied for specific questions. A study by Melmer et al. analysed the retention times of 2-AB labeled glycans from fetuin, RNase-B and a mAb in HILIC, IP-RPC and PGC chromatography, respectively. An example of the main glycan variants in mAB HexNAc4Hex3dHex1, HexNAc4Hex4dHex1(two isomers), and HexNAc4Hex5dHex1 is shown in Fig. 5.4 (Melmer et al. 2011).

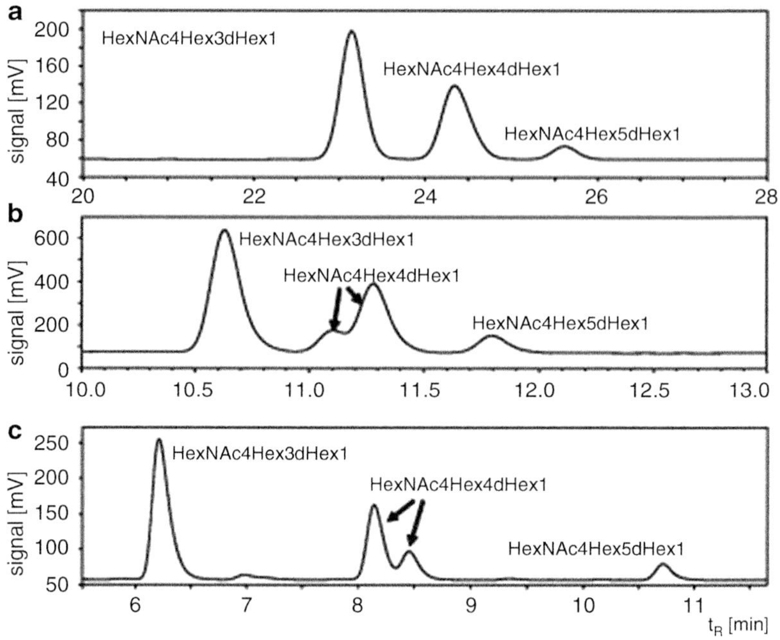

Fig. 5.4 Separation of mAb N-glycans by (**a**) reversed-phase, (**b**) PGC, and (**c**) HILIC chromatography exemplifying the peak patterns of the main N-glycan variants HexNAc4Hex3dHex1, HexNAc4Hex4dHex1 (two isomers), and HexNAc4Hex5dHex1. (Reproduced from Melmer et al. (2011) with permission of the publisher)

5.4 Ion Mobility MS: Gas-Phase Separation of Glycoconjugates

Next to liquid-based separation technologies, it is also possible to use gas-phase separation for the determination of glycoconjugate structures. For this purpose, ion mobility-mass spectrometry (IM-MS) has been gaining more and more attention over the past years. IM-MS measures the mass-to-charge ratio (*m/z*) of ions as well as the time these molecules need to traverse a cell that is filled with an inert gas (May and McLean 2015; Manz and Pagel 2018). In addition, the molecules are under the influence of a weak electric field when traversing the gas filled cell. This approach simultaneously uses various physical–chemical properties of these molecules for their separation, mass, charge, size and shape, subsequently enabling the differentiation of structure isomers. In addition the measured drift time can be converted into a collision cross section (CCS) value that is a molecular property and can be universally compared and used for the structure description.

Recent studies from two independent groups demonstrated that intact glycopeptide isomers can be successfully differentiated using IM-MS (Guttman and Lee 2016; Hinneburg et al. 2016). Hinneburg and co-workers employed synthetic

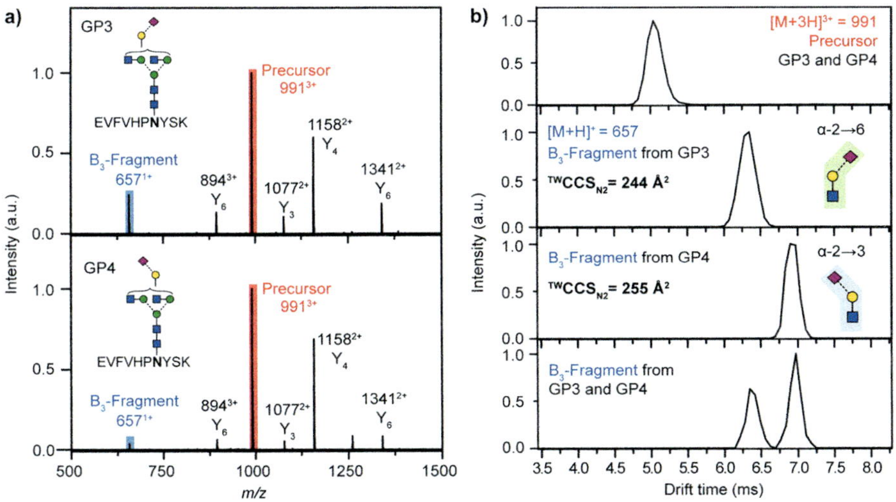

Fig. 5.5 Differentiation of N-acetylneuraminic acid (NeuAc) linkage isomers using CID fragmentation and subsequent IM-MS analysis. Two isomeric glycopeptides, which either carry α2–6 (GP3)- or α2–3 (GP4)-linked NeuAc, were analysed. (**a**) Both peptides exhibit identical MS/MS spectra, as shown for the triply protonated precursor ion (red). (**b**) When analysed as mixture the intact glycopeptide ions cannot be separated by IM-MS (m/z 991, red). B_3-trisaccharide fragments (m/z 657, blue) directly cleaved from the glycopeptide by CID on the other hand show characteristic drift times depending on the regiochemistry of the NeuAc linkage, which allows unambiguous identification of α2–3- and α2–6-linked isoforms. (Reproduced from Hinneburg et al. (2016) with permission of the publisher)

glycopeptides designed based on the human protein C fragment ^{284}EVFVHP**N**YSK^{293} (UniProt entry P04070) carrying either an α2–6 or α2–3 monosialylated, biantennary *N*-glycan. While the intact glycopeptides just differing in the type of sialic acid linkage did not show sufficient different molecular properties in the gas phase that would have enabled their separation by IM-MS, this could easily be achieved when oligosaccharide-only fragment ions produced by CID fragmentation were subjected to IM-MS. The *m/z* 657 B_3 type oxonium ion fragment (trisaccharide consisting of Gal, GlcNAc and NeuAc) showed significantly different drift times that were dependent on the type of NeuAc linkage, with the α2–6 fragment exhibiting considerably shorter drift times compared to the α2–3 equivalent (Fig. 5.5) (Nilsson 2016). The isomeric NeuAc-containing trisaccharide fragments showed excellent baseline separation even in the simultaneous presence of both glycopeptides, thus providing a comparably easy opportunity to perform in-depth glycoproteomics to study sialic acid linkage in a site- and protein-specific manner. In addition, the travelling wave collision cross sections measured in nitrogen drift gas ($^{TW}CCS_{N2}$) differed significantly for these two trisaccharides: 236 Å^2 for the α2–6-linked NeuAc and 246 Å^2 for the α2–3-linked NeuAc-containing fragments (Hofmann et al. 2014). These values were highly diagnostic for the regiochemistry of the underlying NeuAc linkage and could thus be used to gain site-specific information on important glycan structural features directly from individual glycopeptides in a single experiment. Recently Barroso and co-workers evaluated the capacity of travelling wave IM-MS to separate isomeric glycoconjugates at three different levels (Barroso et al. 2018): as an intact glycoprotein, after digestion into glycopeptides and just the released glycans. They put particular focus on the ability to differentiate different types of sialic acid linkage (i.e. α2–3 and α2–6). In agreement with previous studies, isomer separation was achieved for glycans (without fragmentation) and for glycopeptides (after fragmentation as described by Hinneburg et al. and Guttmann et al.). Under native MS conditions, no baseline isomer separation of intact glycoproteins was observed. However, the drift time of various glyco-isoforms increases with an increase in the carbohydrate fraction, i.e. complexity and branching of the glycoforms (Barroso et al. 2018).

5.5 Capillary Electrophoresis-Electrospray Ionisation-Mass Spectrometry

Capillary electrophoresis (CE) has been the primary methodology used for separating and analysing short tandem repeat alleles in forensic DNA laboratories worldwide. The separation of analyte molecules is based upon their electrophoretic mobility on an applied voltage. As the majority of glycans is neutral, often derivatisation of glycans is required for their analysis. Honda et al. explored the possibility of CE-based analysis of derivatised mono- and oligosaccharides using amino pyridine. The separation of pyridylamino sugar derivatives was achieved by the in situ anionic borate complex (Honda et al. 1989). In 1996 Guttmann and co-workers introduced the fluorophore 1-aminopyrene-3,6,8-trisulformate (APTS) as a novel glycan derivatisation agent (Guttman et al. 1996a). APTS having three negatively charged groups also makes otherwise neutral glycans highly negative, and thus they can easily be separated by CE. Following this work, this reagent has also been employed for glycan analysis by capillary gel electrophoresis (CGE) technique offering the opportunity to separate various glycan isomers (Guttman et al. 1996b; Guttman and Pritchett 1995; Chen and Evangelista 1998). In CGE the separation of labelled glycans is based on their difference in hydrodynamic volume to charge ratio (Guttman 2013). Consequently, methods for automated glycan profiling using a single capillary column instrument or a multiplexed capillary array system have been described for glycan analysis purposes (Szigeti and Guttman 2017). Over the past decade, Rapp and co-workers developed high-throughput glycomics workflows based on multiplexed capillary gel electrophoresis coupled with laser-induced fluorescence (LIF) detection, also now known as xCGE-LIF (Callewaert et al. 2004; Ruhaak et al. 2010; Hennig et al. 2015; Schwarzer et al. 2008). Here, glycan identification and characterisation are based on normalised retention times (achieved by using internal standards) and using these retention times to match these with database entries related to the respective structure (Behne et al. 2013; Huffman et al. 2014). If necessary, and to confirm structures not in the database, further extended structural analysis can be achieved by using an array of exoglycosidases in combination with xCGE-LIF-based glycoprofiling.

CE analysis can also be coupled with other detection technologies such as mass spectrometry, significantly increasing the analytical power of this separation methodology. When analysing glycoconjugates this approach provides high selectivity for isomer separation but also the opportunity for tandem MS structure elucidation. Gennaro and Salas-Solano reported an approach for CE-MS with online LIF detection to identify *N*-glycans from therapeutic antibodies. A major advantage of this glycan profiling approach is the ability to identify and discriminate co-migrating isomeric glycan species (Gennaro and Salas-Solano 2008). Gahoul and co-workers (Gahoual et al. 2013) developed a CE-MS/MS method based on the sheathless CE-ESI-MS (CESI) platform (Moini 2007) allowing a fast and precise characterisation of a monoclonal antibody digest. This allowed them to achieve 100% sequence coverage for both heavy and light chain in a single analytical experiment including the glycopeptides from 100 fmol of protein digest. Recently, Giorgetti and co-workers demonstrated that similar glycosylation profiles were obtained when analysing *N*-glycans by CE-ESI-MS and by HILIC of 2-aminobenzamide labelled *N*-glycans, tested on ten different monoclonal antibodies (mAbs) (Giorgetti et al. 2018). CE-ESI-MS analysis has the advantage that it is not restricted to glycan or glycopeptide analyses but can also be used for the overall characterisation of mAbs. Khatri et al. (2017) adopted a microfluidics-based CE-MS system to analyse released glycans, glycopeptides and monosaccharides using standard glycoproteins such as bovine ribonuclease B, human transferrin and human alpha-1-acid glycoprotein. They explored the use of TMT-labelling of glycans to improve the electrophoretic migration and also to enable multiplexed quantitation by tandem MS. They also employed sialic acid derivatisation using the method developed by the Wuhrer group (Reiding et al. 2014) to differentiate sialic acid linkage by MS only. At the same time, the Wuhrer group developed a high-resolution separation platform based on capillary electrophoresis–mass spectrometry (CE-MS) for selective differentiation of α2–3- and α2–6-sialylated glycopeptides without any sample pretreatment (Kammeijer et al. 2017) (Fig. 5.6).

5.6 Differentiation of Isobaric Glycans Using MS

During MALDI analysis, sialylated glycans usually undergo decomposition to give rise to focused (in-source fragmentation) and unfocused (post-source fragmentation) ion peaks when measured in the reflector-time-of-flight detectors (Harvey 1999). This is due to the presence of a labile carboxylic proton. Thus, several strategies have been developed to stabilise and neutralise sialic acid residues and make them more suitable for MALDI analyses. This can be achieved by permethylation (Ciucanu and Kerek 1984), methyl esterification or derivatisation with acetohydrazide (Toyoda et al. 2008). In late 2000, Harvey and co-workers demonstrated a method for stabilising sialic acids and discriminating α2–3 and α2–6 isomers. Here

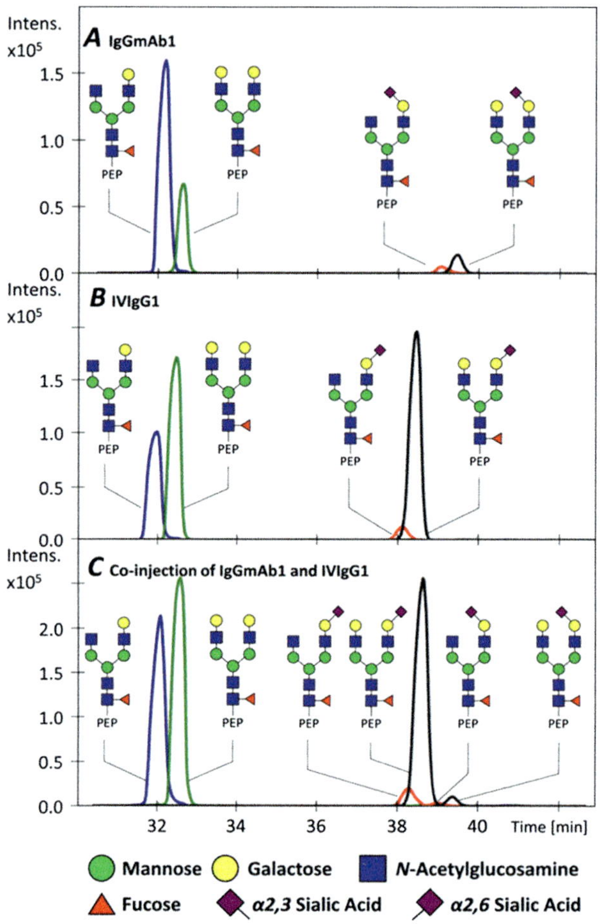

Fig. 5.6 Extracted ion electropherograms (EIEs) of IgGmAb1 and IVIgG1 glycopeptides obtained with CE–ESI-MS after targeted alignment. (**a**) EIEs of IgGmAb1 glycopeptides derived from CHO cells, (**b**) EIEs of IVIgG1 retrieved from human plasma and (**c**) EIEs of a co-injection of IgGmAb1 and IVIgG1. The "PEP" label illustrates the tryptic peptide sequence EEQYNSTYR to which the glycan is attached. (Figure taken from Kammeijer et al. (2017) with permission from publisher)

the glycans were treated with 4-(4,6-dimethoxy-1,3,5-triazin-2-yl)-4-methylmorpholinium chloride (DMT-MM) in methanol converting α2–6-linked sialic acids to methyl esters (+14 Da), and the α2–3-linked sialic acids formed lactones (−18 Da) (Wheeler et al. 2009). In 2014, Reiding et al. developed a simplified procedure for the derivatisation and discrimination of sialic acids using 1-ethyl-3-(3-(dimethylamino)propyl)carbodiimide (EDC) and 1-hydroxybenzotriazole (HOBt) as activators in ethanolic solution. In contrast to the previous approach described by Harvey, this reaction converts α2–6-linked sialic acid to

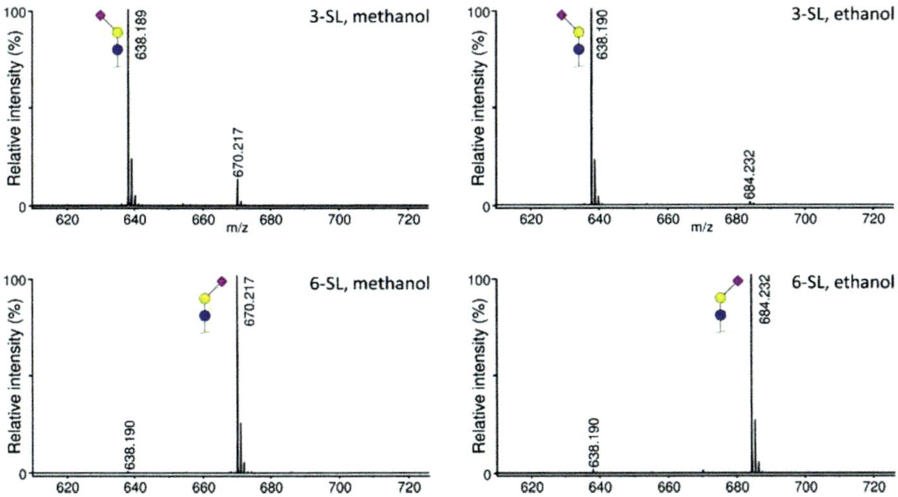

Fig. 5.7 Fig. 5.7 MALDI-TOF-MS spectra of 3′- and 6′-sialyllactose when alkyl esterified for 1 h at 37 °C with EDC + HOBt in methanol or ethanol. The lactonized reaction product is 638.190 Da [M + Na]+, with the esterified masses being 670.217 and 684.232 Da for methanol and ethanol, respectively. Figure taken from Reiding et al 2014 with permission from publisher

dimethylamide (+28 Da) and α2–3-linked sialic acid to a cyclic lactone with the adjacent galactose (−18 Da). For example, Fig. 5.7 shows a RP MALDI-TOF-MS spectra of 3′- and 6′-sialyllactose after alkyl esterification for 1 h at 37 °C with EDC + HOBt in either methanol or ethanol. The lactonized reaction product is visible at 638.190 Da [M + Na]+, with the esterified masses being 670.217 and 684.232 Da for methanol and ethanol, respectively. This particular approach has successfully been applied to differentiate α2–3- and α2–6-linked sialic acids at both glycan (Reiding et al. 2014) and glycopeptide level (de Haan et al. 2015). Though this approach offers an easy and convenient approach to stabilise and differentiate sialic acids and their linkage, other structure isomers will not be resolved by this approach.

5.7 Concluding Remarks

Determination of the in vivo functions of glycoproteins is an arduous task. One of the major analytical challenges in MS-based glycomics is the propensity of glycans to form many different isomers from a comparably limited set of building blocks (Laine 1994). In relation to other OMICS areas such as proteomics and genomics, glycomics requires multi-methodological approaches to analytically capture glycoconjugate structures and subsequently use this information to understand their functional significance. Recent advances in MS instrumentation, availability of synthetic standards and the development of novel methodologies have pushed this field towards the large-scale OMICS technologies. This will open entirely novel opportunities to study glycoconjugates and understand their functional significance in health and disease.

References

Abrahams JL, Packer NH, Campbell MP (2015) Relative quantitation of multi-antennary N-glycan classes: combining PGC-LC-ESI-MS with exoglycosidase digestion. Analyst 140(16):5444–5449

Abrahams JL, Campbell MP, Packer NH (2018) Building a PGC-LC-MS N-glycan retention library and elution mapping resource. Glycoconj J 35(1):15–29

Adamczyk B et al (2014) Comparison of separation techniques for the elucidation of IgG N-glycans pooled from healthy mammalian species. Carbohydr Res 389:174–185

Adamczyk B et al (2018) Sample handling of gastric tissue and O-glycan alterations in paired gastric cancer and non-tumorigenic tissues. Sci Rep 8(1):242

Alagesan K, Khilji SK, Kolarich D (2017) It is all about the solvent: on the importance of the mobile phase for ZIC-HILIC glycopeptide enrichment. Anal Bioanal Chem 409(2):529–538

Almeida A, Kolarich D (2016) The promise of protein glycosylation for personalised medicine. Biochim Biophys Acta 1860(8):1583–1595

Anugraham M et al (2014) Specific glycosylation of membrane proteins in epithelial ovarian cancer cell lines: glycan structures reflect gene expression and DNA methylation status. Mol Cell Proteomics 13(9):2213–2232

Anugraham M et al (2015) A platform for the structural characterization of glycans enzymatically released from glycosphingolipids extracted from tissue and cells. Rapid Commun Mass Spectrom 29(7):545–561

Anugraham M et al (2017) Tissue glycomics distinguish tumour sites in women with advanced serous adenocarcinoma. Mol Oncol 11(11):1595–1615

Arnold JN et al (2004) The glycosylation of human serum IgD and IgE and the accessibility of identified oligomannose structures for interaction with mannan-binding lectin. J Immunol 173(11):6831–6840

Arnold JN et al (2005) Human serum IgM glycosylation: identification of glycoforms that can bind to mannan-binding lectin. J Biol Chem 280(32):29080–29087

Ashline DJ et al (2007) Carbohydrate structural isomers analyzed by sequential mass spectrometry. Anal Chem 79(10):3830–3842

Barroso A et al (2018) Evaluation of ion mobility for the separation of glycoconjugate isomers due to different types of sialic acid linkage, at the intact glycoprotein, glycopeptide and glycan level. J Proteome 173:22–31

Behne A et al (2013) glyXalign: high-throughput migration time alignment preprocessing of electrophoretic data retrieved via multiplexed capillary gel electrophoresis with laser-induced fluorescence detection-based glycoprofiling. Electrophoresis 34(16):2311–2315

Bielik AM, Zaia J (2011) Multistage tandem mass spectrometry of chondroitin sulfate and dermatan sulfate. Int J Mass Spectrom 305(2–3):131–137

Bleckmann C, Geyer H, Geyer R (2011) Nanoelectrospray-MS(n) of native and permethylated glycans. Methods Mol Biol 790:71–85

Callewaert N et al (2004) Noninvasive diagnosis of liver cirrhosis using DNA sequencer-based total serum protein glycomics. Nat Med 10(4):429–434

Campbell MP et al (2014) Validation of the curation pipeline of UniCarb-DB: building a global glycan reference MS/MS repository. Biochim Biophys Acta 1844(1 Pt A):108–116

Carvalho S et al (2016) Preventing E-cadherin aberrant N-glycosylation at Asn-554 improves its critical function in gastric cancer. Oncogene 35(13):1619–1631

Chaudhury NM et al (2016) Reduced Mucin-7 (Muc7) sialylation and altered saliva rheology in Sjogren's syndrome associated oral dryness. Mol Cell Proteomics 15(3):1048–1059
Chen FT, Evangelista RA (1998) Profiling glycoprotein n-linked oligosaccharide by capillary electrophoresis. Electrophoresis 19(15):2639–2644
Ciucanu I, Costello CE (2003) Elimination of oxidative degradation during the per-O-methylation of carbohydrates. J Am Chem Soc 125(52):16213–16219
Ciucanu I, Kerek F (1984) A simple and rapid method for the permethylation of carbohydrates. Carbohydr Res 131(2):209–217
Costello CE, Contado-Miller JM, Cipollo JF (2007) A glycomics platform for the analysis of permethylated oligosaccharide alditols. J Am Soc Mass Spectrom 18(10):1799–1812
Cummings RD (2009) The repertoire of glycan determinants in the human glycome. Mol BioSyst 5(10):1087–1104
de Graaf M, Fouchier RA (2014) Role of receptor binding specificity in influenza A virus transmission and pathogenesis. EMBO J 33(8):823–841
de Haan N et al (2015) Linkage-specific sialic acid derivatization for MALDI-TOF-MS profiling of IgG glycopeptides. Anal Chem 87(16):8284–8291
Deshpande N et al (2010) GlycoSpectrumScan: fishing glycopeptides from MS spectra of protease digests of human colostrum sIgA. J Proteome Res 9(2):1063–1075
Duus J, Gotfredsen CH, Bock K (2000) Carbohydrate structural determination by NMR spectroscopy: modern methods and limitations. Chem Rev 100(12):4589–4614
Dwek RA (1995) Glycobiology: "towards understanding the function of sugars". Biochem Soc Trans 23(1):1–25
Everest-Dass AV et al (2013) Structural feature ions for distinguishing N- and O-linked glycan isomers by LC-ESI-IT MS/MS. J Am Soc Mass Spectrom 24(6):895–906
Flowers SA et al (2013) Selected reaction monitoring to differentiate and relatively quantitate isomers of sulfated and unsulfated core 1 O-glycans from salivary MUC7 protein in rheumatoid arthritis. Mol Cell Proteomics 12(4):921–931
Flowers SA et al (2017) Lubricin binds cartilage proteins, cartilage oligomeric matrix protein, fibronectin and collagen II at the cartilage surface. Sci Rep 7(1):13149
Fukuda M (1996) Possible roles of tumor-associated carbohydrate antigens. Cancer Res 56(10):2237–2244
Furukawa K et al (2001) Novel functions of complex carbohydrates elucidated by the mutant mice of glycosyltransferase genes. Biochim Biophys Acta 1525(1–2):1–12
Fuster MM, Esko JD (2005) The sweet and sour of cancer: glycans as novel therapeutic targets. Nat Rev Cancer 5(7):526–542
Gahoual R et al (2013) Rapid and multi-level characterization of trastuzumab using sheathless capillary electrophoresis-tandem mass spectrometry. MAbs 5(3):479–490
Gao WN et al (2015) Microfluidic chip-LC/MS-based glycomic analysis revealed distinct N-glycan profile of rat serum. Sci Rep 5:12844
Garner B et al (2001) Structural elucidation of the N- and O-glycans of human apolipoprotein(a): role of o-glycans in conferring protease resistance. J Biol Chem 276(25):22200–22208
Gennaro LA, Salas-Solano O (2008) On-line CE-LIF-MS technology for the direct characterization of N-linked glycans from therapeutic antibodies. Anal Chem 80(10):3838–3845
Giorgetti J et al (2018) Monoclonal antibody N-glycosylation profiling using capillary electrophoresis – mass spectrometry: assessment and method validation. Talanta 178:530–537
Gustafsson OJ et al (2015) MALDI imaging mass spectrometry of N-linked glycans on formalin-fixed paraffin-embedded murine kidney. Anal Bioanal Chem 407(8):2127–2139
Guttman A (2013) Capillary electrophoresis in the N-glycosylation analysis of biopharmaceuticals. TrAC-Trends Anal Chem 48:132–143
Guttman M, Lee KK (2016) Site-specific mapping of sialic acid linkage isomers by ion mobility spectrometry. Anal Chem 88(10):5212–5217
Guttman A, Pritchett T (1995) Capillary gel electrophoresis separation of high-mannose type oligosaccharides derivatized by 1-aminopyrene-3,6,8-trisulfonic acid. Electrophoresis 16(10):1906–1911

Guttman A et al (1996a) High-resolution capillary gel electrophoresis of reducing oligosaccharides labeled with 1-aminopyrene-3,6,8-trisulfonate. Anal Biochem 233(2):234–242
Guttman A, Chen FT, Evangelista RA (1996b) Separation of 1-aminopyrene-3,6,8-trisulfonate-labeled asparagine-linked fetuin glycans by capillary gel electrophoresis. Electrophoresis 17(2):412–417
Haltiwanger RS (2002) Regulation of signal transduction pathways in development by glycosylation. Curr Opin Struct Biol 12(5):593–598
Harvey DJ (1999) Matrix-assisted laser desorption/ionization mass spectrometry of carbohydrates. Mass Spectrom Rev 18(6):349–450
Hennig R et al (2015) N-Glycosylation Fingerprinting of Viral Glycoproteins by xCGE-LIF. Methods Mol Biol 1331:123–143
Hinneburg H et al (2016) Distinguishing N-acetylneuraminic acid linkage isomers on glycopeptides by ion mobility-mass spectrometry. Chem Commun (Camb) 52(23):4381–4384
Hinneburg H et al (2017) Unlocking cancer glycomes from histopathological formalin-fixed and paraffin-embedded (FFPE) tissue microdissections. Mol Cell Proteomics 16(4):524–536
Hofmann J et al (2014) Estimating collision cross sections of negatively charged N-glycans using traveling wave ion mobility-mass spectrometry. Anal Chem 86(21):10789–10795
Honda S et al (1989) Simultaneous determination of reducing monosaccharides by capillary zone electrophoresis as the borate complexes of N-2 pyridylglycamines. Anal Biochem 176(1):72–77
Hua S et al (2011) Comprehensive native glycan profiling with isomer separation and quantitation for the discovery of cancer biomarkers. Analyst 136(18):3663–3671
Hua S et al (2013) Isomer-specific LC/MS and LC/MS/MS profiling of the mouse serum N-glycome revealing a number of novel sialylated N-glycans. Anal Chem 85(9):4636–4643
Huang Y et al (2017) LC-MS/MS isomeric profiling of permethylated N-glycans derived from serum haptoglobin of hepatocellular carcinoma (HCC) and cirrhotic patients. Electrophoresis 38(17):2160–2167
Huffman JE et al (2014) Comparative performance of four methods for high-throughput glycosylation analysis of immunoglobulin G in genetic and epidemiological research. Mol Cell Proteomics 13(6):1598–1610
Jensen PH et al (2012) Structural analysis of N- and O-glycans released from glycoproteins. Nat Protoc 7(7):1299–1310
Kammeijer GSM et al (2017) Sialic acid linkage differentiation of glycopeptides using capillary electrophoresis – electrospray ionization - mass spectrometry. Sci Rep 7(1):3733
Karlsson NG et al (2004a) Negative ion graphitised carbon nano-liquid chromatography/mass spectrometry increases sensitivity for glycoprotein oligosaccharide analysis. Rapid Commun Mass Spectrom 18(19):2282–2292
Karlsson NG, Schulz BL, Packer NH (2004b) Structural determination of neutral O-linked oligosaccharide alditols by negative ion LC-electrospray-MSn. J Am Soc Mass Spectrom 15(5):659–672
Karlsson H, Halim A, Teneberg S (2010) Differentiation of glycosphingolipid-derived glycan structural isomers by liquid chromatography/mass spectrometry. Glycobiology 20(9):1103–1116
Khatri K et al (2017) Microfluidic capillary electrophoresis-mass spectrometry for analysis of monosaccharides, oligosaccharides, and glycopeptides. Anal Chem 89(12):6645–6655
Knox JH, Kaur B, Millward GR (1986) Structure and performance of porous graphitic carbon in liquid-chromatography. J Chromatogr 352:3–25
Kolarich D et al (2006) Comprehensive glyco-proteomic analysis of human alpha1-antitrypsin and its charge isoforms. Proteomics 6(11):3369–3380
Kolarich D et al (2008) Glycoproteomic characterization of butyrylcholinesterase from human plasma. Proteomics 8(2):254–263
Kolarich D, Lepenies B, Seeberger PH (2012) Glycomics, glycoproteomics and the immune system. Curr Opin Chem Biol 16(1–2):214–220
Kolarich D et al (2015) Isomer-specific analysis of released N-Glycans by LC-ESI MS/MS with porous graphitized carbon. Methods Mol Biol 1321:427–435

Kuberan B et al (2002) Analysis of heparan sulfate oligosaccharides with ion pair-reverse phase capillary high performance liquid chromatography-microelectrospray ionization time-of-flight mass spectrometry. J Am Chem Soc 124(29):8707–8718
Laine RA (1994) A calculation of all possible oligosaccharide isomers both branched and linear yields 1.05 x 10(12) structures for a reducing hexasaccharide: the Isomer Barrier to development of single-method saccharide sequencing or synthesis systems. Glycobiology 4(6):759–767
Lee A et al (2010) The lectin riddle: glycoproteins fractionated from complex mixtures have similar glycomic profiles. OMICS 14(4):487–499
Lee A et al (2011) Liver membrane proteome glycosylation changes in mice bearing an extra-hepatic tumor. Mol Cell Proteomics 10(9):M900538MCP200
Mancera-Arteu M et al (2017) Zwitterionic-hydrophilic interaction capillary liquid chromatography coupled to tandem mass spectrometry for the characterization of human alpha-acid-glycoprotein N-glycan isomers. Anal Chim Acta 991:76–88
Manz C, Pagel K (2018) Glycan analysis by ion mobility-mass spectrometry and gas-phase spectroscopy. Curr Opin Chem Biol 42:16–24
Marino K et al (2010) A systematic approach to protein glycosylation analysis: a path through the maze. Nat Chem Biol 6(10):713–723
May JC, McLean JA (2015) Ion mobility-mass spectrometry: time-dispersive instrumentation. Anal Chem 87(3):1422–1436
Melmer M et al (2011) Comparison of hydrophilic-interaction, reversed-phase and porous graphitic carbon chromatography for glycan analysis. J Chromatogr A 1218(1):118–123
Miller RL et al (2017) Versatile separation and analysis of heparan sulfate oligosaccharides using graphitized carbon liquid chromatography and electrospray mass spectrometry. Anal Chem 89(17):8942–8950
Moginger U et al (2018) Alterations of the human skin N- and O-glycome in basal cell carcinoma and squamous cell carcinoma. Front Oncol 8:70
Moini M (2007) Simplifying CE-MS operation. 2. Interfacing low-flow separation techniques to mass spectrometry using a porous tip. Anal Chem 79(11):4241–4246
Muramatsu T (1993) Carbohydrate signals in metastasis and prognosis of human carcinomas. Glycobiology 3(4):291–296
Nilsson J (2016) Liquid chromatography-tandem mass spectrometry-based fragmentation analysis of glycopeptides. Glycoconj J 33(3):261–272
Ohtsubo K, Marth JD (2006) Glycosylation in cellular mechanisms of health and disease. Cell 126(5):855–867
Pabst M, Altmann F (2008) Influence of electrosorption, solvent, temperature, and ion polarity on the performance of LC-ESI-MS using graphitic carbon for acidic oligosaccharides. Anal Chem 80(19):7534–7542
Pabst M et al (2007) Mass + retention time = structure: a strategy for the analysis of N-glycans by carbon LC-ESI-MS and its application to fibrin N-glycans. Anal Chem 79(13):5051–5057
Pabst M et al (2009) Comparison of fluorescent labels for oligosaccharides and introduction of a new postlabeling purification method. Anal Biochem 384(2):263–273
Pabst M et al (2012) Isomeric analysis of oligomannosidic N-glycans and their dolichol-linked precursors. Glycobiology 22(3):389–399
Packer NH et al (1998) A general approach to desalting oligosaccharides released from glycoproteins. Glycoconj J 15(8):737–747
Pang PC et al (2011) Human sperm binding is mediated by the sialyl-Lewis(x) oligosaccharide on the zona pellucida. Science 333(6050):1761–1764
Pucic M et al (2011) High throughput isolation and glycosylation analysis of IgG-variability and heritability of the IgG glycome in three isolated human populations. Mol Cell Proteomics 10(10):M111 010090
Rath CB et al (2018) Flagellin glycoproteomics of the periodontitis associated pathogen Selenomonas sputigena reveals previously not described O-glycans and rhamnose fragment rearrangement occurring on the glycopeptides. Mol Cell Proteomics 17(4):721–736

Reiding KR et al (2014) High-throughput profiling of protein N-glycosylation by MALDI-TOF-MS employing linkage-specific sialic acid esterification. Anal Chem 86(12):5784–5793
Rini JM, Esko JD (2015) Glycosyltransferases and glycan-processing enzymes. In: Varki A, Cummings RD, Esko JD et al (eds) Essentials of glycobiology, 3rd edn. Cold Spring Harbor, New York, pp 65–75
Royle L et al (2003) Secretory IgA N- and O-glycans provide a link between the innate and adaptive immune systems. J Biol Chem 278(22):20140–20153
Royle L et al (2008) HPLC-based analysis of serum N-glycans on a 96-well plate platform with dedicated database software. Anal Biochem 376(1):1–12
Rudd PM et al (2001) Glycosylation and the immune system. Science 291(5512):2370–2376
Ruhaak LR et al (2010) Optimized workflow for preparation of APTS-labeled N-glycans allowing high-throughput analysis of human plasma glycomes using 48-channel multiplexed CGE-LIF. J Proteome Res 9(12):6655–6664
Saeland E, van Kooyk Y (2011) Highly glycosylated tumour antigens: interactions with the immune system. Biochem Soc Trans 39(1):388–392
Saphire E et al (2003) Crystal structure of an Intact human IgG: antibody asymmetry, flexibility, and a guide for HIV-1 vaccine design. In: Axford J (ed) Glycobiology and medicine. Springer, New York, pp 55–66
Schultz MJ, Swindall AF, Bellis SL (2012) Regulation of the metastatic cell phenotype by sialylated glycans. Cancer Metastasis Rev 31(3–4):501–518
Schwarzer J, Rapp E, Reichl U (2008) N-glycan analysis by CGE-LIF: profiling influenza A virus hemagglutinin N-glycosylation during vaccine production. Electrophoresis 29(20):4203–4214
Seeberger PH (2015) Monosaccharide diversity. In: Varki A, Cummings RD, Esko JD et al (eds) Essentials of glycobiology, 3rd edn. Cold Spring Harbor, New York, pp 19–30
Stadlmann J et al (2008) Analysis of immunoglobulin glycosylation by LC-ESI-MS of glycopeptides and oligosaccharides. Proteomics 8(14):2858–2871
Stavenhagen K, Kolarich D, Wuhrer M (2015) Clinical glycomics employing graphitized carbon liquid chromatography-mass spectrometry. Chromatographia 78(5–6):307–320
Sumer-Bayraktar Z et al (2011) N-glycans modulate the function of human corticosteroid-binding globulin. Mol Cell Proteomics 10(8):M111 009100
Sumer-Bayraktar Z et al (2012) Micro- and macroheterogeneity of N-glycosylation yields size and charge isoforms of human sex hormone binding globulin circulating in serum. Proteomics 12(22):3315–3327
Szigeti M, Guttman A (2017) Automated N-glycosylation sequencing of biopharmaceuticals By capillary electrophoresis. Sci Rep 7(1):11663
Thanawiroon C et al (2004) Liquid chromatography/mass spectrometry sequencing approach for highly sulfated heparin-derived oligosaccharides. J Biol Chem 279(4):2608–2615
Theodoratou E et al (2016) Glycosylation of plasma IgG in colorectal cancer prognosis. Sci Rep 6:28098
Tomiya N et al (1988) Analyses of N-linked oligosaccharides using a two-dimensional mapping technique. Anal Biochem 171(1):73–90
Toyoda M et al (2008) Quantitative derivatization of sialic acids for the detection of sialoglycans by MALDI MS. Anal Chem 80(13):5211–5218
Varki A (1993) Biological roles of oligosaccharides: all of the theories are correct. Glycobiology 3(2):97–130
Varki A (2017) Biological roles of glycans. Glycobiology 27(1):3–49
Varki A, Gagneux P (2015) Biological functions of glycans. In: Varki A, Cummings RD, Esko JD et al (eds) Essentials of glycobiology, 3rd edn. Cold Spring Harbor, New York, pp 77–88
Varki A, Lowe JB (2009) Biological roles of glycans. In: Varki A et al (eds) Essentials of glycobiology. Cold Spring Harbor, New York
Wassarman PM (2011) The sperms sweet tooth. Science 333(6050):1708–1709
West C, Elfakir C, Lafosse M (2010) Porous graphitic carbon: a versatile stationary phase for liquid chromatography. J Chromatogr A 1217(19):3201–3216

Wheeler SF, Domann P, Harvey DJ (2009) Derivatization of sialic acids for stabilization in matrix-assisted laser desorption/ionization mass spectrometry and concomitant differentiation of alpha(2 --> 3)- and alpha(2 --> 6)-isomers. Rapid Commun Mass Spectrom 23(2):303–312

Wilson NL et al (2002) Sequential analysis of N- and O-linked glycosylation of 2D-PAGE separated glycoproteins. J Proteome Res 1(6):521–529

Wongtrakul-Kish K et al (2013) Characterization of N- and O-linked glycosylation changes in milk of the tammar wallaby (Macropus eugenii) over lactation. Glycoconj J 30(5):523–536

Wuhrer M et al (2004) Nano-scale liquid chromatography-mass spectrometry of 2-aminobenzamide-labeled oligosaccharides at low femtomole sensitivity. Int J Mass Spectrom 232(1):51–57

Wuhrer M, de Boer AR, Deelder AM (2009) Structural glycomics using hydrophilic interaction chromatography (HILIC) with mass spectrometry. Mass Spectrom Rev 28(2):192–206

Zhuo Y, Bellis SL (2011) Emerging role of alpha2,6-sialic acid as a negative regulator of galectin binding and function. J Biol Chem 286(8):5935–5941

Chapter 6
Synchrotron-Radiation Vacuum-Ultraviolet Circular-Dichroism Spectroscopy for Characterizing the Structure of Saccharides

Koichi Matsuo and Kunihiko Gekko

Abstract Circular-dichroism (CD) spectroscopy is a powerful tool for analyzing the structures of chiral molecules and biomolecules. The development of CD instruments using synchrotron radiation has greatly expanded the utility of this method by extending the spectra to the vacuum-ultraviolet (VUV) region below 190 nm and thereby yielding information that is unobtainable by conventional CD instruments. This technique is especially advantageous for monitoring the structure of saccharides that contain hydroxy and acetal groups with high-energy transitions in the VUV region. Combining VUVCD spectra with theoretical calculations provides new insight into the contributions of anomeric hydroxy groups and rotational isomers of hydroxymethyl groups to the dynamics, intramolecular hydrogen bonds, and hydration of saccharides in aqueous solution.

Keywords Circular dichroism · Glycoprotein · Glycosaminoglycan · Hydration · Intramolecular hydrogen bond · Molecular dynamics simulation · Saccharide · Solution structure · Structural dynamics · Synchrotron radiation · Time-dependent density functional theory · Vacuum ultraviolet

Abbreviations

AGP	α_1-acid glycoprotein
CD	circular dichroism
GG	gauche-gauche
GT	gauche-trans

K. Matsuo (✉) · K. Gekko
Hiroshima Synchrotron Radiation Center, Hiroshima University, Higashi-Hiroshima, Hiroshima, Japan
e-mail: pika@hiroshima-u.ac.jp

Y. Yamaguchi, K. Kato (eds.), *Glycobiophysics*, Advances in Experimental Medicine and Biology 1104, https://doi.org/10.1007/978-981-13-2158-0_6

MD	molecular dynamics
methyl α-D-Glc	methyl α-D-glucopyranoside
methyl β-D-Glc	methyl β-D-glucopyranoside
NMR	nuclear magnetic resonance
SR	synchrotron radiation
TDDFT	time-dependent density functional theory
TG	trans-gauche
UV	ultraviolet
VUV	vacuum ultraviolet
VUVCD	vacuum-ultraviolet circular dichroism

6.1 Introduction

Saccharides perform various biological functions in a living body such as energy storage (e.g., glycogen and starch), the constituents of plant and crustacean/arthropod exoskeletons (e.g., cellulose and chitin), the molecular recognition between proteins and cell membranes (e.g., mannose and sialic acid), and the protection against damage from cold and dry conditions (e.g., trehalose) (Sharon and Lis 1989; Schneegurt et al. 1994; Ravi Kumar 2000; Jagdale et al. 2005). They also have various physical properties such as gelation (e.g., carrageenan) and retrogradation (e.g., amylose) (Miles et al. 1985; Morris 1986; Collins and Birkinshaw 2007). Spectroscopic techniques such as X-ray crystallography and nuclear magnetic resonance (NMR) spectroscopy have been used to characterize the structures of saccharides in attempts to understand these biological functions and physical properties, but the structure–function relationships remain controversial due to the complicated contributions made by intramolecular and solvent interactions.

Circular dichroism (CD) is defined as the difference between the absorption of left- and right-handed circularly polarized light, and it sensitively reflects the steric structures of optically active materials such as biomolecules. While X-ray crystallography and NMR spectroscopy provide information at atomic spatial resolutions, whereas CD does not, this technique has been widely used for analyzing the equilibrium structures and conformations of biomolecules such as saccharides because CD spectra can be measured in samples at low concentrations for molecules of any sizes under various solvent conditions and without requiring crystallization. The principles, techniques, and applications of CD spectroscopy have been comprehensively reviewed in many books (Fasman 1996; Berova et al. 2000; Wallace and Janes 2009).

CD spectra of saccharides can be roughly divided into three wavelength regions. The two most common substituents (acetamido and carboxyl groups) display CD bands associated with the n–π* transitions at 200–240 nm and the π–π* transitions at 180–200 nm in the far-ultraviolet (far-UV) region, whereas the n–σ* transitions of acetal and hydroxy groups produce bands at 140–180 nm in the vacuum-ultraviolet (VUV) region. In particular, the CD of saccharides in the VUV region can yield basic structural information related to their biological functions and physical properties because acetal and hydroxy groups are common constituents of the saccharide backbone (Johnson Jr 1987) (the different configurations in glucose and mannose affect the carbohydrate-lectin binding constant (Farina and Wilkins 1980; Chervenak and Toone 1995), and the equatorial hydroxyl group is effective for stabilization of protein (Gekko and Koga 1983)). However, technical difficulties associated with the light source and optical device mean that conventional CD spectroscopy is not available for the high-energy transitions of chromophores such as hydroxy and acetal groups. This situation has prompted efforts aimed at developing vacuum-ultraviolet circular-dichroism (VUVCD) spectrometers.

The first VUVCD spectrometer was developed using a Hinteregger hydrogen lamp at the beginning of the 1970s (Schnepp et al. 1970), and this has been applied for the structural analysis of biomaterials, predominantly saccharides (Johnson Jr 1971, 1987; Pysh 1976; Brahms et al. 1977; Stevens 1978, 1986). These pioneering studies demonstrated that the CD spectra in the VUV region can provide important information that is unobtainable in the far-UV region, but the insufficient power of the light source restricted VUVCD measurements in solution and the principle of pairwise relationships between the configurations and the CD spectra of saccharides remain unclear for the solution structures in equilibrium between the α- and β-anomers and between the gauche (G) and trans (T) configurations. Since the 1980s, synchrotron radiation (SR) has been used as the light source of VUVCD spectrometers because a synchrotron can provide an excellent high-flux source of photons that is three to six orders of magnitude more intense than that from xenon lamps in the VUV wavelength region around 180 nm (Wallace and Janes 2009; Snyder and Rowe 1980; Sutherland et al. 1982, 1986; Wallace 2000; Ojima et al. 2001; Wallace et al. 2011). In 2017 there were more than 15 VUVCD beamlines worldwide, including at the National Synchrotron Light Source (USA), Aarhus Storage Ring (Denmark), Synchrotron Radiation Source (UK), Hiroshima Synchrotron Radiation Center (Japan) (Matsuo and Gekko 2013a), Beijing Synchrotron Radiation Facility (China), National Synchrotron Radiation Research Center (Taiwan), Synchrotron SOLEIL (France), TERAS (Japan), BESSYII (Germany), ANKA (Germany), and Diamond Light Source (UK), and at present 9 of these beamlines are operational (Wallace et al. 2011).

The VUVCD spectra of saccharides obtained using SR have successfully disclosed the contributions of configurations of hydroxy and acetal groups to the structural dynamics, intramolecular hydrogen bonds, and hydration of saccharides with an aid of theoretical calculation of the spectra using the time-dependent density function theory (TDDFT) and the molecular dynamics (MD) methods (Fig. 6.1), thereby opening a new field in glycostructural biology. This review focuses on the recent development of structural analyses of saccharides (unsubstituted and substituted saccharides) using SR-VUVCD spectroscopy, primarily based on our own work (Matsuo and Gekko 2013a, b).

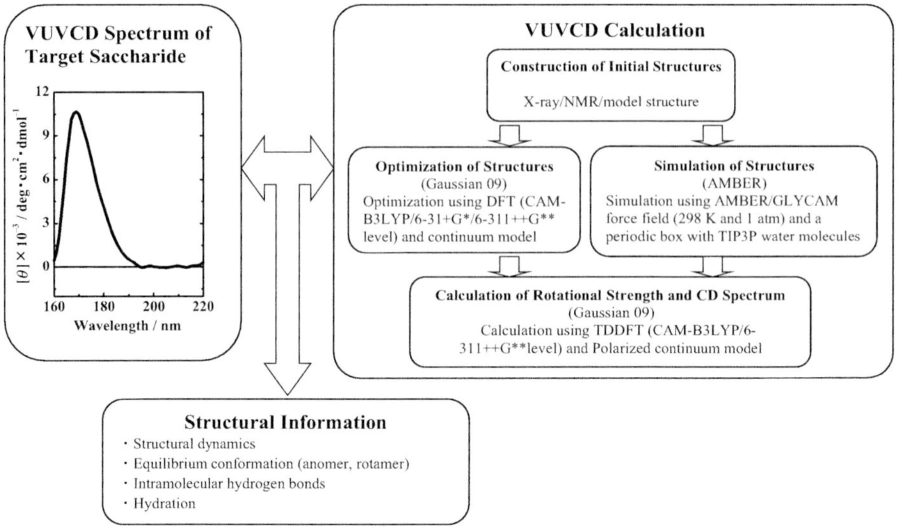

Fig. 6.1 Scheme for the structural analysis of saccharide by VUVCD spectroscopy in combination with theoretical calculation of the spectrum

6.2 Basic Theory and Calculation of Circular Dichroism

CD is defined as the difference (ΔA) between the absorbance of left- and right-handed circularly polarized light, each of which follows the Beer-Lambert law at a given wavelength:

$$\Delta A = A_{\rm L} - A_{\rm R} = \left(\varepsilon_{\rm L} - \varepsilon_{\rm R}\right) lC = \Delta\varepsilon lC \tag{6.1}$$

where $\varepsilon_{\rm L}$ and $\varepsilon_{\rm R}$ are the molar absorptions of the sample for left- and right-handed circularly polarized light, respectively, C is the molar concentration of the sample (in M), and l is the path length of the optical cell (in cm). CD instruments usually represent the CD in $\Delta\varepsilon$ or a molar ellipticity, $[\theta]$, with the latter being widely used for saccharides and proteins. $[\theta]$ is related to $\Delta\varepsilon$ by

$$\left[\theta\right] = 3298\Delta\varepsilon \tag{6.2}$$

which is in units of deg.·cm^2·dmol^{-1} (Fasman 1996; Berova et al. 2000). For polysaccharides (and other polymers such as proteins) the molar concentration unit is based on the mean residue weight, which corresponds to the molecular weight of the constituent monosaccharide or disaccharide units.

CD is induced by the interaction between electric and magnetic dipole transition moments of chromophores, and its intensity is related to the rotational strength that is theoretically defined by

$$R_{0a} = \mathrm{Im}\left\{\Psi_0 \middle| \hat{\mu} \middle| \Psi_a \cdot \Psi_a \middle| \hat{\mathbf{m}} \middle| \Psi_0\right\} \tag{6.3}$$

where R_{0a} is the rotational strength of the electric transition from the "0" to "a" states, $\hat{\mu}$ and $\hat{\mathbf{m}}$ are the electric and magnetic dipole moments, respectively, and Im{ } is the imaginary part of a complex number.

The final CD spectrum can be calculated using the following equations:

$$R_i = 1.23 \times 10^{-42} \frac{[\theta]_i \Delta\lambda}{\lambda_i} \tag{6.4}$$

$$[\theta](\lambda) = \sum_i [\theta]_i \exp\left[-\left(\frac{\lambda - \lambda_i}{\Delta\lambda}\right)^2\right] \tag{6.5}$$

where λ_i is the wavelength of the ith transition and $\Delta\lambda$ is the half bandwidth of a spectrum calculated assuming that it conforms to a Gaussian distribution (Berova et al. 2000).

Figure 6.1 shows a general scheme for the theoretical calculation of the CD spectrum. The initial structure of a target molecule (e.g., a monosaccharide) is obtained from X-ray crystallography or NMR spectroscopy, or from the model structure constructed with the standard molecular parameters. This initial structure is optimized by the density functional theory taking into consideration solvent effects or simulated by MD method such as Amber and Gromacs in explicit water molecules. The rotational strength and CD spectrum for the optimized or simulated structure are calculated with Eqs. 6.4 and 6.5 using TDDFT. The associated software and programs (e.g., Gaussian 09) can nowadays be installed on a standard workstation computer, which makes it possible to easily compare a calculated spectrum with an experimentally observed one and also identify the electric transition responsible for producing the spectrum and estimate the intact structure of the molecule including the effects of hydration.

6.3 Structural Analysis of Unsubstituted Saccharides

6.3.1 Monosaccharides

Monosaccharides exist in a complicated equilibrium structure. Figure 6.2a shows the chemical structure of D-glucose, in which there are two anomeric configurations (α and β) of the hydroxy group at C-1, three rotamer conformations of the hydroxymethyl group at C-5, and two chair conformations ($^4C_1$ and $^1C_4$). The gauche-gauche (GG), gauche-trans, (GT), and trans-gauche (TG) rotamers differ in the positions of the hydroxy oxygen at C-6 relative to the ring oxygen (Fig. 6.2b). In order to minimize conformational problems, the saccharides that preferentially adopt the $^4C_1$ conformation of the pyranose form have often been used in CD spectroscopy

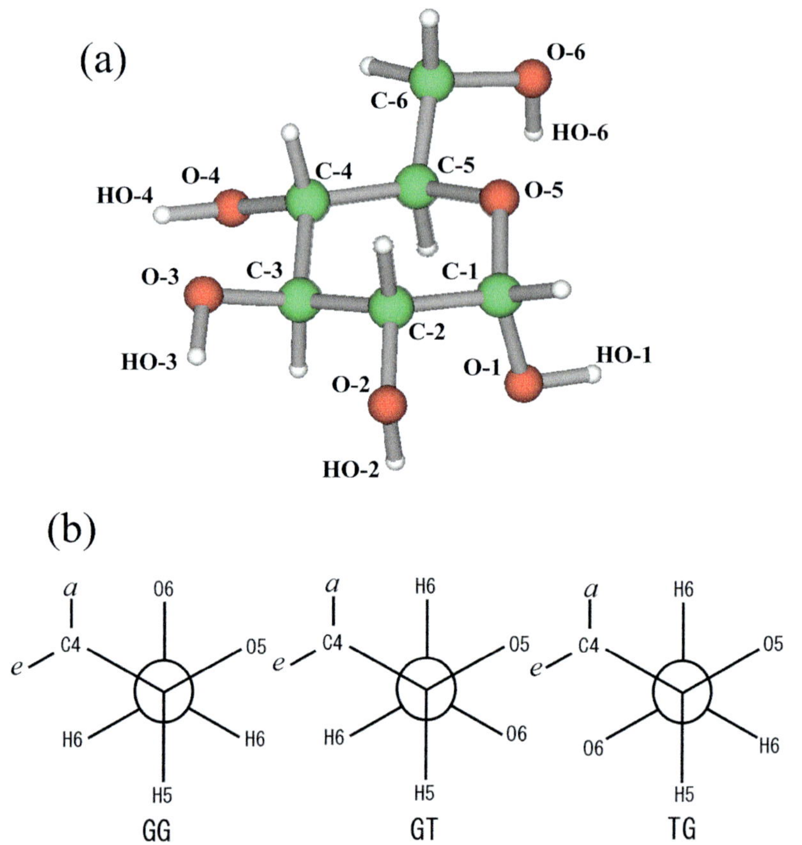

Fig. 6.2 (**a**) Chemical structure of D-glucose with an α-anomer configuration and a GT conformation. (**b**) Three possible staggered conformations (GG, GT, and TG) of the hydroxymethyl group at C-5. The C-4 hydroxy is axial (*a*) for D-galactose and equatorial (*e*) for D-glucose. The view is looking down the bond from C-5 to C-6

studies. Monosaccharides are basically composed of six conformers: α-GG, α-GT, α-TG, β-GG, β-GT, and β-TG.

The CD spectra of many monosaccharides and methyl aldopyranosides have been measured down to 165 nm in water and D_2O, and to 140 nm in thin solid films (Johnson Jr 1987; Stevens 1996). Film CD spectra provide important information on the originating orbital and energy levels (state assignments), but the pairwise relationships between CD spectra and structure in aqueous solution have not been explicitly determined due to the complexity of the equilibrium structures. These relationships have been investigated by comparing the observed SR-VUVCD spectra with those calculated by a TDDFT method (Matsuo and Gekko 2004; Matsuo et al. 2012, 2015; Kanematsu et al. 2015).

Figure 6.3 shows the VUVCD spectra of five monosaccharides in aqueous solution (Matsuo and Gekko 2004), which were obtained using a VUVCD

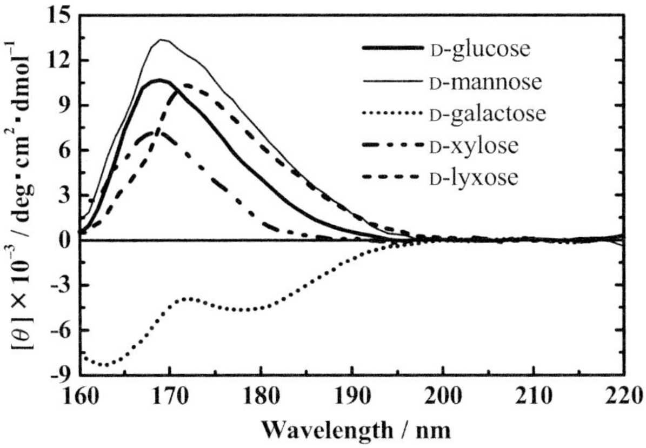

Fig. 6.3 VUVCD spectra of monosaccharides in aqueous solution at 25 °C. (Reproduced from Matsuo and Gekko 2004)

spectrophotometer and an SR light source. These monosaccharides have very similar structures but exhibit markedly different spectra. For example, D-glucose and D-galactose differ only in the configuration of the hydroxy group at C-4 (equatorial type for glucose and axial type for galactose; see Fig. 6.2), but glucose has a positive CD peak around 170 nm, while galactose produces two negative CD peaks around 165 and 180 nm; this illustrates meaning that VUVCD is highly sensitive to the structural differences of saccharides. The CD bands around 160–180 nm predominantly arise from the electronic transitions (n–σ*) of the ring oxygen (Matsuo et al. 2012; Listowsky and Englard 1968), which would be affected by the nearby hydroxy group at C-1 and the hydroxymethyl group at C-5.

It is difficult to estimate the contribution of each anomeric configuration at C-1 and each rotamer conformation at C-5 separately, since they exist in equilibrium in solution. However, deconvolution of the VUVCD spectra of D-glucose, D-mannose, and D-galactose into six independent Gaussian components (α-GG, α-GT, α-TG, β-GG, β-GT, and β-TG) using their compositions determined by NMR spectroscopy suggested that the GG and GT conformers produce positive and negative peaks, respectively, and that the negative peaks around 165 and 177 nm for D-galactose are attributable to the GT and β-TG conformers, respectively, as shown in Fig. 6.4a (Matsuo and Gekko 2004).

These component spectra were also estimated from the theoretical calculation of the spectra for α-GG, α-GT, and α-TG rotamers of methyl α-D-glucopyranoside (methyl α-D-Glc) using a TDDFT method and MD simulation (Matsuo et al. 2012). The calculated GG and GT spectra of both molecules respectively showed positive and negative CD peaks around 170 nm, which were similar to the respective component spectra obtained by deconvolution analyses (Fig. 6.4b). The same characteristic spectra were observed in the theoretical spectra of β-GG and β-GT rotamers of methyl β-D-glucopyranoside (methyl β-D-Glc), showing that the method combining TDDFT and MD is useful for characterizing monosaccharide structures (Matsuo

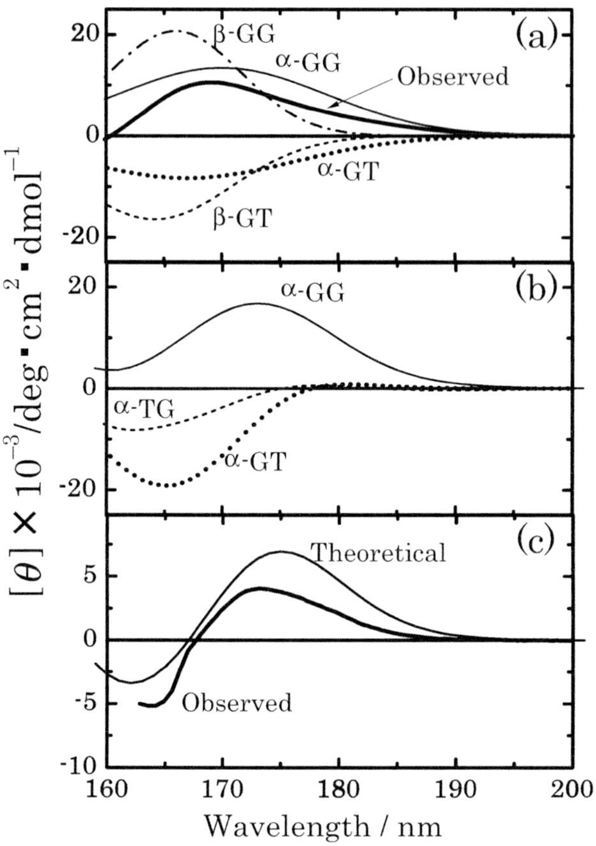

Fig. 6.4 (**a**) Component VUVCD spectra of D-glucose (α-GG, α-GT, β-GG, and β-GT). (**b**) Component VUVCD spectra of methyl α-D-Glc calculated using the TDDFT method (α-GG, α-GT, and α-TG). (**c**) Theoretical and experimental VUVCD spectra of methyl α-D-Glc. (Reproduced from Matsuo and Gekko 2004 and Matsuo et al. 2012)

et al. 2015). Further, the comparisons between the component spectra of methyl α- and β-D-Glc revealed that the α- and β-anomer configurations contribute negatively and positively to the CD around 160 nm, respectively. This prediction is supported by the finding that methyl α- and β-D-xylopyranosides with no hydroxymethyl group at C-5 exhibit negative and positive VUVCD around 163 nm, respectively (Matsuo et al. 2015). A linear combination of component spectra of three rotamers in the α- and β-anomeric forms, which differ markedly from each other, essentially reproduced the experimentally observed spectra of methyl α- and β-D-Glc (Fig. 6.4c shows the results for methyl α-D-Glc) (Matsuo et al. 2012, 2015). These results indicate that the VUVCD spectra of monosaccharides are dominantly influenced by the configurations of the hydroxy group (α/β) at C-1 and the hydroxymethyl group (GG/GT/TG) at C-5.

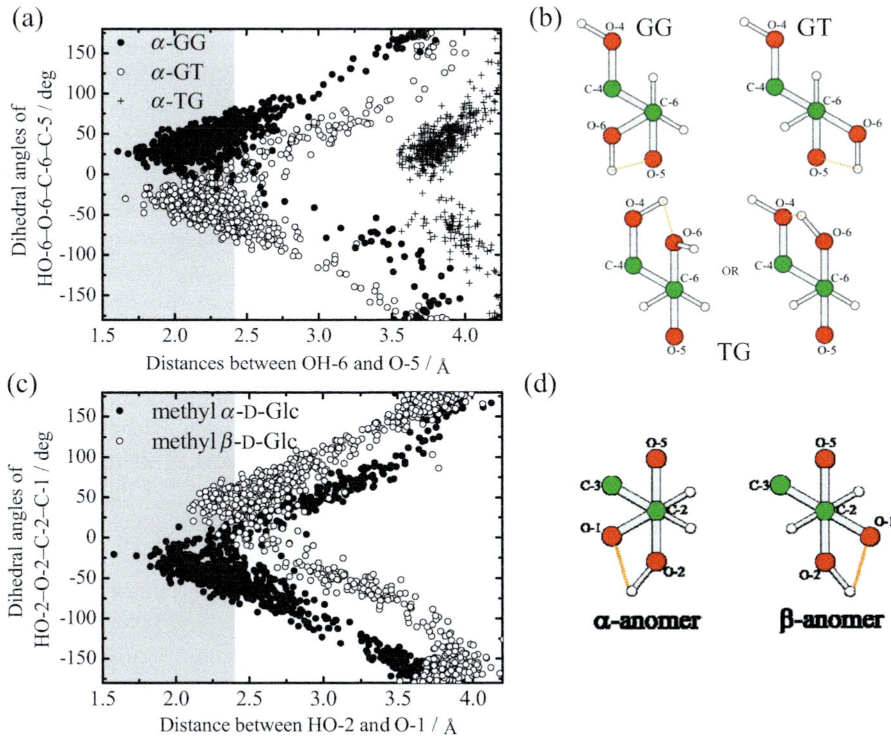

Fig. 6.5 (**a**) Plots of dihedral angles of HO-6–O-6–C-6–C-5 versus the distance between HO-6 and O-5 atoms in the α-GT, α-GG, and α-TG rotamers for methyl α-D-Glc. (**b**) Speculative steric configurations around the ring oxygen and the hydroxymethyl group of the GT, GG, and TG rotamers in methyl α- and β-D-Glc in aqueous solution. Dotted lines indicate hydrogen bonds. The view in each panel is looking down the bond from C-6 to C-5. (**c**) Plots of dihedral angles of HO-2–O-2–C-2–C-1 versus the distance between HO-2 and O-1 atoms for methyl α- and β-D-Glc. (**d**) Speculative steric configurations around the methoxy oxygen of methyl α- and β-D-Glc in aqueous solution. The view in each panel is looking down the bond from C-2 to C-1. (Reproduced from Matsuo et al. 2012, 2015)

The theoretical calculation of the VUVCD of a saccharide using the TDDFT method and MD simulation is also powerful for obtaining structural information such as the structural dynamics and intermolecular hydrogen bonds of GG, GT, and TG rotamers and α- and β-anomers. Figure 6.5a shows plots of the dihedral angles of HO-6–O-6–C-6–C-5 versus the distance between the HO-6 and O-5 atoms in the GT and GG rotamers of methyl α-D-Glc. The dihedral angles of HO-6–O-6–C-6–C-5 were around −50° and 50° for the GT and GG rotamers, respectively, within an interatomic distance of 2.4 Å, where the hydrogen atom (HO-6) could form a hydrogen bond with the ring oxygen (O-5) in both rotamers (Fig. 6.5b). On the other hand, in the TG rotamer, the dihedral angles of HO-4–O-4–C-4–C-3 converged to about 50 ° within a distance of 2.4 Å between the HO-6 and O-4 atoms and to about

−150° within a distance of 2.4 Å between the O-6 and HO-4 atoms (Matsuo et al. 2012), implying the formation of two types of hydrogen bonds (O-6–HO-4 and O-4–HO-6) in the TG rotamer (Fig. 6.5b). The differences in GT, GG, and TG spectra were therefore ascribed to the hydroxymethyl group at C-5 and the hydroxy group at C-4, which strongly affect the orientations of intramolecular hydrogen bonds around the ring oxygen (Matsuo et al. 2012, 2015).

These calculation methods can be used to characterize the α- and β-anomer configurations. Figure 6.5c shows plots of the dihedral angles of the hydroxy group of HO-2–O-2–C-2–C-1 versus the distance between the HO-2 and O-1 atoms for three rotamers of methyl α-D-Glc and methyl β-D-Glc. It is evident that the dihedral angles of HO-2–O-2–C-2–C-1 in all three rotamers were around −50° for the α-anomer and around 50° for the β-anomer, within an interatomic distance of 2.4 Å, where the hydrogen atom (HO-2) could form a hydrogen bond with the methoxy oxygen (O-1) (Fig. 6.5d), although the hydrogen bond would be considerably more unstable in the β-anomer than in the α-anomer. The differences in the α- and β-anomers could therefore be ascribed to the different dynamics of the hydroxy group at C-2 and the methoxy group at C-1, which strongly affect the orientations of intramolecular hydrogen bonds around the methoxy oxygen (Matsuo et al. 2015).

The effects of hydration are reflected in the different VUVCD spectra of methyl α-D-Glc in H_2O and D_2O, where all hydrogen atoms of the hydroxy groups are replaced by deuterium in the latter. The observed isotope differences in the peak position (blue shift) and the intensity of spectra were compared with those theoretically calculated using multicomponent quantum mechanics, which incorporates the quantum deviation of hydrogen nuclei from equilibrium geometry to provide information about the conformation of the hydroxy groups and water molecules along the solvation surface (Kanematsu et al. 2015). It was found that the modification of the solvation surface is essential for reproducing the observed isotope effect on the

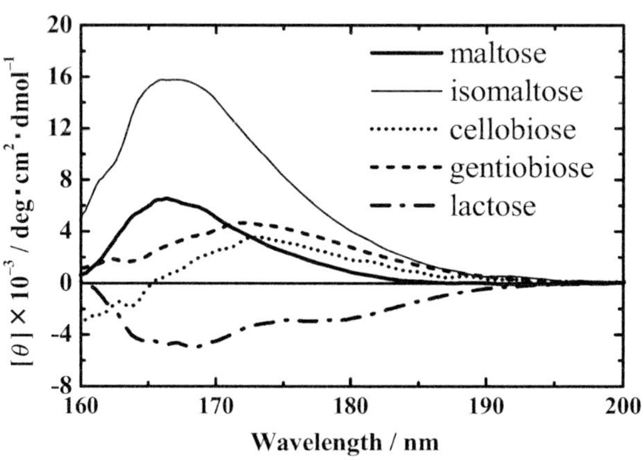

Fig. 6.6 VUVCD spectra of disaccharides in aqueous solution at 25 °C. (Reproduced from Matsuo and Gekko 2004)

spectrum; that is, the isotope differences are strongly dependent on solute–solvent interactions. These observations indicate that the VUVCD spectrum can offer insight into not only the conformations but also the hydration of saccharides.

6.3.2 *Disaccharides*

Figure 6.6 shows the SR-VUVCD spectra of various disaccharides (Matsuo and Gekko 2004). While these spectra qualitatively resemble those of the monomer components, it is apparent that the spectra of the constituent monomers are not additive since the configuration of the anomeric hydroxy group that links two monomers is fixed and the resulting ether linkages ($1 \rightarrow 2$, $1 \rightarrow 3$, $1 \rightarrow 4$, and $1 \rightarrow 6$) result in different CD. The deconvolution analysis of VUVCD spectra might not be applicable to disaccharides because they comprise a larger number of independent components compared with monosaccharides. However, comparisons of the spectra for disaccharides allowed us to predict the contributions of the glycosidic linkage to the disaccharide CD spectra. Isomaltose and maltose, which consist of two glucose molecules joined by α-1,6- and α-1,4-glycosidic linkages, respectively, both exhibit a positive CD peak around 170 nm, which is larger for isomaltose. Gentiobiose and cellobiose, which consist of two D-glucose molecules joined by β-1,6- and β-1,4-glycosidic linkages, respectively, also both show a positive CD peak around 170 nm, with this being larger for gentiobiose. It is therefore expected that the 1,6-glycosidic linkage has a greater potential to increase the CD than does the 1,4-glycosidic linkage. The CD peaks of gentiobiose and cellobiose are red shifted relative to those of isomaltose and maltose, suggesting that the α-glycosidic linkage is a higher-energy bond than the β-glycosidic linkage in both 1,4- and 1,6-linkages. Lactose, which is a disaccharide consisting of D-galactose and D-glucose joined by a β-1,4-glycosidic linkage, exhibits two successive negative CD peaks around 177 and 168 nm, which would arise mainly from the constituent galactose unit that produces negative CD peaks. These predictions should be confirmed by theoretical calculations using TDDFT in order to further improve our understanding of the VUVCD spectra of disaccharides.

6.3.3 *Oligosaccharides and Polysaccharides*

Few SR-VUVCD spectra of oligosaccharides and polysaccharides have been measured, but the measured VUVCD spectra of unsubstituted oligosaccharides have interesting features. The CD spectra of three malto-oligosaccharides with an α-1,4-glycoside linkage between glucose units (maltose, maltotriose, and maltoheptaose) exhibited negative CD around 190 nm and positive CD around 170 nm, and this changed markedly as the chain length increased. However, there were no indications of a chain-length dependence of CD attributable to the formation of a helical structure (Lewis and Johnson Jr 1978), although amylose—which is a polysaccharide composed of maltose units—is known to have a helical conformation in solution (Wu and Sarko 1978). The CD spectra of an isomalto-oligosaccharide series with an α-1,6-glycoside linkage between glucose units (dextran) were measured in aqueous solution down to 175 nm and in films down to 150 nm over average molecular weights ranging from 410 to 303,000 (Stipanovic et al. 1980). Although the solution spectra did not exhibit the positive CD around 167 nm that was observed in film spectra, the ellipticity at 177 nm increased slightly (by less than 10%) with the molecular weight, but without showing a critical point around a molecular weight of 2,000 that was observed in various solution properties (Stipanovic et al. 1980). The ability to measure the SR-VUVCD spectra of oligosaccharides in solution would yield useful information on the roles of intra- and intermolecular interactions in the conformational transition from an oligomer to a polymer, which is an important step toward understanding the structures and properties of polysaccharides.

A considerable amount of VUVCD data is available for polysaccharides, most of which have been reported by Stevens and coworkers (Stevens 1996; Stipanovic et al. 1980; Liang et al. 1979; Stipanovic and Stevens 1980, 1981). All of the studied D-glucans consisting of (1 → 3)- and (1 → 4)-linkages exhibit positive bands for the α-linkage (e.g., amylose) and negative bands for the β-linkage (e.g., cellulose) at 164–172 nm, but no such correlation with anomeric configuration is observed for (1 → 6) glucans such as dextran and pustulan, with the CD being positive in both cases. The band around 180–190 nm is assigned to the ether oxygen atom of the linkage and so may reflect the local flexibility of the polysaccharide chain. The CD spectrum of pustulan with a β-1,6-glycoside linkage exhibited a positive peak around 177 nm but a negative peak around 185 nm, with the pustulan solution turning into a gel over time at 10°C (Stipanovic and Stevens 1981). X-ray diffraction confirmed that the time dependence of this negative peak was attributable to the helical formation of pustulan in solution. An agarose solution shows a positive band at 180 nm, which decreases in intensity and shifts to longer wavelengths due to melting of the gel as the temperature increases (Stipanovic and Stevens 1981). Comparing the VUVCD spectra of polysaccharide gels with those theoretically calculated for the cross-linking unit structures would give new insights into the gel-formation mechanism.

6.4 Substituted Saccharides

Many biologically important saccharides include various substituents such as carboxyl, acetamido, and sulfate groups, some of which exhibit absorption in the far-UV region. VUVCD measurements in aqueous solutions have revealed that *N*-acetylglucosamine has a negative band around 210 nm, a positive band around 180 nm, and a shoulder around 190 nm, while *N*-acetylgalactosamine shows two negative bands around 210 and 170 nm and a positive band around 190 nm (Bush and Ralapati 1981; Matsuo et al. 2009a). D-Glucuronic acid shows a positive band around 210 nm and a negative band around 185 nm, both of which are strongly dependent on the pH value. The electronic transitions of these chromophores are influenced in a complex manner by the solvent conditions, the surrounding substituents, and the ring conformation. The VUVCD spectra of the chitin series of oligosaccharides with β-1,4-glycoside linkages between *N*-acetylglucosamine units all exhibit a negative CD band at about 210 nm and a positive band at about 192 nm. There is a small progressive increase in the molar ellipticity per monosaccharide unit for both the positive and negative bands as the chain length increases (Coduti et al. 1977). Carboxymethyl dextran shows very different CD spectra depending on the content (degree of substitution), neutralization, and counterions of the carboxyl group due to its modified n–π* transition (Gekko 1979).

Glycosaminoglycans, which are typical substituted polysaccharides, are linear polymers composed of repeating disaccharide units of hexosamine (glucosamine or galactosamine) and uronic acid (glucuronic or iduronic acid). These disaccharide units have similar structures, while glycosaminoglycans exhibit very different VUVCD spectra in aqueous solution, as shown in Fig. 6.7 (Matsuo et al. 2009a; Chakrabarti 1981). The differences in the CD spectra are larger in the VUV region than in the far-UV region, indicating that VUVCD spectroscopy provides much more information about the structures of glycosaminoglycans in aqueous solution. The configuration of the carboxyl group and the number and position of sulfate group definitely influence the spectra in the VUV region. Theoretically assigning the bands in the VUV region is difficult due to the large number of overlapping electronic transitions (n–π*, π–π*, and n–σ*) and the modified ring conformation. However, the contributions of the substituents may be estimated by comparing the spectra of polysaccharides and the constituent monomers such as *N*-acetylglucosamine and D-glucuronic acid.

As described above, saccharides exhibit complex VUVCD spectra due to their structural diversity. However, the high sensitivity of VUVCD spectra to the

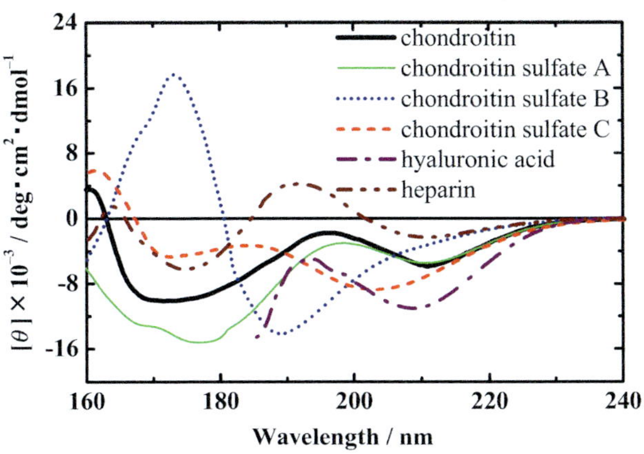

Fig. 6.7 VUVCD spectra of glycosaminoglycans in aqueous solution at 25 °C. (Reproduced from Matsuo et al. 2009a)

electronic transitions of chromophores can provide important and new information about the conformation, interaction, and hydration of saccharides in solution. Combining VUVCD spectroscopy with theoretical calculations is therefore a powerful technique for elucidating the structural biology of glycoconjugates.

6.5 Application to Glycoprotein

Glycoproteins are difficult targets for structural analysis because they are barely crystallized and exceed the size measurable by NMR spectroscopy. VUVCD spectroscopy can detect the CD spectra of sugar moieties in the VUV region below 200 nm, thereby improving the predictive accuracy of the secondary structures.

Human α_1-acid glycoprotein (AGP) is the major acute-phase protein that exhibits the ability to bind to neutral drugs and steroid hormones. This protein consists of 183 amino acid residues, while its 5 glycan chains constitute about 40% of the total mass (36 kDa), and so hence its tertiary structure was unknown until Schönfeld et al. (2008) successfully applied X-ray crystallography to the recombinant unglycosylated protein (Schönfeld et al. 2008). Also, Nishi et al. (2011) studied the differences in drug-binding selectivity based on the crystal structures (Nishi et al. 2011). Although the constituent carbohydrate groups are generally thought to have little effect on the structure and function of glycoproteins, the tertiary structure of intact AGP (its glycosylated form) remains unknown. Its native state exhibited a characteristic VUVCD spectrum (down to 160 nm) of β-strand-rich proteins, as shown in Fig. 6.8 (Matsuo et al. 2009b). The VUVCD spectrum of the protein moiety was obtained by subtracting the contributions of the constituent sugars (L-fucose, D-mannose, D-galactose, *N*-acetylglucosamine, and *N*-acetylneuraminic acid) (inset

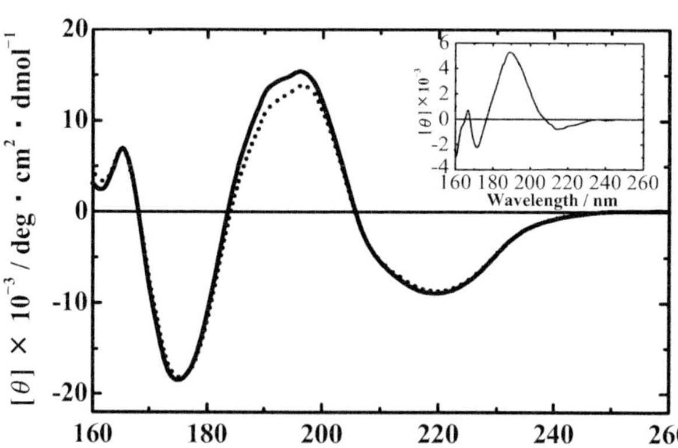

Fig. 6.8 VUVCD spectra of AGP at 25 °C. Dotted line shows the spectrum of the protein moiety obtained by subtracting the spectrum of the glycan moiety (inset) from the spectrum of AGP (solid line). (Reproduced from Matsuo et al. 2009b)

of Fig. 6.8). The glycan chains make small but non-negligible contributions to the secondary structure. The spectrum of the protein moiety indicates that it comprises 37.7% β-strands (ten segments) and 14.4% α-helices (three segments), which is almost consistent with the structure predicted by homology modeling due to the Modeller package, but differs considerably from that based on the X-ray structure: 39.3% β-strands (eight segments) and 25.7% α-helices (four segments) (Schönfeld et al. 2008). This large difference between the VUVCD estimate and X-ray data suggests that the constituent glycan chains—if they have any structural effects—can affect the tertiary structure in the N- and C-terminal regions, since the effects of crystal packing might not be disregarded (Schönfeld et al. 2008). These observations indicate that VUVCD spectroscopy is useful for predicting the intact structure of glycoproteins, which is essential for understanding their structure–function relationships.

6.6 Conclusions

VUVCD spectroscopy using SR has yielded novel detailed information about the structures of saccharides in solution that are unobtainable by conventional CD spectroscopy. Although VUVCD spectroscopy, unlike X-ray crystallography or NMR spectroscopy, does not provide information at atomic spatial resolutions, it can be used to analyze the equilibrium conformation associated with the structural dynamics, intramolecular hydrogen bonds, and hydration of monosaccharides in combination with theoretical calculation of the spectra, expanding its applicability to solution chemistry. Since the structures of numerous glycoconjugates are difficult to analyze by X-ray crystallography or NMR spectroscopy, the role of SR-VUVCD spectroscopy is very likely to increase in structural glycobiology.

Acknowledgments The authors sincerely thank Professor Masaki Taniguchi and staff members of Hiroshima Synchrotron Radiation Center, and Dr. Tomoyuki Fukazawa of JASCO Corporation for constructing the SR-VUVCD spectrometer. The authors are indebted to many collaborators for their helpful technical assistance and discussions. This work was supported by a Grant-in-Aid for Scientific Research from the Ministry of Education, Science, Sports, and Culture of Japan (No. 15K07028 to K.M.) and by a grant from Kurata Memorial Hitachi Science and Technology Foundation.

References

Berova N, Nakanishi K, Woody RW (eds) (2000) Circular dichroism: principles and applications, 2nd edn. Wiley-VCH, New York

Brahms S, Brahms J, Spach G, Brack A (1977) Identification of beta, beta-turns and unordered conformations in polypeptide chains by vacuum ultraviolet circular dichroism. Proc Natl Acad Sci USA 74:3208–3212

Bush CA, Ralapati S (1981) Vacuum UV circular dichroism spectroscopy of acetamido sugars. In: Brant DA (ed) Solution properties of polysaccharides. American Chemical Society, Washington, DC, pp 293–302
Chakrabarti B (1981) Carboxyl and amide transitions in the circular dichroism of glycosaminoglycans. In: Brant DA (ed) Solution properties of polysaccharides. American Chemical Society, Washington, DC, pp 275–292
Chervenak MC, Toone EJ (1995) Calorimetric analysis of the binding of lectins with overlapping carbohydrate-binding ligand specificities. Biochemistry 34:5685–5695
Coduti PL, Gordon EC, Bush CA (1977) Circular dichroism of oligosaccharides containing N-acetyl amino sugars. Anal Biochem 78:9–20
Collins MN, Birkinshaw C (2007) Comparison of the effectiveness of four different crosslinking agents with hyaluronic acid hydrogel films for tissue-culture applications. J Appl Polym Sci 104:3183–3191
Farina RD, Wilkins RG (1980) Kinetics of interaction of some α- and β-D-monosaccharides with concanavalin A. Biochim Biophys Acta 631:428–438
Fasman GR (ed) (1996) Circular dichroism and the conformational analysis of biomolecules. Plenum, New York
Gekko K (1979) Circular dichroism study on polyelectrolytic properties of carboxymethyldextran. Biopolymers 18:1989–2003
Gekko K, Koga S (1983) Increased thermal stability of collagen in the presence of sugars and and polyols. J Biochem 94:199–205
Jagdale GB, Grewal PS, Salminen SO (2005) Both heat-shock and cold-shock influence trehalose metabolism in an entomopathogenic nematode. J Parasitol 91:988–994
Johnson WC Jr (1971) A circular dichroism spectrometer for the vacuum ultraviolet. Rev Sci Instrum 42:1283–1286
Johnson WC Jr (1987) The circular dichroism of carbohydrates. Adv Carbohydr Chem Biochem 45:73–124
Kanematsu Y, Kamiya Y, Matsuo K, Gekko K, Kato K, Tachikawa M (2015) Isotope effect on the circular dichroism spectrum of methyl α-D-glucopyranoside in aqueous solution. Sci Rep 5:17900
Lewis DG, Johnson WC Jr (1978) Optical properties of sugars. VI Circular dichroism of amylose and glucose oligomers. Biopolymers 17:1439–1449
Liang JN, Stevens ES, Morris ER, Rees DA (1979) Spectroscopic origin of conformation-sensitive contributions to polysaccharide optical activity: vacuum-ultraviolet circular dichroism of agarose. Biopolymers 18:327–333
Listowsky I, Englard S (1968) Characterization of the far ultraviolet optically active absorption bands of sugars by circular dichroism. Biochem Biophys Res Commun 30:329–332
Matsuo K, Gekko K (2004) Vacuum-ultraviolet circular dichroism study of saccharides by synchrotron radiation spectroscopy. Carbohydr Res 339:591–597
Matsuo K, Gekko K (2013a) Construction of a synchrotron-radiation vacuum-ultraviolet circular-dichroism spectrophotometer and its application to the structural analysis of biomolecules. Bull Chem Soc Jpn 86:675–689
Matsuo K, Gekko K (2013b) Circular-dichroism and synchrotron-radiation circular-dichroism spectroscopy as tools to monitor protein structure in a lipid environment. Methods Mol Biol 974:151–176
Matsuo K, Namatame H, Taniguchi M, Gekko K (2009a) Vacuum-ultraviolet circular dichroism analysis of glycosaminoglycans by synchrotron-radiation spectroscopy. Biosci Biotechnol Biochem 73:557–561
Matsuo K, Namatame H, Taniguchi M, Gekko K (2009b) Membrane-induced conformational change of alpha1-acid glycoprotein characterized by vacuum-ultraviolet circular dichroism spectroscopy. Biochemistry 48:9103–9111
Matsuo K, Namatame H, Taniguchi M, Gekko K (2012) Vacuum-ultraviolet electronic circular dichroism study of methyl α-D-glucopyranoside in aqueous solution by time-dependent density functional theory. J Phys Chem A 116:9996–10003

Matsuo K, Namatame H, Taniguchi M, Gekko K (2015) Solution structures of methyl aldopyranosides revealed by vacuum-ultraviolet electronic circular-dichroism spectroscopy. Biomed Spectrosc Imaging 4:269–282
Miles MJ, Morris VJ, Orford PD, Ring SG (1985) The roles of amylose and amylopectin in the gelation and retrogradation of starch. Carbohydr Res 135:271–281
Morris VJ (1986) Gelation of polysaccharides. In: Mitchell JR, Ledward DA (eds) Functional properties of food macromolecules. Elsevier Applied Science Publishers, New York, pp 121–170
Nishi K, Ono T, Nakamura T, Fukunaga N, Izumi M, Watanabe H, Suenaga A, Maruyama T, Yamagata Y, Curry S, Otagiri M (2011) Structural insights into differences in drug-binding selectivity between two forms of human alpha1-acid glycoprotein genetic variants, the A and F1*S forms. J Biol Chem 286:14427–14434
Ojima N, Sakai K, Matsuo K, Matsui T, Fukazawa T, Namatame H, Taniguchi M, Gekko K (2001) Vacuum-ultraviolet circular dichroism spectrophotometer using synchrotron radiation: optical system and on-line performance. Chem Lett 30:522–523
Pysh ES (1976) Optical activity in the vacuum ultraviolet. Annu Rev Biophys Bioeng 5:63–75
Ravi Kumar MNV (2000) A review of chitin and chitosan applications. React Funct Polym 46:1–27
Schneegurt MA, Sherman DM, Nayar S, Sherman LA (1994) Oscillating behavior of carbohydrate granule formation and dinitrogen fixation in the cyanobacterium Cyanothece sp. strain ATCC 51142. J Bacteriol 176:1586–1597
Schnepp O, Allen S, Peason EF (1970) The measurement of circular dichroism in the vacuum ultraviolet. Rev Sci Instrum 40:1136–1141
Schönfeld DL, Ravelli RBG, Mueller U, Skerra A (2008) The 1.8-Å crystal structure of alpha1-acid glycoprotein (Orosomucoid) solved by UV RIP reveals the broad drug-binding activity of this human plasma lipocalin. J Mol Biol 384:393–405
Sharon N, Lis H (1989) Lectins as cell recognition molecules. Science 246:227–234
Snyder PA, Rowe EM (1980) The first use of synchrotron radiation for vacuum ultraviolet circular dichroism measurements. Nucl Inst Methods 172:345–349
Stevens ES (1978) Far (vacuum) ultraviolet circular dichroism. Methods Enzymol 49:214–221
Stevens ES (1986) Vacuum UV circular dichroism of polysaccharides. Photochem Photobiol 44:287–293
Stevens ES (1996) Carbohydrate. In: Fasman GR (ed) Circular dichroism and the conformational analysis of biomolecules. Plenum Press, New York, pp 501–530
Stipanovic AJ, Stevens ES (1980) Vacuum ultraviolet circular dichroism of (1→6)-β-D-glucan. Int J Biol Macromol 2:209–212
Stipanovic AJ, Stevens ES (1981) Vacuum UV dichroism of D-glucans. In: Brant DA (ed) Solution properties of polysaccharides. American Chemical Society, Washington, DC, pp 303–315
Stipanovic AJ, Stevens ES, Gekko K (1980) Vacuum ultraviolet circular dichroism of dextran. Macromolecules 13:1471–1473
Sutherland JC, Keck PC, Griffin KP, Takacs PZ (1982) Simultaneous measurement of absorption and circular dichroism in a synchrotron spectrometer. Nucl Inst Methods Phys Res 195:375–379
Sutherland JC, Lin B, Mugavero JA, Trunk J, Tomasz M, Santella R, Marky L, Breslauer KJ (1986) Vacuum ultraviolet circular dichroism of double stranded nucleic acids. Photochem Photobiol 44:295–301
Wallace BA (2000) Conformational changes by synchrotron radiation circular dichroism spectroscopy. Nat Struct Biol 7:708–709
Wallace BA, Janes RW (eds) (2009) Modern techniques for circular dichroism and synchrotron radiation circular dichroism spectroscopy. IOS, Amsterdam
Wallace BA, Gekko K, Hoffmann SV, Lin YH, Sutherland JC, Tao J, Wien F, Janes RW (2011) Synchrotron radiation circular dichroism (SRCD) spectroscopy: an emerging method in structural biology for examining protein conformations and protein interactions. Nucl Instrum Methods Phys Res A 649:177–178
Wu HCH, Sarko A (1978) The double-helical molecular structure of crystalline b-amylose. Carbohydr Res 61:7–25

Chapter 7
Biophysical Analyses for Probing Glycan-Protein Interactions

Masamichi Nagae and Yoshiki Yamaguchi

Abstract Glycan-protein interactions occur at many physiological events, and the analyses are of considerable importance for understanding glycan-dependent mechanisms. Biophysical approaches including 3D structural analysis are essential for revealing glycan-protein interactions at the atomic level. The inherent diversity of glycans suits them to function as identification tags, e.g., distinguish self from the nonself components of pathogens. However, the complexity of glycans and poor affinities for interaction partners limit the usefulness of conventional analyses. To cope with such troublesome glycans, a logical sequence of biophysical analyses need to be developed. In this chapter, we introduce a workflow of glycan-protein interaction analysis consisting of six steps: preparation of lectin and glycan, screening of glycan ligand, determination of binding epitope, quantitative interaction analysis, 3D structural analysis, and molecular dynamics simulation. Our increasing knowledge and understanding of lectin-glycan interactions will hopefully lead to the design of glyco-based medicines and vaccines.

Keywords Glycan microarray · Nuclear magnetic resonance · X-ray crystallography · Isothermal titration calorimetry · Surface plasmon resonance · Molecular dynamics simulation · Frontal affinity chromatography

Abbreviations

DIS	Deuterium-induced isotope shift
EM	Electron microscopy
FAC	Frontal affinity chromatography
GlcNAc	*N*-Acetylglucosamine

M. Nagae
Graduate School of Pharmaceutical Sciences, The University of Tokyo, Tokyo, Japan

Y. Yamaguchi (✉)
Synthetic Cellular Chemistry Laboratory, RIKEN Cluster for Pioneering Research, Wako, Saitama, Japan
e-mail: yyoshiki@riken.jp

Y. Yamaguchi, K. Kato (eds.), *Glycobiophysics*, Advances in Experimental Medicine and Biology 1104, https://doi.org/10.1007/978-981-13-2158-0_7

ITC	Isothermal titration calorimetry
MD	Molecular dynamics
NMR	Nuclear magnetic resonance
SPR	Surface plasmon resonance
STD-NMR	Saturation transfer difference NMR
trNOE	Transferred nuclear Overhauser effect

7.1 Introduction

Glycan-protein interactions are crucial for numerous biological processes such as quality control of nascent proteins, cell-cell interaction, and pathogenic recognition. Of special interest is the breadth of glycan-mediated recognition events, which comes from the extremely high information-coding capacity of glycans. The diversity of oligosaccharides is inherent in their configurational variability, created by the broad range of monosaccharide species, anomeric configurations, multiple linkages, and branching. Glycan complexity is potentially suitable for coding biological information, and interest is growing to "decode" the sugar codes. To this end, investigating glycan-protein interaction at the atomic level is a necessity as only then will we understand the detailed mechanisms of these glycan-mediated biological activities. Recent techniques introduced include glycan microarray analysis and other high-throughput screening methods, and now we can begin to have a more systematic approach to analyzing glycan-protein interactions. We here introduce biophysical approaches to analyze the structure and function of glycan-protein interactions.

7.2 A Flowchart for Analyzing Glycan-Protein Interactions by Biophysical Methods

Biophysical investigation of glycan-protein interactions is typically divided into six steps in our laboratory:

1. Large-scale preparation of protein and glycan
2. Screening of glycan ligand
3. Determination of binding epitope (glycotope)
4. Quantitative glycan-protein interaction analysis
5. 3D structural study on protein-glycan interaction
6. Molecular dynamics simulation of glycan-protein interaction

Crystallographic and nuclear magnetic resonance (NMR) analyses are direct methods for obtaining atomic details of glycan-protein interactions if the binding epitope is known. In those cases where the binding epitopes are unknown, biophysical approaches such as listed above are required for structural studies. A computational approach combined with experiments is especially effective to investigate the flexible nature of glycans.

7.3 Large-Scale Preparation of Protein and Glycan

Biophysical studies require large amounts of protein and glycan. In general, much effort needs to be paid for their preparation. Examples of our achievements are shown below.

7.3.1 Preparation of Protein

Carbohydrate recognition domain (lectin domain) or glyco-enzymes directly bind glycan, and it is convenient to use the isolated domain for biophysical interaction analysis. Normally, 5–10 mg of purified protein is required. Bacterial expression systems are frequently used to express lectin domains in recombinant form (Fig. 7.1a), but obtaining large amounts of pure lectin in a soluble form is often difficult using bacterial expression systems. For instance, C-type lectin and C-type lectin-like domains contain three conserved disulfide bridges, and the cysteine residues hamper soluble expression with correct folding. Refolding of protein from inclusion bodies may be needed to produce functional lectin domains. The protein is solubilized and refolded under controlled conditions. This refolding procedure has been effective for the purification of mammalian C-type lectins and C-type lectin-like domains (Nagae et al. 2013b, 2014a, b, 2016b).

Fusion of a tag protein can improve the solubility of target proteins. In our laboratory, several fusion tags have been used, including protein disulfide isomerase (PDI) and B1 domain of streptococcal protein G (GB1). Single chain Fv (scFv) originating from anti-carbohydrate MLS128 antibody was successfully expressed as a soluble form with PDI fusion (Subedi et al. 2012), while the other fusion tags such as thioredoxin or disulfide isomerase DsbC were not sufficient to produce soluble scFv (Fig. 7.1b). The expressed MLS128-scFv retains full binding activity toward synthetic carbohydrate antigens. The lectin domain of a C-type lectin-like receptor Dectin-1 was expressed as a GB1-fused protein (Dulal et al. 2016). The domain is correctly folded, is a monomer, and specifically binds a β-glucan ligand. The addition of the GB1 tag to the lectin domain greatly increased both yield (five-fold) and solubility (14-fold) (Fig. 7.1c).

Bacterial expression systems are the first choice for protein expression but are not always successful for glycosylated proteins. In these cases, mammalian expression systems may be necessary to express correctly folded proteins with posttranslational modifications. HEK293 cells, CHO cells, and their variant cells are frequently used as hosts. For example, protein *O*-mannosyl kinase (POMK), which phosphorylates the mannose residue on α-dystroglycan, has three *N*-glycosylation

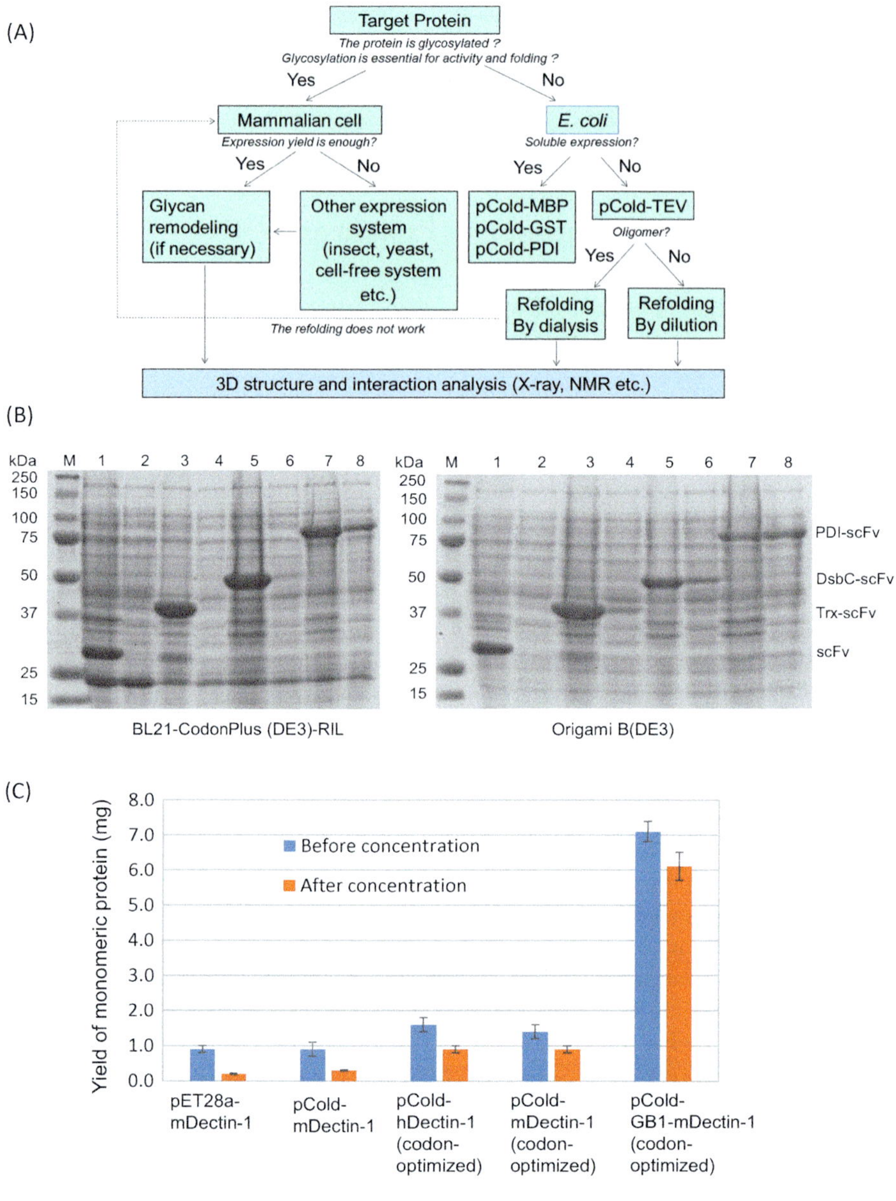

Fig. 7.1 Large-scale preparation of recombinant proteins (**a**) A flowchart of large-scale preparation of lectins for 3D structure and interaction studies (**b**) Small-scale expression of differently fused MLS128-scFv in BL21 CodonPlus (left) and Origami B (right) strains analyzed by SDS-PAGE. M: Molecular weight of marker proteins; Lanes 1–8, odd number lanes are composition of total cell lysates, and even number lanes are composition of soluble cell lysates. Lanes 1 and 2 for scFv without fusion tag, Lanes 3 and 4 for thioredoxin (Trx)-fused scFv, Lanes 5 and 6 for DsbC-fused scFv, and Lanes 7 and 8 for PDI-fused scFv. The figure was adapted from a previous paper (Subedi et al. 2012) with permission (**c**) Comparison of expression yield of Dectin-1 with and without GB1 fusion tag before and after concentration (to 1 mL) of purified sample. Results are presented as mean ± SD (n = 3) for pET28a murine Dectin-1, pCold murine Dectin-1, pCold-human Dectin-1 (codon-optimized), pCold murine Dectin-1 (optimized), and pCold GB1-fused murine Dectin-1 (codon-optimized). The figure was adapted from the paper (Dulal et al. 2016) with permission

sites within the catalytic domain, and two sites out of three are partially *N*-glycosylated. To abolish the heterogeneity originating from partial *N*-glycan occupancy, two asparagine residues were mutated to glutamine and the mutated POMK expressed in HEK293 variant cells lacking *N*-acetylglucosaminyl transferase I (GnT-I), to produce uniformly glycosylated protein ($Man_5GlcNAc_2$-glycan attached), which was successfully crystallized and the atomic structure elucidated (Nagae et al. 2017b).

7.3.2 Preparation of Glycan

It is another challenge to obtain sufficient amounts of glycan and glycoconjugates suitable for biophysical assays. Prior to interaction analysis, it is necessary to design and prepared a suitable glycan ligand. Ligand design is an extremely important step. The design may arise from information in the literature, or else the binding specificity needs to be ascertained experimentally (Steps 2 and 3). Several methods can be used to prepare glycan ligands: (1) purification of a free natural glycan or liberation of a specific glycan from natural glycoconjugates (glycoprotein, glycolipid, etc.), (2) chemical synthesis, and (3) chemoenzymatic synthesis. Liberation of glycan from natural glycoprotein can yield large amounts of glycan provided that there is a plentiful supply of the natural glycoprotein. *N*-glycans attached onto glycoproteins can be liberated by chemical methods (e.g., hydrazinolysis) or by enzymatic methods (e.g., PNGase F) and then labeled with a fluorescent tag 2-aminopyridine for detection. Pyridylamino derivatives of oligosaccharides can be purified by HPLC, and their chemical structures confirmed by solution NMR spectroscopy and mass spectrometry. For example, a biantennary *N*-glycan bearing bisecting GlcNAc was successfully purified with a yield of 120 μg from 200 mg of IgG and used for the analysis of a plant lectin E4-PHA (Nagae et al. 2014c, 2016b).

Glycan ligands "designed" through synthetic approaches are of great importance in structural studies. *N*-glycan core units in particular cannot be prepared easily by enzymatic approaches, and chemical synthesis plays a significant role (Hanashima et al. 2014b). Synthetic bisected hexasaccharide has been used as ligand for the C-type lectin receptor murine DCIR2, the legume lectin PHA-E, and the plant lectin Calsepa. The use of designed synthetic ligands has been invaluable for revealing the ligand recognition mode of these lectins by X-ray crystallography and NMR.

Recombinant and/or chemoenzymatic methods have been developed to obtain sufficient amounts of glycopeptides for structural analysis. It is known that several lectins recognize not only the glycan part but also the aglycon part (Nagae and Yamaguchi 2015), and here it is best to use the appropriate glycoconjugate (e.g., glycopeptide, glycolipid, etc.) as the ligand molecule. A C-type lectin receptor CLEC-2 binds to *O*-glycosylated podoplanin, and binding requires both disialyl core 1 glycan (NeuAcα2-3Galβ1–3[NeuAcα2–6]GalNAc) and neighboring amino acid residues. The *O*-glycosylated podoplanin glycopeptide was prepared by overexpression using engineered yeast cells and in vitro sialylation (Kato et al. 2008).

The substrate of POMK was chemoenzymatically synthetized, composed of core M3 trisaccharide unit (GalNAcβ1–3GlcNAcβ1-4Man) and α-dystroglycan peptide. Mono-mannosylated peptide was chemically synthesized in advance, and then additional disaccharide units were successively introduced by two glycosyltransferases (Nagae et al. 2017b).

7.4 Screening of Glycan Ligand

It is very important to understand how glycans function in recognition and signaling systems within an organism and the responses with microbes and pathogens (Cummings and Pierce 2014). Heterogeneous glycans are potentially a vast source of information, and glycan microarrays embedding such diverse glycans are now becoming one of the key technologies to provide information on recognition determinants (glycotopes) of carbohydrate-binding proteins (Feizi and Chai 2004; Geissner and Seeberger 2016).

7.4.1 Glycan Microarray

When doing the glycan microarray analysis, it is essential to immobilize a series of glycans onto the plate. The immobilization is one of the critical issues in the development of a glycan microarray. There are several ways to immobilize carbohydrate probes onto a suitable surface with retention of function. The chemically modified or unmodified glycans can be non-covalently or covalently attached onto the surface (Fig. 7.2a). Neoglycolipid (NGL) is an artificial glycolipid used to construct glycan microarrays via non-covalent interaction. In NGL technology, lipid-conjugated oligosaccharides prepared from natural sources or chemically synthesized are immobilized onto nitrocellulose-coated microarray slides (Palma et al. 2014). In covalent immobilization, various chemical reactions have been developed (Hyun et al. 2017). Among them, the conjugation reaction using amine-linked glycans is most widely used, and the Consortium for Functional Glycomics uses this chemistry for a glycan microarray platform (Fig. 7.2b). Several different functional groups which react with amines are utilized such as cyanuric chloride, *N*-hydroxysuccinimide ester (NHS), and epoxide. Measuring lectin binding requires labeling of the lectin with a fluorescent probe or the use of an antibody. Even so carbohydrate-lectin interaction is often too weak to be detected. In such cases, oligomerization of the lectin is a strategy to enhance inherent weak binding. Interactions may be detected by introducing dimeric glutathione *S*-transferase (GST) into the lectin as a fusion tag, as in the case of a mammalian lectin ZG16p-binding immobilized glycans (Kanagawa et al. 2014; Hanashima et al. 2015). Alternatively, it is possible to enhance binding by increasing the amount of surface glycans (Hanashima et al. 2015). However, it should be noted that the density of immobilized glycans may affect the specificity

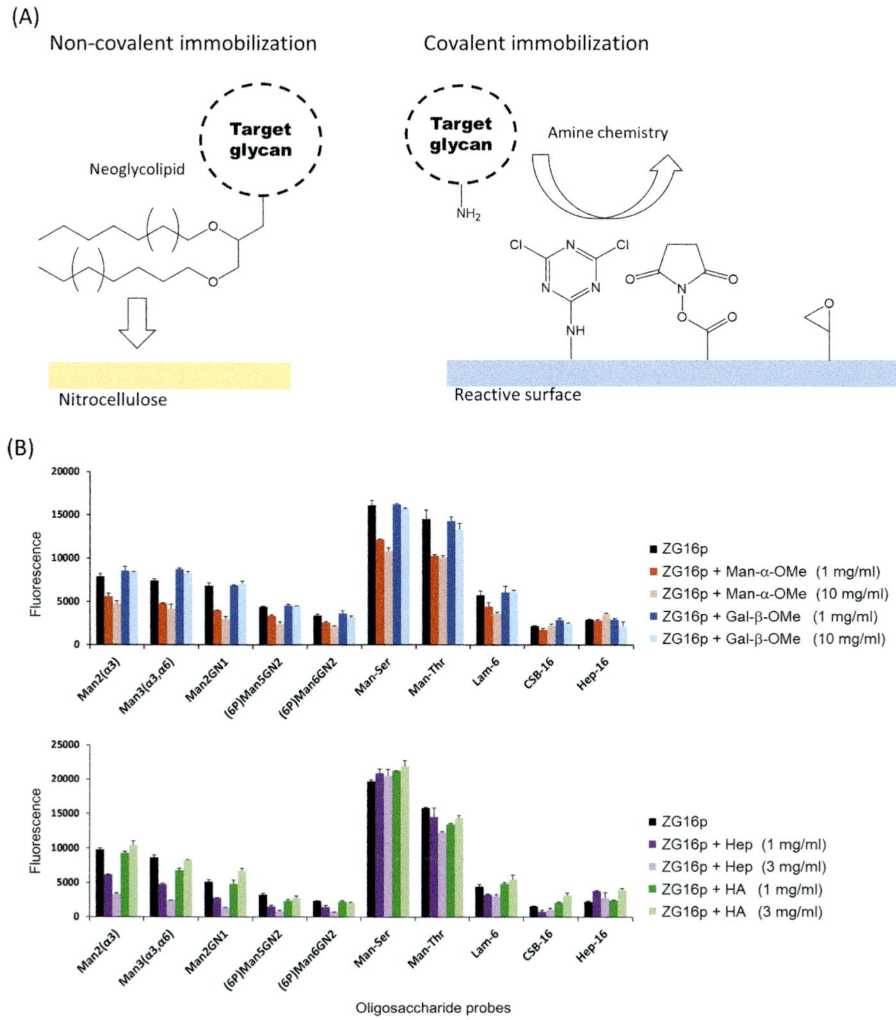

Fig. 7.2 Glycan microarray analysis for the lectin specificity
(**a**) Non-covalent and covalent glycan immobilization for glycan microarray analysis
(**b**) On-array inhibition assays for analyzing the binding of GST-fused human ZG16p to selected ligands which are non-covalently immobilized on the plate as neoglycolipids. Man-α-OMe, Gal-β-OMe (upper panel), oligosaccharide fractions (>20-mer) of heparin (Hep), and hyaluronic acid (HA) (lower panel) were used as inhibitors. This figure was adapted from a previous paper (Kanagawa et al. 2014) with permission

of lectins (Horan et al. 1999). Instead of fluorescence detection, label-free glycan biosensors have been developed to push the limit of detection (Hushegyi and Tkac 2014).

Glycan microarrays have various applications. Microarrays usually simply display relative binding affinity, but quantitative parameters (e.g., dissociation constant, K_D) can be obtained by applying various concentrations (Liang et al. 2007). The estimated K_D values are comparable to those obtained by other methods such as surface plasmon resonance (SPR). Moreover, effects of heterogeneous glycan mixing can be evaluated by immobilizing a mixture of glycans onto one spot (Liang et al. 2011). On-array inhibition assays can show if two different ligands bind competitively to the lectin (sharing a common binding site) or independently (each using a different binding site) (Fig. 7.2b) (Kanagawa et al. 2014). Glycan microarrays can also evaluate direct binding of mammalian cells and viruses (Nimrichter et al. 2004; Song et al. 2011), as well as biologically active glycans, if cell signaling can be linked to a fluorescent probe (Pai et al. 2016). This could result in rapid progress toward elucidating pathway-specific glycans.

7.4.2 Other Screening Methods to Evaluate Lectin-Glycan Interaction

Glycan microarray analysis often may overestimate the strength of weak binding, and weak binding is often seen in carbohydrate-lectin interactions. In addition, nonspecific binding is usually present, which contributes a background signal. Therefore it is preferable to include other screening methods to validate the results. Several methods have already been developed to screen glycans which bind target lectins. The thermal shift assay is widely used for ligand screening. In this approach, melting temperature is measured in the absence and the presence of ligands and any shifts assessed (Vedadi et al. 2006). A temperature shift is indicative of increased stability acquired upon ligand binding. The purified protein is mixed with ligand in the presence of a fluorescent probe, such as SYPRO Orange, that binds to the denatured state. This method is simple and can be used for rapid ligand screening. Frontal affinity chromatography (FAC) can quantitatively evaluate weak interactions between lectins and glycans, as explained in the following section. The FAC method can also be used to screen for ligand binding. NMR is an excellent technique to determine the binding epitope of a lectin-glycan interaction. When the binding epitope is determined, a suitable ligand of minimum size is designed for use in X-ray crystallography. NMR can also be applied to high-throughput ligand screening (Meyer and Peters 2003) and be advantageously used with a mixture of binder and non-binders.

7.5 Determination of Binding Epitope

The binding epitope of a glycan (glycotope) can be deduced from the results of glycan microarray and FAC analysis using a series of glycans. More directly, NMR spectroscopy can provide the glycotope at atomic level. NMR spectroscopy has demonstrated its potential with a wide range of applications for studies of structure and function of biological molecules. The unique potential of NMR spectroscopy is its ability to generate high-resolution data not only on structure but also on dynamics and binding interactions. There are many types of NMR experiments that can be applied for very different experimental conditions both for the sample and the NMR parameters. NMR experiments provide various parameters such as chemical shifts, spin-spin couplings (*J* couplings), nuclear Overhauser effect (NOEs), and relaxation times. These parameters provide the atomic information on chemical environment, dihedral angles, distance, and mobility, respectively. Here two NMR techniques, saturation transfer difference NMR (STD-NMR) and deuterium-induced isotope shift (DIS), are described below.

7.5.1 Saturation Transfer Difference NMR in Lectin–Glycan Interactions

Saturation transfer difference NMR (STD-NMR) spectroscopy is a method for studying protein-ligand interactions by observing signals emanating from the ligand (Mayer and Meyer 2001; Haselhorst et al. 2009). When the target protein is saturated with selective pulses, the saturation is transferred to a bound ligand through intermolecular ^{1}H-^{1}H cross relaxation (Fig. 7.3a). Protons which are spatially close to the protein surface receive a higher degree of saturation compared with protons further away, and those on a non-binding ligand do not receive any saturation from the protein and are therefore not attenuated in the *on*-resonance spectrum. The subtraction of two spectra, one with protein resonance saturation (*on*-resonance) and the other without saturation of the protein resonance (*off*-resonance), results in a final STD-NMR spectrum showing only signals from bound ligands. For binding epitope analyses, the signal intensities of each peak are evaluated from an amplification factor (AF) = $(I_{\mathrm{off}} - I_{\mathrm{on}})/I_{\mathrm{off}}$, where I_{on} is the *on*-resonance signal intensity and I_{off} is *off*-resonance signal intensity. Association constants can be estimated from a plot of AF against ligand concentration or initial growth rates of STD AF (Hanashima et al. 2010; Angulo et al. 2010). STD-NMR requires only a small amount of native protein, and no expensive protein labeling is necessary. STD-NMR spectroscopy can be utilized to detect binding of ligands with K_{D} values in the 10^{-2}–10^{-8} M range. Often screening for optimum STD-NMR conditions is required, such as probe temperature and optimal ligand-to-protein ratio (Fig. 7.3b). D_2O solvent is preferred over H_2O, to avoid direct and water-mediated saturation artifacts. High-affinity ligands typically reside longer within the protein binding site, undergo slow chemical exchange, and are thus not detectable by STD-NMR spectroscopy.

1D STD-NMR analysis is often hampered by severe overlapping of non-anomeric signals. For instance, the ring proton signals from the two branches (α1-3 and α1-6 branches) of bisected biantennary *N*-glycan largely overlap in 1D STD-NMR analysis. In contrast, the methyl proton signals from GlcNAc and core fucose are sharp

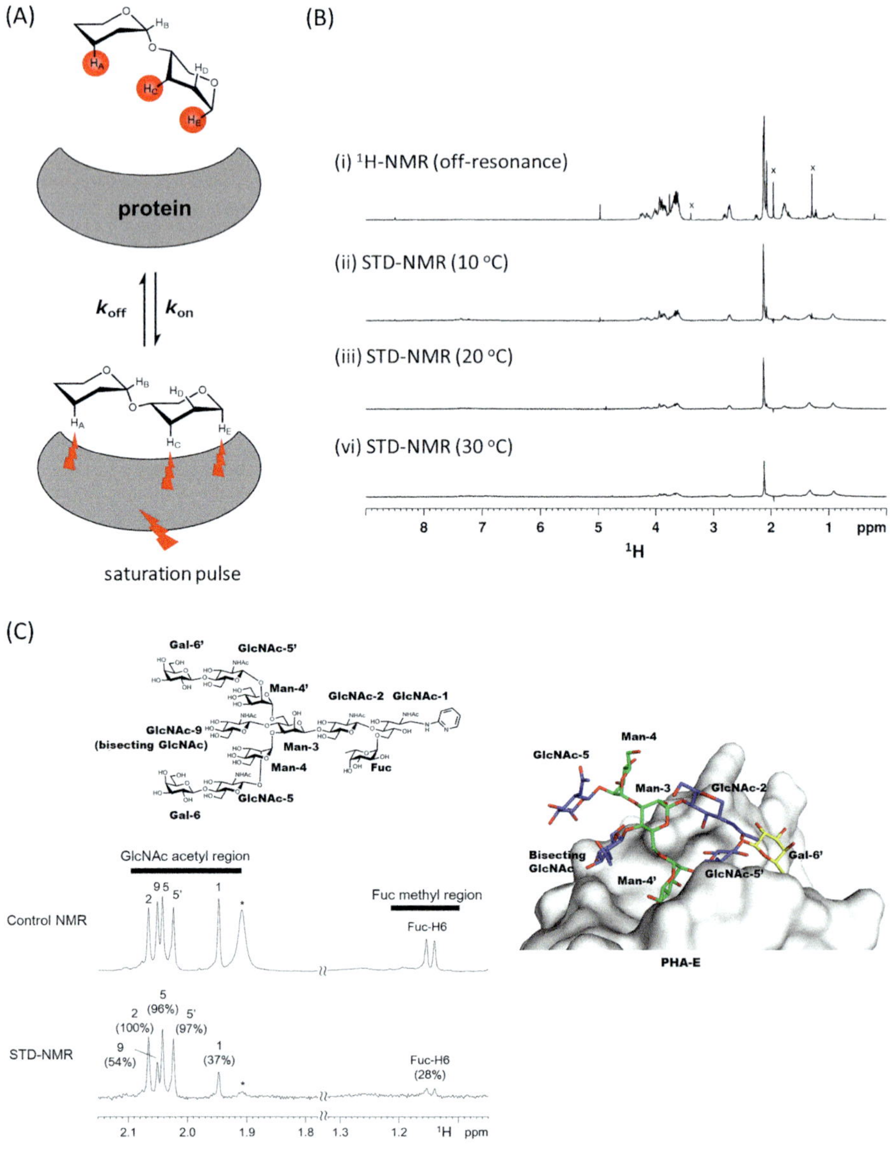

Fig. 7.3 Saturation transfer difference NMR (STD-NMR) analysis for group epitope mapping
(**a**) Basic principle of saturation transfer difference NMR (STD-NMR). The ligand protons which receive saturation from the protein are highlighted in red. From the degree of ligand saturation, one can map the group epitope which is recognized by the protein
(**b**) STD-NMR spectra of $(Neu5Ac)_6$ with an anti-polysialic acid antibody 12E3 at 10 °C (ii), 20 °C (iii), and 30 °C (iv) with the control ^{1}H-NMR spectrum (off-resonance) at 10 °C (i). STD-NMR spectra were collected by irradiating iteratively at 7 ppm (on resonance)/ 40 ppm (off resonance). x; the signals from low-molecular-weight impurities, which are nearly null in STD-NMR spectra. This figure was adapted from a previous paper (Hanashima et al. 2013) with permission
(**c**) STD-NMR spectrum (lower) and reference ^{1}H-NMR spectrum (upper) of the di-galactosylated biantennary glycan bearing bisecting GlcNAc and core fucose (top panel) in the presence of PHA-E (glycan: subunit molar ratio is 1:2). Normalized amplification factor (GlcNAc-2 signal is set to 100%) is shown in parenthesis. The signal indicated with an asterisk originates from PHA-E protein. This figure was adapted from a previous paper (Nagae et al. 2014c) with permission. Crystal structure of PHA-E in complex with bisected *N*-glycan (PDB code: 5AVA) is shown in right panel

and well separated for certain glycans. Hence these signals can be used as a reliable probe for branch-specific epitope mapping as exemplified by an analysis of the interaction between biantennary bisected glycan and a legume lectin PHA-E (Nagae et al. 2014c). The methyl proton signals of GlcNAc from outer branches and inner chitobiose unit show almost comparable intensities, indicating that the entire glycan contributes equally to binding with PHA-E (Fig. 7.3c). Consistent with this result, the crystal structure of bisected glycan-PHA-E complex showed that most sugar residues of the glycan ligand are recognized by PHA-E (Fig. 7.3c, right panel). Another approach is the use of 2D ^{1}H-^{13}C STD-heteronuclear single quantum coherence (HSQC) to avoid the severe signal overlap in 1D NMR spectra. 2D STD- spectra have been collected to investigate the interaction between oligosialic acid and anti-oligosialic acid antibodies, 12E3 and A2B5 (Hanashima et al. 2013). STD-NMR is also applicable to analyze multiple binding modes of a unique ligand in a single binding site (Angulo and Nieto 2011). The duplicate sugar-binding modes of C-type lectin, DC-SIGN (Angulo et al. 2008), and anti-HIV antibody, 2G12 (Enríquez-Navas et al. 2012), have been analyzed in this way.

7.5.2 *Application of ^{13}C-NMR Using Deuterium-Induced Isotope Shift*

In solution NMR experiments, it is difficult to directly observe the hydroxyl protons of glycans because of their inherently fast exchange with water protons. Deuterium secondary isotope shifts on ^{13}C-chemical shifts (deuterium-induced ^{13}C isotope shifts, DIS) have been observed at the geminal ^{13}C-signal of exchangeable hydroxyl protons (Pfeffer et al. 1979). In the presence of H_2O and D_2O, the line shape of the ^{13}C-OH/D signal is highly dependent on the H/D exchange rate (Fig. 7.4a). In a very slow H/D exchange environment, a deuterium-induced ^{13}C isotope shift (DIS) is observed in corresponding isotopomers, ^{13}C-OH and ^{13}C-OD, with a 0.09–0.15 ppm difference, whereas in the case of fast exchange, the equilibrium makes each isotopomer indistinguishable and gives a singlet signal at an averaged chemical shift.

This NMR technique has been applied in the weak carbohydrate-carbohydrate interaction of LewisX (Hanashima et al. 2011). LewisX, a trisaccharide of galactose, fucose and GlcNAc, is often found at the termini of glycolipids and glycoproteins and initiates cell-cell interactions by forming a LewisX-Ca^{2+} complex (Eggens et al. 1989). Calcium-dependent intermolecular interactions of the LewisX-Ca^{2+} complex are observed, via the proton exchange rates of each hydroxyl group in an equimolar H_2O/D_2O solution (Fig. 7.4b). The conformation of the glycan ligand can also be analyzed by this method, including that of β(1,3)-glucan dependent on the degree of polymerization (DP) (Hanashima et al. 2014a). With β-glucan, the exchange of ^{13}C-OH/D is dramatically retarded as DP extends, suggesting changes in quaternary structure (Fig. 7.4c). Recently a new two-dimensional ^{1}H-^{13}C NMR method, named INTOXSY, was developed to estimate the H/D exchange rate constants of OH groups based on DIS (Battistel et al. 2017).

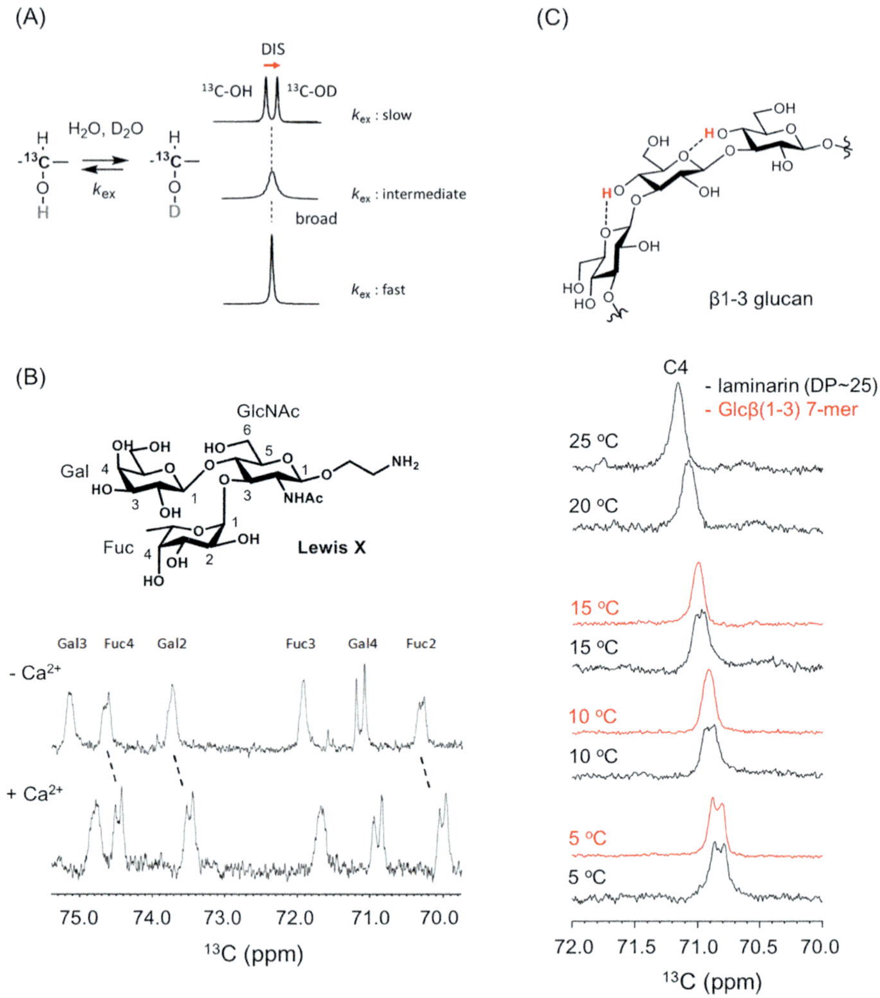

Fig. 7.4 Hydrogen exchange analysis using deuterium-induced isotope shift
(**a**) Deuterium-induced ^{13}C isotope shifts for analyzing the proton exchange rate of sugar hydroxyl groups. k_{ex}, exchange rates of hydroxyl protons with water. The chemical shift difference (isotope shift) is ~0.15 ppm, which is not dependent on the magnetic field
(**b**) ^{13}C-NMR spectra of 40 mM LewisX tetrasaccharide with expanded secondary carbon area. LewisX at 5 °C (top), and LewisX with 1.0 M calcium chloride at 5 °C (bottom). The sample was dissolved in 10 mM sodium acetate buffer (pH 6.0) composed of H_2O:D_2O = 1:1
(**c**) ^{13}C-NMR spectra of short β1–3-glucans DP7 (red) and laminarin (black) at 5, 10, 15, 20, and 25 °C. ^{13}C-NMR signal focuses on the C4 position. All ligands were dissolved in the same buffer shown in (B). The figures were adapted from previous papers (Hanashima et al. 2011, 2014a) with modifications

7.6 Quantitative Lectin-Glycan Interaction Analysis

Like other protein-ligand interactions, lectin-glycan interactions are characterized by several biophysical properties, such as the dissociation constant K_d, thermodynamics parameters such as enthalpy ΔH and entropy ΔS, and kinetics (e.g., k_{on} and k_{off} rate constants). Many quantitative biophysical methods exist to analyze lectin-glycan interactions such as frontal affinity chromatography (FAC) analysis, isothermal titration calorimetry (ITC), and surface plasmon resonance (SPR). All these methods can provide K_d; however, a suitable method needs to be selected for each purpose. Among them, ITC can be performed without labeling or immobilization of lectin or glycan. A lot of sample (typically 10 mmol for ligand and protein) is usually needed. In FAC analysis, immobilization of lectin is required, and the glycan must be labeled for detection. SPR also requires immobilization of lectin or glycan, but amount of sample is typically small. Combining complementary methods provides more insights than a single method.

7.6.1 *Frontal Affinity Chromatography (FAC)*

Frontal affinity chromatography (FAC) is a powerful method to assess lectin-glycan interactions as it is applicable to weak interactions (Tateno et al. 2007). In FAC analysis, glycans are applied to a lectin-immobilized column and the elution profile monitored. Dissociation constants for each glycan are calculated from the equation:

$$V - V_0 = \frac{B_t}{K_d + [A]_0}$$

where V and V_0 are the elution volumes of the column and negative control, respectively. B_t is the effective lectin content expressed in mole, and $[A]_0$ is the initial glycan molar concentration. The amount of analyte needed depends on the affinity of analyte and ligand: higher affinities require smaller amounts.

An example of FAC analysis is that of the binding of various *N*-glycans to a mannose-binding Jacalin-related lectin Calsepa. The lectin was immobilized on the column, and a panel of pyridylaminated (PA) or *p*-nitrophenyl (*p*NP) glycans were successively passed through the column. In this way, Calsepa was found to specifically bind bisected *N*-glycans (Nagae et al. 2017a) (Fig. 7.5a). The specificity is explained by the crystal structure of bisected glycan-Calsepa complex (Fig. 7.5a, right). The ligand is in a flipped-back conformation, which bisected glycan prefers, and the unique ligand conformation enables the 2:1 sandwich binding mode.

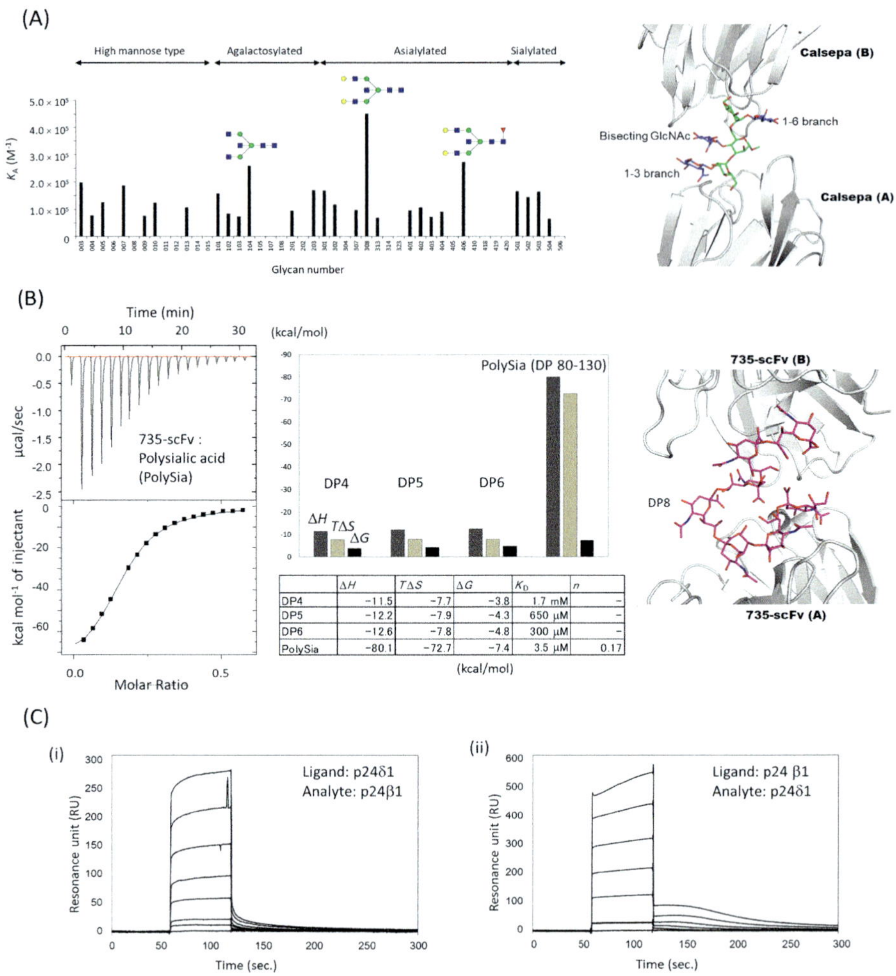

Fig. 7.5 Quantitative analyses of lectin-glycan interaction
(**a**) FAC analysis of a mannose-binding Jacalin-related lectin Calsepa toward *N*-glycans. Association constant (K_A) of Calsepa lectin for each glycan is shown. Monosaccharide symbols follow the SNFG (symbol nomenclature for glycans) system (Varki et al. 2015). The FAC data is adapted from a paper (Nagae et al. 2017a) with permission. Crystal structure of Calsepa in complex with bisected glycan (corresponding to #104) is shown in right panel (PDB code: 5AV7)
(**b**) Isothermal titration calorimetry (ITC) analysis of scFv735 using polysialic acid (PolySia, DP 80–130) (left). Bar graph and table of thermodynamic parameters for DP4, DP5, DP6, and PolySia are shown in middle panel. Crystal structure of scFv735 in complex with octa-sialic acid (DP8) is shown in the right panel (PDB code: 3WBD)
(**c**) Surface plasmon resonance (SPR) analyses of the protein-protein interaction between p24β1 and p24δ1 GOLD domains. (i) Sensorgram showing the interaction between immobilized p24δ1 GOLD domain and analyte p24β1 GOLD domain. p24β1 GOLD domain aliquots (0, 5, 25, 50, 100, 200, and 400 μM) were injected for 60 s at a flow rate of 30 μl/min. Subtracted sensorgrams with blank lanes are shown. (ii) Sensorgram showing the interaction between immobilized p24β1 GOLD domain and analyte p24δ1 GOLD domain. The dilution series of p24δ1 GOLD domain are the same as those of p24β1 GOLD domain shown in (i). The SPR data was adapted from a paper (Nagae et al. 2016a) with permission

7.6.2 Isothermal Titration Calorimetry (ITC)

Isothermal titration calorimetry (ITC) is a technique that directly determines thermodynamic parameters of protein-ligand interactions (Dam et al. 2016). In a single experiment, ITC measures n (number of binding sites on the protein), ΔH (enthalpy of binding), and K_A (association constant). From the K_A value, the free energy of binding (ΔG) and the entropy of binding (ΔS) can be calculated from the following equation:

$$\Delta G = \Delta H - T\Delta S = -RT \ln K_A$$

where ΔG, ΔH, and ΔS are the changes in free energy, enthalpy, and entropy of binding, respectively. T is the absolute temperature and R is the gas constant (8.314 J/mol/K).

An example is the ITC analysis of an anti-polysialic acid antibody 735. This antibody shows high affinity toward α2–8-linked polysialic acid as the degree of polymerization (DP) increases. To investigate the DP-dependent affinity enhancement mechanism, a series of oligo (DP4–6) and polysialic acids (DP 80–130) and single chain Fv originating from antibody 735 (scFv735) were evaluated by ITC (Nagae et al. 2013a) (Fig. 7.5b). As DP increases so does the affinity of scFv735. Thermodynamic parameters revealed that a large entropy loss upon interaction ($T\Delta S = -72.7$ kcal/mol for polysialic acid) is compensated for by a larger enthalpy gain ($\Delta H = -80.1$ kcal/mol). Importantly, the crystal structure of scFv735-ligand complex indicates that antibody 735 can recognize internal residues of polysialic acids (Fig. 7.5b, right). Therefore, increasing DP may provide a rapid cycle of dissociation-association and thereby enhance the affinity (Nagae and Yamaguchi 2014).

Typical lectin-glycan interactions are exothermic, but occasionally an endothermic reaction is observed, e.g., interaction between heparan sulfate and heparin-binding hemagglutinin (Huang et al. 2017). Kinetic parameters can also be obtained from ITC experiments (Burnouf et al. 2012; Vander Meulen et al. 2016).

7.6.3 Surface Plasmon Resonance (SPR)

Surface plasmon resonance (SPR) is one of the most widely used analytical methods for detecting biomolecular interactions and kinetics. It monitors interactions between an immobilized ligand on a thin metal layer and a soluble-free analyte by detecting changes in the resonance angle due to an increase in concentration of analyte at the surface. SPR can monitor real-time interaction with biosensors by measuring the resonance units against time.

Sensorgrams can be utilized for the determination of kinetic parameters, association/dissociation constants (K_A/K_D), on/off rates (k_{on}/k_{off}), as well as thermodynamics parameters. There are two major methods to calculate kinetic parameters, first is by directly fitting the curve sensorgram to ideal binding models (more applicable to slow association/dissociation binding) and the second is steady-state analysis (for fast association/dissociation interactions). Maximum resonance values of a series of analyte concentrations (R_{eq}) are plotted against analyte concentration. In simple 1:1 steady-state binding model, the plot is fitted to the following equation:

$$R_{eq} = C \times R_{max} / (C + K_D)$$

where C is concentration of analyte, R_{max} is maximum binding response, and K_D is the equilibrium dissociation constant. For example, the heterophilic interaction of p24 Golgi dynamics (GOLD) domains is shown between p24β1 and p24δ1. These GOLD domains interact with each other to form a hetero p24 family protein complex which is involved in the transport of glycosylphosphatidylinositol (GPI)-anchored proteins. The sensorgram shows fast association/dissociation binding and was analyzed by a 1:1 steady-state model (Nagae et al. 2016a) (Fig. 7.5c). To avoid immobilization artifacts, two experiments were performed: first p24β1 was immobilized and p24δ1 injected and then the vice versa experiment performed. The estimated K_D values of the two runs are comparable (1.2×10^{-4} and 1.5×10^{-4} M). SPR is widely used in lectin-glycan interactions, either alone or combined with other methods such as the enzyme-linked immunosorbent assay (ELISA), ITC, and thermal shift assays (Wesener et al. 2015; Ribeiro et al. 2016; Machon et al. 2017). When the lectin is immobilized and glycan injected, the SPR detection limit needs to be confirmed in advance. To overcome the sensitivity problem, biotinylated carbohydrates are immobilized onto streptavidin or neutravidin-coated sensor chips. Diluted lectins are injected and affinities calculated.

7.7 3D Structural Studies on Lectin-Glycan Interactions

Biophysical data of lectin-glycan interaction is strengthened when combined with structural data obtained from X-ray crystallography and NMR spectroscopy. X-ray crystallography is the favored method to determine the 3D structures of protein-ligand complexes at atomic resolution. However, glycans are highly flexible in solution, and the electron density is sometimes missing even in the presence of lectin. In contrast, solution NMR analyses such as titration assays and transferred nuclear Overhauser effects (trNOE) can give a more realistic picture under physiological conditions. Titration assays provide affinities, while trNOEs give intramolecular distance constraints of the bound glycan. This chapter focuses on NMR analysis of lectin-glycan interaction.

7.7.1 NMR Titration Study

For any binding scheme, a ligand exchanges between free and bound states with rate constant k (expressed per second). In NMR spectra, resonance frequencies of bound and free ligand are different, with $\Delta\nu$ representing the chemical shift difference (expressed in Hz). In a fast exchange regime, $k >> \Delta\nu$, the signal will appear in a population-weighted position between free and bound resonances. This allows one to monitor binding using the chemical shift change. In ^{1}H-^{15}N HSQC experiments, the dissociation constant K_D can be obtained by plotting the weighted chemical shift change, $\Delta\delta$, as a function of the carbohydrate/protein molar ratio. $\Delta\delta$ of each concentration is calculated with the following equation:

$$\Delta\delta = \left[\left(\Delta\delta_H\right)^2 + \left(\alpha \times \Delta\delta_N\right)^2 \right]^{1/2}$$

where $\Delta\delta_H$ and $\Delta\delta_N$ are the observed chemical shift changes (ppm) of ^{1}H and ^{15}N and α is a scaling factor (typically 0.1~0.2). Then the dissociation constants is estimated by fitting the curve:

$$\Delta\delta = \Delta\delta_{max} \left[[L]_T + [P]_T + K_D - \sqrt{\left([L]_T + [P]_T + K_D\right)^2 - 4[L]_T [P]_T} \right] \times \left(\frac{1}{2[P]_T} \right)$$

where $\Delta\delta_{max}$ is the maximum chemical shift change, $[L]_T$ the total ligand concentration, $[P]_T$ the total protein concentration, and K_D the dissociation constant for a 1:1 binding equilibrium.

In a slow exchange regime, $k << \Delta\nu$, two distinct resonances are observed, corresponding to the free and bound states. Increasing concentrations of carbohydrate increases the signal of the bound form and decreases that of the free. When the exchange rate is comparable to the frequency difference between the two states (intermediate exchange regime), $k \sim \Delta\nu$, extensive line broadening occurs, and sometimes the signals are beyond detection. Occasionally, it may be possible to escape this undesirable intermediate exchange situation by changing experimental conditions, aiming to either accelerate or slow the exchange.

A first example is the interaction of human soluble lectin ZG16p and phosphatidylinositol mannosides (PIMs) (Hanashima et al. 2015). PIM2 was titrated into a ^{15}N-labeled ZG16p solution, and signal perturbations in the ^{1}H-^{15}N HSQC spectra were recorded (Fig. 7.6a). The binding is a fast exchange process, since each set of specific signals featured a gradual chemical shift change. The binding interface of PIM2 can be inferred from a map of the hot spots on the 3D structure of ZG16p, and the dissociation constant for PIM2 was 3.0 mM from a fit of the titration curve.

The second example is the interaction of a C-type lectin receptor mDCIR2 and bisected *N*-glycan (Nagae et al. 2013b). Bisected and non-bisected *N*-glycans

(glycans ***a*** and ***b***) were analyzed. Concentration-dependent spectral shifts in the ^{1}H-NMR spectra were only observed in the presence of bisected glycan (Fig. 7.6b). The interaction involved slow exchange in terms of chemical shift, and the dissociation constant was estimated to be 3.0×10^{-5} M, a ligand concentration at which there is an equal amount of ligand-free and ligand-bound lectin. The specificity is well explained from the crystal structure of mDCIR2-bisected glycan complex, in which the bisected GlcNAc residue is directly interacting with the protein (Fig. 7.6b, right).

7.7.2 Transferred Nuclear Overhauser Effects (trNOE)

The nuclear Overhauser effect (NOE) describes the phenomenon of magnetization transfer from one spin to another. This process proceeds through cross relaxation originating from dipolar interaction between two spins that are spatially close (less than 5 Å apart). Roughly, the NOE is related to the distance between the two protons and shows a $1/r^6$ dependence.

If the glycan ligand binds to lectin with slow exchange, both intermolecular lectin-glycan NOEs and intramolecular NOEs of the glycan are important in elucidating the bound conformation of the ligand. However, in most lectin-glycan interactions, the exchange is fast because the affinity is weak. In such cases, transferred-NOESY (tr-NOESY) is the suitable measurement for conformational analysis of bound glycan ligand. trNOE analysis is generally valid for ligands that bind in the μM-mM range (Fig. 7.6c).

TrNOE experiments were applied to the conformational analysis of bisected *N*-glycan bound to a lectin, Calsepa (Nagae et al. 2016c). The NOE build-up curve shows linearity up to 200 ms for the glycan-Calsepa mixture and to 500 ms for glycan alone (Fig. 7.6d). The mixing time was then chosen as 200 ms for the trNOE experiments. In the presence of Calsepa lectin, long-range inter-residual trNOEs were observed between core β-mannose (Man-3) and GlcNAc-5′ on the outer branch (Fig. 7.6e). The presence of this long-range trNOE indicates that the glycan assumes a compact back-fold conformation when bound to Calsepa lectin. In a second example, the interaction between ZG16p and phosphatidylinositol mannoside glycans (PIM1 and PIM2) was investigated by trNOE (Hanashima et al. 2015). In the presence of ZG16p, several inter- and intra-residual NOE correlations were observed in PIM1 (Fig. 7.6f). There were none with the inactive ZG16p mutant D151N. From these correlations, the conformation of PIM1 in ZG16p-bound state was elucidated.

The structural information obtained from trNOE is usually limited by low proton density and by the inherently flexible nature of carbohydrate chains. Thus, obtained distances are typically further evaluated by molecular dynamics simulations and/or X-ray crystallographic analysis.

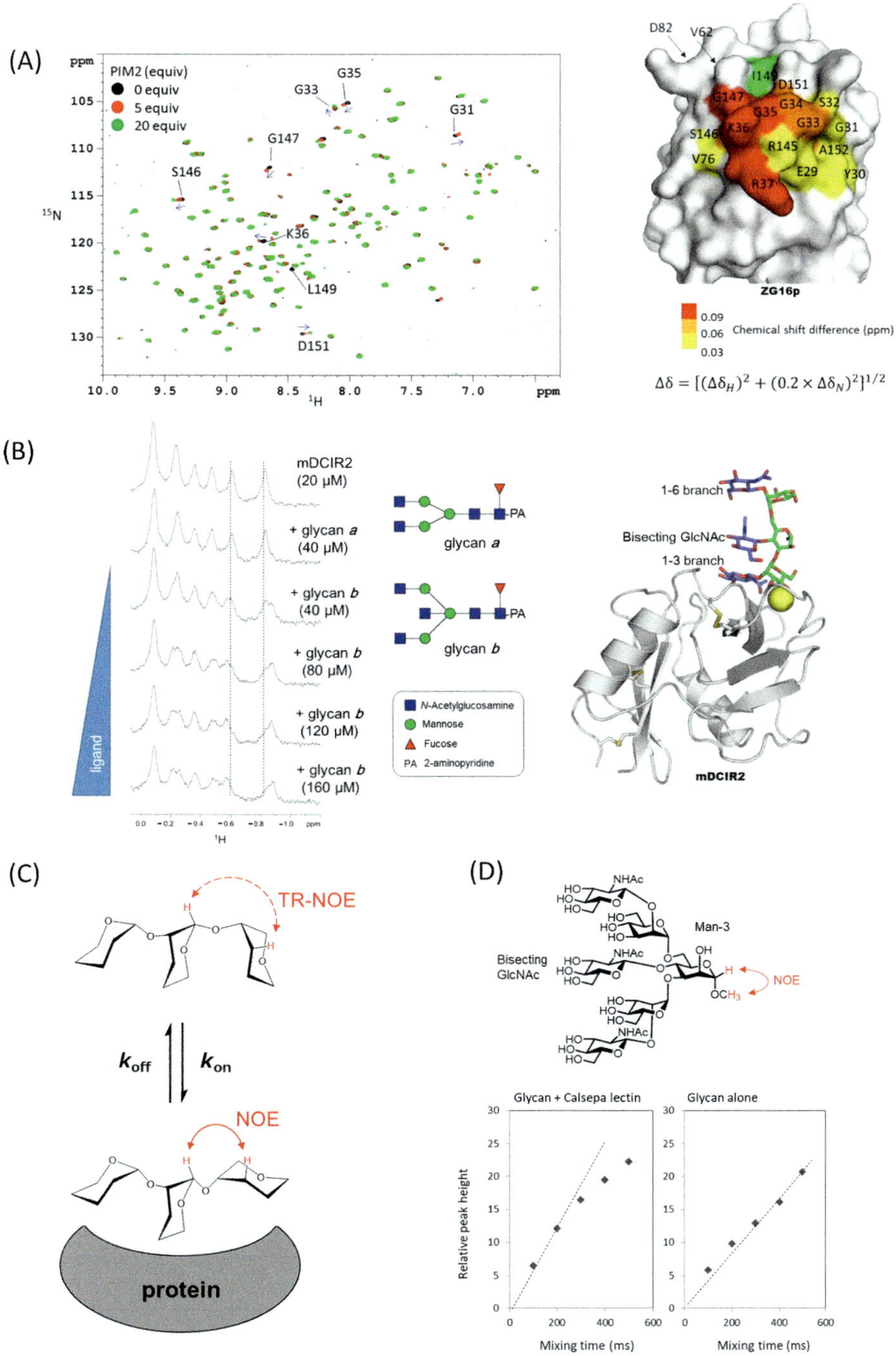

Fig. 7.6 NMR titration studies and trNOE-based conformational analyses
(a) NMR titration study of the interaction between ZG16p and PIM2. (Left panel) ^{1}H-^{15}N HSQC spectra of uniformly ^{15}N-labeled ZG16p in titration with PIM2 glycan (Black, 0 equiv.; red, 5 equiv.; and green, 20 equiv. of PIM2). Blue arrows indicate directions of the chemical shift changes. The figure was adapted from a paper (Hanashima et al. 2015) with permission.

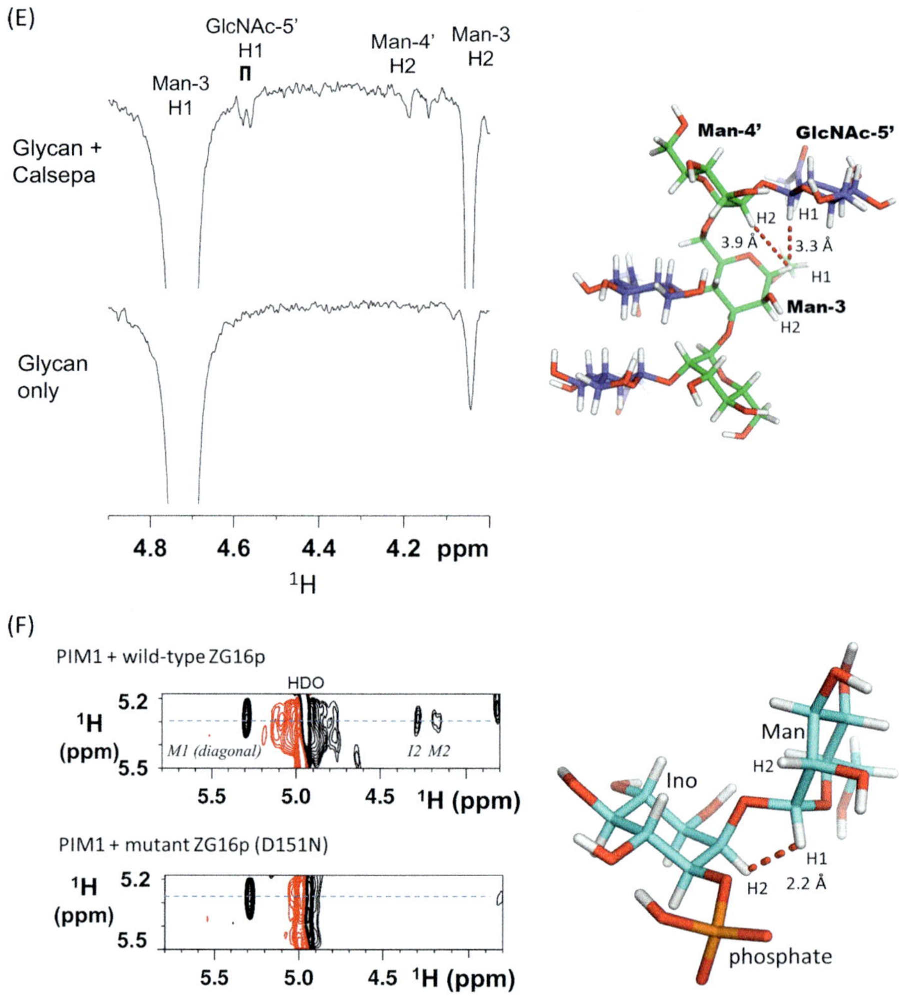

Fig. 7.6 (continued) (Right panel) Mapping surface residues in the PIM2 glycan interaction on the crystal structure of human ZG16p (PDB ID; 3APA). The signal from I149 (green) was broadened upon PIM2 binding

(**b**) NMR titration study of the interaction between a C-type lectin receptor mDCIR2 CRD and glycans. The pyridylaminated bisected and non-bisected *N*-glycans (glycans a and b) used in this study are shown in middle panel. A part of ^{1}H-NMR spectra of 20 μM mDCIR2 CRD in the absence of glycan (top), in the presence of 40 μM glycan ***a*** (second), and 40–120 μM glycan ***b*** (third to the last). The figure was adapted and modified from a previous paper (Nagae et al. 2013b) with permission. Crystal structure of mDCIR2 CRD in complex with bisected glycan corresponding to glycan ***b*** is shown in right panel (PDB code: 3VYK)

(**c**) Basic principle of transferred NOE (trNOE). If the glycan ligand is bound to protein for a sufficiently short time (in fast exchange), one can observed the bound NOEs (reflecting the interproton distances in the bound state) by using the free ligand signals

(**d**) NOE build up curves of Man-3 H1 and Man-3 *O*-methyl proton signals in the presence (left) and absence (right) of Calsepa lectin. Man-3 H1 signal was selectively inverted using a 180° rectangular pulse with 40-ms duration. This figure was adapted from a paper (Nagae et al. 2016c) with permission

(**e**) 1D selective NOESY spectra of bisected glycan in the presence (upper) and absence (lower) of Calsepa. Strong intra-residue trNOE was observed from Man-3 H1 to Man-3 H2, and long-range

7.7.3 Other NMR Techniques

Residual dipolar coupling (RDC) measured in liquid crystalline alignment medium offer a viable alternative to traditional NOE-based approaches for structural analysis. RDC provides a way to constrain the relative orientation of two molecules in complex with each other by aligning their independently determined order tensors (Jain 2009). Proton-proton and proton-carbon RDC of heparin tetrasaccharide has been measured (Jin et al. 2009), which, when combined with molecular dynamics simulations, clarified ring conformer and global shape of the glycan.

Paramagnetic probes attached to the reducing ends of oligosaccharides cause paramagnetic relaxation enhancement (PRE) and/or pseudocontact shift (PCS) which can resolve the peak overlap problem (Kato and Yamaguchi 2015). These spatial perturbations can be sources of long-range atomic distance information (~10 Å), which complements the local conformational information derived from *J*-couplings and NOEs (~5 Å).

7.8 Molecular Dynamics Simulation Studies of Lectin-Glycan Interactions

Theoretical approaches such as docking models and molecular dynamics (MD) simulation complement experimental information and have great possibilities for visualizing the highly mobile glycans. Docking simulations of lectin-glycan complexes provide a theoretical foundation and reduce ambiguities in experimental results. We validated the accuracy of docking simulation by applying it to a crystal structure of POMK-ligand complex, in which the electron density of ligand was partially observed (Fig. 7.7a). The position of the trisaccharide ligand in the docking model coincides well with the fragmented electron density in the crystal structure, underscoring the usefulness of docking and MD simulations in determining bound structures (Nagae et al. 2017b).

MD simulation is also useful for validation of possible binding modes with the estimation of binding free energy. Orysata lectin, one of the mannose-binding

Fig. 7.6 (continued) trNOEs were also observed from Man-3 H1 to GlcNAc-5′ (α1–6 branch) H1 and to Man-4' H2 signals (Right panel). Proton-proton distances were indicated between Man-3 H1 and GlcNAc-5' H1 and between Man-3 H1 and Man-4' H2. Structures of back-fold conformations are derived from Calsepa complex (PDB code: 5AV7). Hydrogen atoms are generated with PyMOL. This figure was adapted and modified from a paper (Nagae et al. 2016c) with permission

(**f**) 2D ^{1}H-^{1}H NOESY spectrum of PIM1 glycan (4.6-fold excess) in the presence of wild-type ZG16p collected with a mixing time of 250 ms at 10 °C (upper panel). Key inter-residual correlation was provided between Man-H1 and Ino-H2 and intra-residual correlations (Man-H1 and Man-H2) as negative NOEs. The atomic distance of inter-residue Man-H1-Ino-H2 was determined as 2.2 Å from the relative intensity of the signal. No correlation was observed when using the inactive ZG16p mutant (D151N) (lower panel). This figure is reproduced from the paper (Hanashima et al. 2015) with permission

Jacalin-related lectins, shows a preference for complex-type glycans with a α1-6 branch extension rather than a α1-3 branch. It was not easy to understand the structural basis for the preference for the α1-6 branch from the "snap-shot" crystal structure of Orysata in complex with biantennary glycan. MD simulations yield the dynamic behavior of the glycan attached to Orysata lectin (Fig. 7.7b). Several docking models of Orysata in complex with a series of *N*-glycans were constructed, and the calculated binding free energies suggest that multiple binding modes exist in the solution, and the binding preference seems somehow to originate from an average of multiple binding modes that favors α1-6 branch binding (Fig. 7.7c). MD simulations are especially useful for analyzing such multiple ligand-binding modes.

7.9 Future Perspective: Application of Cryo-electron Microscopy Analysis to Investigate Lectin-Glycan Interaction

Recent improvement in cryo-electron microscopy (cryo-EM) is a great breakthrough for structural biology (Cheng et al. 2015). EM analysis has an inherent advantage for investigating the global conformations of macromolecules, rather than atomic details of local interfaces. The structural changes of glycans on glycoproteins affect the physiological functions of carrier glycoproteins. This has been clearly exemplified by EM analysis. For example, EM has shown that mutation or elimination of *N*-glycans affects the conformational equilibrium of α5β1 integrin (Li et al. 2017). Likewise, the inter-domain angle of LDL-receptor-related protein 6 (LRP6) is regulated by the evolutionarily conserved *N*-glycan (Matoba et al. 2017).

In X-ray crystallography, the trimming of surface glycans on glycoproteins by glycosidase is usually necessary to produce diffraction quality crystals (Chang et al. 2007). Mutations at *N*-glycosylation sites can also produce highly uniform proteins (Nagae et al. 2017b). However, such treatments prevent visualization of the atomic structure of glycans, although some native glycoproteins provide good glycan electron density due to extensive intramolecular interactions between glycan and carrier glycoproteins (Nagae and Yamaguchi 2012). In contrast, there is no need to eliminate the surface glycans of glycoproteins in single particle analysis of cryo-EM. Thus, cryo-EM has great potential to provide the atomic structure of native glycans. Actually, cryo-EM structures of glycoproteins, such as γ-secretase and human coronavirus, often include the coordinates of attached glycans (Bai et al. 2015; Walls et al. 2016) (Fig. 7.8a and b). Even tetra-antennary *N*-glycan structures are visible in the EM map of the HIV-1 envelope glycoprotein trimer (Env) (Lee et al. 2016) (Fig. 7.8c).

Due to limitation of molecular weight, it is normally difficult to directly visualize lectin-glycan interactions using single particle analysis by cryo-EM. However, population analysis of many particles is the great advantage of EM analysis. Although the flexibility of the glycan part remains a serious problem for visualization (averaging

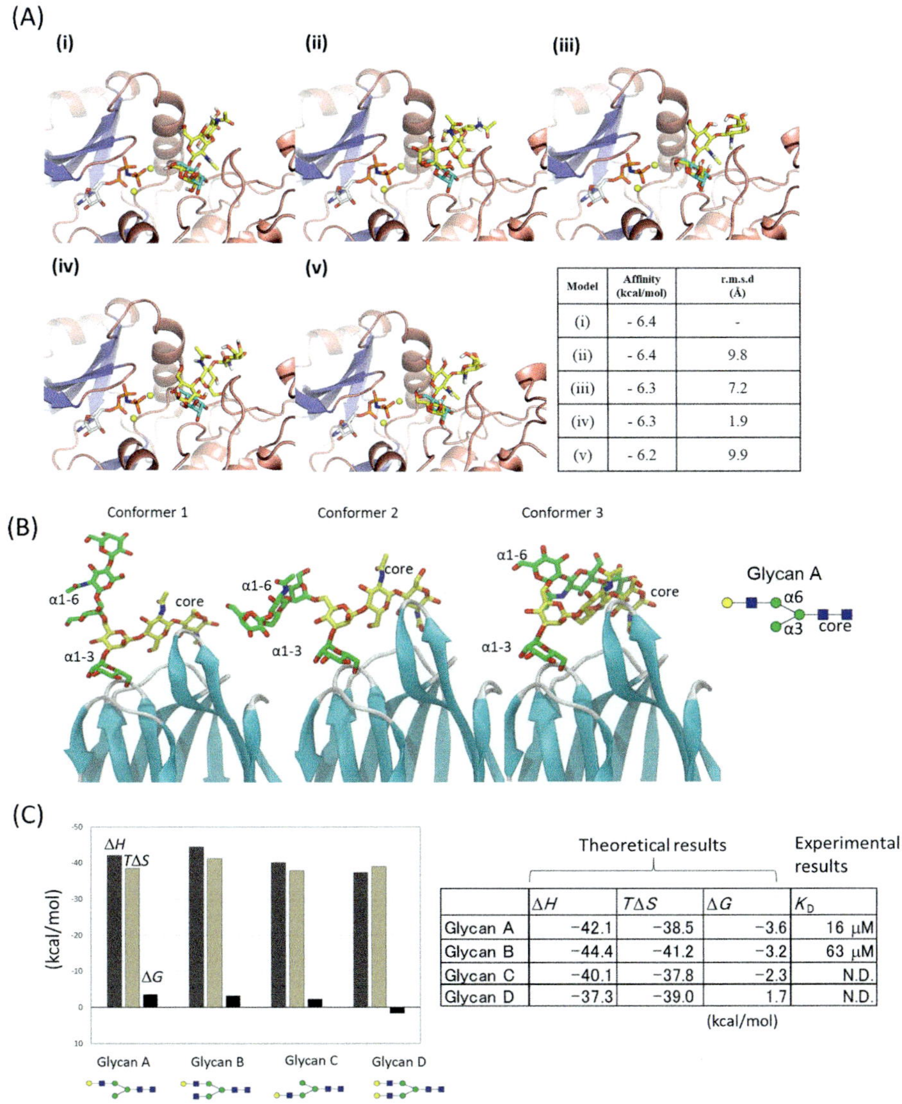

Model	Affinity (kcal/mol)	r.m.s.d (Å)
(i)	- 6.4	-
(ii)	- 6.4	9.8
(iii)	- 6.3	7.2
(iv)	- 6.3	1.9
(v)	- 6.2	9.9

	ΔH	$T\Delta S$	ΔG	K_D
Glycan A	−42.1	−38.5	−3.6	16 μM
Glycan B	−44.4	−41.2	−3.2	63 μM
Glycan C	−40.1	−37.8	−2.3	N.D.
Glycan D	−37.3	−39.0	1.7	N.D.

Fig. 7.7 Molecular dynamics simulations of protein-glycan complexes
(**a**) Top five docking poses of the core M3 trisaccharide (GalNAcβ1–3GlcNAcβ1-4Man, shown in yellow) in POMK binding site (i-v) by AutoDock Vina. The mannose residue assigned in the crystal structure (PDB code: 5GZ9) is overlaid in cyan. The table contains docking results for each model (i)-(v). Calculated binding affinity (in kcal/mol) together with root mean square deviation (r.m.s.d. in Å) from top scoring docking pose (i) is shown for all five models. This figure is adapted from the paper (Nagae et al. 2017b) with permission
(**b**) Three major conformations (Conformers 1, 2, and 3) of the α1–6 branch of glycan A bound to Orysata lectin in MD simulations. The simulation was performed using the crystal structure of Orysata in complex with glycan A (PDB code: 5XFH). This figure is adapted from the paper (Nagae et al. 2017a) with permission
(**c**) Bar graph and table of calculated average enthalpy (ΔH), entropy ($T\Delta S$), and total binding free energy (ΔG) for the biantennary glycans A–D. Experimental dissociation constant ($K_D = 1/K_A$) determined by FAC analysis is also indicated. N.D.: K_D was not determined due to weak or no binding

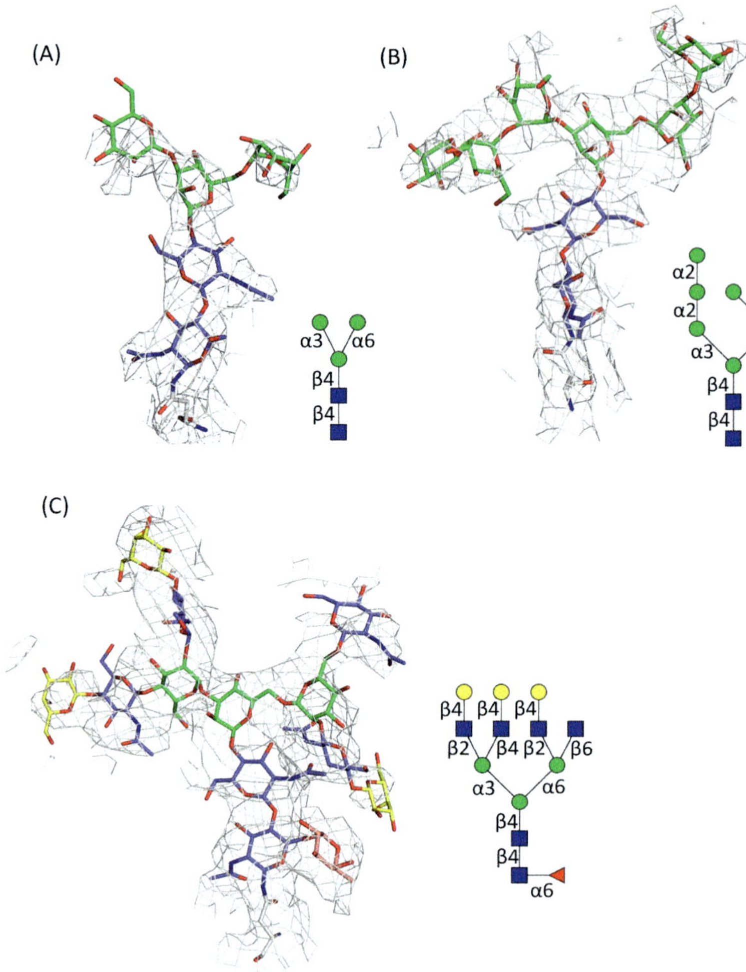

Fig. 7.8 Electron microscopic images of carbohydrates covalently attached onto proteins
(**a**) *N*-glycan core attached onto N55 of human nicastrin, a component of γ-secretase complex ((Bai et al. 2015), PDB code: 5A63, EMDB: 3061). The density map is depicted at 5.0 σ level cutoff. Schematic representation of the observed glycan is shown in right panel
(**b**) High-mannose-type *N*-glycan attached onto N426 of human coronavirus NL63 (HCoV-NL63) ((Walls et al. 2016), PDB code, 5SZS; EMDB, 8331). The density map is contoured at 7.0 σ level
(**c**) Tetra-antennary glycan attached onto N637 of HIV-1 Env trimer ((Lee et al. 2016), PDB code, 5FUU; EMDB, 3308). The density map is depicted in cyan mesh contoured at 5.5 σ level

problem), we can expect that cryo-EM will become the method of choice for analyzing the structure of native glycan on glycoproteins. The quaternary structural analyses of elaborate complex systems have recently been reported, such as human MHC-I peptide-loading complex (PLC), which includes TAP1, TAP2, tapasin, MHC-I heavy chain, β2m, calreticulin, and ERp57 in the complex (Blees et al. 2017). It is likely that multicomponent glycoprotein complexes and lectin receptor-glycoprotein complexes will be analyzed by cryo-EM at atomic resolution, coupled with other biophysical methods in the very near future.

Acknowledgment Most of the work described here was done with many collaborators as well as our lab members, all of whom we wish to acknowledge greatly. Optimization of recombinant protein expression was conducted by Drs. Ganesh P. Subedi and Hari P. Dulal, glycan microarray analysis was conducted by Dr. Yan Liu and Prof. Ten Feizi (Imperial College London), NMR study by Dr. Shinya Hanashima, FAC analysis by Dr. Hiroaki Tateno (AIST, Japan), ITC analysis by Drs. Masato Kikkawa and Seiji Yamamoto (GE Healthcare, Japan), and MD simulations by Dr. Sushil K. Mishra. SPR analysis was performed at One-Stop Facility Center, the University of Tokyo, by courtesy of Prof. Nobuaki Higashi. We also thank Noriko Tanaka for secretarial assistance.

References

Angulo J, Nieto PM (2011) STD-NMR: application to transient interactions between biomolecules-a quantitative approach. Eur Biophys J 40(12):1357–1369. https://doi.org/10.1007/s00249-011-0749-5

Angulo J, Díaz I, Reina JJ, Tabarani G, Fieschi F, Rojo J, Nieto PM (2008) Saturation transfer difference (STD) NMR spectroscopy characterization of dual binding mode of a mannose disaccharide to DC-SIGN. Chembiochem 9(14):2225–2227. https://doi.org/10.1002/cbic.200800361

Angulo J, Enríquez-Navas PM, Nieto PM (2010) Ligand-receptor binding affinities from saturation transfer difference (STD) NMR spectroscopy: the binding isotherm of STD initial growth rates. Chemistry (Weinheim an der Bergstrasse, Germany) 16(26):7803–7812. https://doi.org/10.1002/chem.200903528

Bai XC, Yan C, Yang G, Lu P, Ma D, Sun L, Zhou R, Scheres SHW, Shi Y (2015) An atomic structure of human γ-secretase. Nature 525(7568):212–217. https://doi.org/10.1038/nature14892

Battistel MD, Azurmendi HF, Freedberg DI (2017) Glycan OH exchange rate determination in aqueous solution: Seeking evidence for transient hydrogen bonds. J Phys Chem B 121(4):683–695. https://doi.org/10.1021/acs.jpcb.6b10594

Blees A, Januliene D, Hofmann T, Koller N, Schmidt C, Trowitzsch S, Moeller A, Tampe R (2017) Structure of the human MHC-I peptide-loading complex. Nature 551(7681):525–528. https://doi.org/10.1038/nature24627

Burnouf D, Ennifar E, Guedich S, Puffer B, Hoffmann G, Bec G, Disdier F, Baltzinger M, Dumas P (2012) kinITC: a new method for obtaining joint thermodynamic and kinetic data by isothermal titration calorimetry. J Am Chem Soc 134(1):559–565. https://doi.org/10.1021/ja209057d

Chang VT, Crispin M, Aricescu AR, Harvey DJ, Nettleship JE, Fennelly JA, Yu C, Boles KS, Evans EJ, Stuart DI, Dwek RA, Jones EY, Owens RJ, Davis SJ (2007) Glycoprotein structural genomics: solving the glycosylation problem. Structure 15(3):267–273. https://doi.org/10.1016/j.str.2007.01.011

Cheng Y, Grigorieff N, Penczek PA, Walz T (2015) A primer to single-particle cryo-electron microscopy. Cell 161(3):438–449. https://doi.org/10.1016/j.cell.2015.03.050

Cummings RD, Pierce JM (2014) The challenge and promise of glycomics. Chem Biol 21(1):1–15. https://doi.org/10.1016/j.chembiol.2013.12.010

Dam TK, Talaga ML, Fan N, Brewer CF (2016) Measuring Multivalent Binding Interactions by Isothermal Titration Calorimetry. Methods Enzymol 567:71–95. https://doi.org/10.1016/bs.mie.2015.08.013

Dulal HP, Nagae M, Ikeda A, Morita-Matsumoto K, Adachi Y, Ohno N, Yamaguchi Y (2016) Enhancement of solubility and yield of a β-glucan receptor Dectin-1 C-type lectin-like domain in *Escherichia coli* with a solubility-enhancement tag. Protein Expr Purif 123:97–104

Eggens I, Fenderson B, Toyokuni T, Dean B, Stroud M, Hakomori S (1989) Specific interaction between Le^x and Le^x determinants. A possible basis for cell recognition in preimplantation embryos and in embryonal carcinoma cells. J Biol Chem 264(16):9476–9484

Enríquez-Navas PM, Chiodo F, Marradi M, Angulo J, Penadés S (2012) STD NMR study of the interactions between antibody 2G12 and synthetic oligomannosides that mimic selected branches of gp120 glycans. Chembiochem 13(9):1357–1365. https://doi.org/10.1002/cbic.201200119

Feizi T, Chai W (2004) Oligosaccharide microarrays to decipher the glyco code. Nat Rev Mol Cell Biol 5(7):582–588. https://doi.org/10.1038/nrm1428

Geissner A, Seeberger PH (2016) Glycan arrays: From basic biochemical research to bioanalytical and biomedical applications. Annu Rev Anal Chem (Palo Alto, Calif) 9(1):223–247. https://doi.org/10.1146/annurev-anchem-071015-041641

Hanashima S, Sato K, Naito Y, Takematsu H, Kozutsumi Y, Ito Y, Yamaguchi Y (2010) Synthesis and binding analysis of unique AG2 pentasaccharide to human Siglec-2 using NMR techniques. Bioorg Med Chem 18(11):3720–3725. https://doi.org/10.1016/j.bmc.2010.03.062

Hanashima S, Kato K, Yamaguchi Y (2011) ^{13}C-NMR quantification of proton exchange at LewisX hydroxyl groups in water. Chem Commun (Camb) 47(38):10800–10802. https://doi.org/10.1039/c1cc13310a

Hanashima S, Sato C, Tanaka H, Takahashi T, Kitajima K, Yamaguchi Y (2013) NMR study into the mechanism of recognition of the degree of polymerization by oligo/polysialic acid antibodies. Bioorg Med Chem 21(19):6069–6076. https://doi.org/10.1016/j.bmc.2013.07.023

Hanashima S, Ikeda A, Tanaka H, Adachi Y, Ohno N, Takahashi T, Yamaguchi Y (2014a) NMR study of short β(1-3)-glucans provides insights into the structure and interaction with Dectin-1. Glycoconj J 31(3):199–207. https://doi.org/10.1007/s10719-013-9510-x

Hanashima S, Korekane H, Taniguchi N, Yamaguchi Y (2014b) Synthesis of *N*-glycan units for assessment of substrate structural requirements of *N*-acetylglucosaminyltransferase III. Bioorg Med Chem Lett 24(18):4533–4537. https://doi.org/10.1016/j.bmcl.2014.07.074

Hanashima S, Götze S, Liu Y, Ikeda A, Kojima-Aikawa K, Taniguchi N, Varon Silva D, Feizi T, Seeberger PH, Yamaguchi Y (2015) Defining the interaction of human soluble lectin ZG16p and mycobacterial phosphatidylinositol mannosides. Chembiochem 16(10):1502–1511

Haselhorst T, Lamerz AC, Itzstein M (2009) Saturation transfer difference NMR spectroscopy as a technique to investigate protein-carbohydrate interactions in solution. Methods Mol Biol 534:375–386. https://doi.org/10.1007/978-1-59745-022-5_26

Horan N, Yan L, Isobe H, Whitesides GM, Kahne D (1999) Nonstatistical binding of a protein to clustered carbohydrates. Proc Natl Acad Sci U S A 96(21):11782–11786

Huang TY, Irene D, Zulueta MM, Tai TJ, Lain SH, Cheng CP, Tsai PX, Lin SY, Chen ZG, Ku CC, Hsiao CD, Chyan CL, Hung SC (2017) Structure of the complex between a heparan sulfate octasaccharide and mycobacterial heparin-binding hemagglutinin. Angew Chem Int Ed Engl 56(15):4192–4196. https://doi.org/10.1002/anie.201612518

Hushegyi A, Tkac J (2014) Are glycan biosensors an alternative to glycan microarrays? Anal Methods 6(17):6610–6620. https://doi.org/10.1039/C4AY00692E

Hyun JY, Pai J, Shin I (2017) The glycan microarray story from construction to applications. Acc Chem Res 50(4):1069–1078. https://doi.org/10.1021/acs.accounts.7b00043

Jain NU (2009) Use of residual dipolar couplings in structural analysis of protein-ligand complexes by solution NMR spectroscopy. Methods Mol Biol 544:231–252. https://doi.org/10.1007/978-1-59745-483-4_15

Jin L, Hricovíni M, Deakin JA, Lyon M, Uhrín D (2009) Residual dipolar coupling investigation of a heparin tetrasaccharide confirms the limited effect of flexibility of the iduronic acid on the molecular shape of heparin. Glycobiology 19(11):1185–1196. https://doi.org/10.1093/glycob/cwp105

Kanagawa M, Liu Y, Hanashima S, Ikeda A, Chai W, Nakano Y, Kojima-Aikawa K, Feizi T, Yamaguchi Y (2014) Structural basis for multiple sugar recognition of Jacalin-related human ZG16p lectin. J Biol Chem 289(24):16954–16965

Kato K, Yamaguchi T (2015) Paramagnetic NMR probes for characterization of the dynamic conformations and interactions of oligosaccharides. Glycoconj J 32(7):505–513. https://doi.org/10.1007/s10719-015-9599-1

Kato Y, Kaneko MK, Kunita A, Ito H, Kameyama A, Ogasawara S, Matsuura N, Hasegawa Y, Suzuki-Inoue K, Inoue O, Ozaki Y, Narimatsu H (2008) Molecular analysis of the pathophysiological binding of the platelet aggregation-inducing factor podoplanin to the C-type lectin-like receptor CLEC-2. Cancer Sci 99(1):54–61. https://doi.org/10.1111/j.1349-7006.2007.00634.x

Lee JH, Ozorowski G, Ward AB (2016) Cryo-EM structure of a native, fully glycosylated, cleaved HIV-1 envelope trimer. Science 351(6277):1043–1048. https://doi.org/10.1126/science.aad2450

Li J, Su Y, Xia W, Qin Y, Humphries MJ, Vestweber D, Cabañas C, Lu C, Springer TA (2017) Conformational equilibria and intrinsic affinities define integrin activation. EMBO J 36(5):629–645. https://doi.org/10.15252/embj.201695803

Liang PH, Wang SK, Wong CH (2007) Quantitative analysis of carbohydrate-protein interactions using glycan microarrays: determination of surface and solution dissociation constants. J Am Chem Soc 129(36):11177–11184. https://doi.org/10.1021/ja072931h

Liang CH, Wang SK, Lin CW, Wang CC, Wong CH, Wu CY (2011) Effects of neighboring glycans on antibody-carbohydrate interaction. Angew Chem Int Ed Engl 50(7):1608–1612. https://doi.org/10.1002/anie.201003482

Machon O, Baldini SF, Ribeiro JP, Steenackers A, Varrot A, Lefebvre T, Imberty A (2017) Recombinant fungal lectin as a new tool to investigate O-GlcNAcylation processes. Glycobiology 27(2):123–128. https://doi.org/10.1093/glycob/cww105

Matoba K, Mihara E, Tamura-Kawakami K, Miyazaki N, Maeda S, Hirai H, Thompson S, Iwasaki K, Takagi J (2017) Conformational freedom of the LRP6 ectodomain is regulated by N-glycosylation and the binding of the Wnt antagonist Dkk1. Cell Rep 18(1):32–40. https://doi.org/10.1016/j.celrep.2016.12.017

Mayer M, Meyer B (2001) Group epitope mapping by saturation transfer difference NMR to identify segments of a ligand in direct contact with a protein receptor. J Am Chem Soc 123(25):6108–6117

Meyer B, Peters T (2003) NMR spectroscopy techniques for screening and identifying ligand binding to protein receptors. Angew Chem Int Ed Engl 42(8):864–890. https://doi.org/10.1002/anie.200390233

Nagae M, Yamaguchi Y (2012) Function and 3D structure of the *N*-glycans on glycoproteins. Int J Mol Sci 13(7):8398–8429. https://doi.org/10.3390/ijms13078398

Nagae M, Yamaguchi Y (2014) Three-dimensional structural aspects of protein-polysaccharide interactions. Int J Mol Sci 15(3):3768–3783. https://doi.org/10.3390/ijms15033768

Nagae M, Yamaguchi Y (2015) Sugar recognition and protein-protein interaction of mammalian lectins conferring diverse functions. Curr Opin Struct Biol 34:108–115

Nagae M, Ikeda A, Hane M, Hanashima S, Kitajima K, Sato C, Yamaguchi Y (2013a) Crystal structure of anti-polysialic acid antibody single chain Fv fragment complexed with octasialic acid: insight into the binding preference for polysialic acid. J Biol Chem 288(47):33784–33796. https://doi.org/10.1074/jbc.M113.496224

Nagae M, Yamanaka K, Hanashima S, Ikeda A, Morita-Matsumoto K, Satoh T, Matsumoto N, Yamamoto K, Yamaguchi Y (2013b) Recognition of bisecting *N*-acetylglucosamine: structural basis for asymmetric interaction with the mouse lectin dendritic cell inhibitory receptor 2. J Biol Chem 288(47):33598–33610

Nagae M, Ikeda A, Kitago Y, Matsumoto N, Yamamoto K, Yamaguchi Y (2014a) Crystal structures of carbohydrate recognition domain of blood dendritic cell antigen-2 (BDCA2) reveal a common domain-swapped dimer. Proteins 82(7):1512–1518. https://doi.org/10.1002/prot.24504

Nagae M, Morita-Matsumoto K, Kato M, Kaneko MK, Kato Y, Yamaguchi Y (2014b) A platform of C-type lectin-like receptor CLEC-2 for binding *O*-glycosylated podoplanin and nonglycosylated rhodocytin. Structure 22(12):1711–1721. https://doi.org/10.1016/j.str.2014.09.009

Nagae M, Soga K, Morita-Matsumoto K, Hanashima S, Ikeda A, Yamamoto K, Yamaguchi Y (2014c) Phytohemagglutinin from *Phaseolus vulgaris* (PHA-E) displays a novel glycan recognition mode using a common legume lectin fold. Glycobiology 24(4):368–378. https://doi.org/10.1093/glycob/cwu004

Nagae M, Hirata T, Morita-Matsumoto K, Theiler R, Fujita M, Kinoshita T, Yamaguchi Y (2016a) 3D structure and interaction of p24β and p24δ Golgi dynamics domains: Implication for p24 complex formation and cargo transport. J Mol Biol 428(20):4087–4099. https://doi.org/10.1016/j.jmb.2016.08.023

Nagae M, Ikeda A, Hanashima S, Kojima T, Matsumoto N, Yamamoto K, Yamaguchi Y (2016b) Crystal structure of human dendritic cell inhibitory receptor C-type lectin domain reveals the binding mode with *N*-glycan. FEBS Lett 590(8):1280–1288. https://doi.org/10.1002/1873-3468.12162

Nagae M, Kanagawa M, Morita-Matsumoto K, Hanashima S, Kizuka Y, Taniguchi N, Yamaguchi Y (2016c) Atomic visualization of a flipped-back conformation of bisected glycans bound to specific lectins. Sci Rep 6:22973

Nagae M, Mishra SK, Hanashima S, Tateno H, Yamaguchi Y (2017a) Distinct roles for each *N*-glycan branch interacting with mannose-binding type Jacalin-related lectins Orysata and Calsepa. Glycobiology 27(12):1120–1133

Nagae M, Mishra SK, Neyazaki M, Oi R, Ikeda A, Matsugaki N, Akashi S, Manya H, Mizuno M, Yagi H, Kato K, Senda T, Endo T, Nogi T, Yamaguchi Y (2017b) 3D structural analysis of protein *O*-mannosyl kinase, POMK, a causative gene product of dystroglycanopathy. Genes Cells 22(4):348–359

Nimrichter L, Gargir A, Gortler M, Altstock RT, Shtevi A, Weisshaus O, Fire E, Dotan N, Schnaar RL (2004) Intact cell adhesion to glycan microarrays. Glycobiology 14(2):197–203. https://doi.org/10.1093/glycob/cwh022

Pai J, Hyun JY, Jeong J, Loh S, Cho EH, Kang YS, Shin I (2016) Carbohydrate microarrays for screening functional glycans. Chem Sci 7(3):2084–2093. https://doi.org/10.1039/c5sc03789a

Palma AS, Feizi T, Childs RA, Chai W, Liu Y (2014) The neoglycolipid (NGL)-based oligosaccharide microarray system poised to decipher the *meta*-glycome. Curr Opin Chem Biol 18:87–94. https://doi.org/10.1016/j.cbpa.2014.01.007

Pfeffer PE, Valentine KM, Parrish FW (1979) Deuterium-induced differential isotope shift ^{13}C NMR.1. Resonance reassignments of mono- and disaccharides. J Am Chem Soc 101(5):1265–1274

Ribeiro JP, Pau W, Pifferi C, Renaudet O, Varrot A, Mahal LK, Imberty A (2016) Characterization of a high-affinity sialic acid-specific CBM40 from *Clostridium perfringens* and engineering of a divalent form. Biochem J 473(14):2109–2118. https://doi.org/10.1042/BCJ20160340

Song X, Yu H, Chen X, Lasanajak Y, Tappert MM, Air GM, Tiwari VK, Cao H, Chokhawala HA, Zheng H, Cummings RD, Smith DF (2011) A sialylated glycan microarray reveals novel interactions of modified sialic acids with proteins and viruses. J Biol Chem 286(36):31610–31622. https://doi.org/10.1074/jbc.M111.274217

Subedi GP, Satoh T, Hanashima S, Ikeda A, Nakada H, Sato R, Mizuno M, Yuasa N, Fujita-Yamaguchi Y, Yamaguchi Y (2012) Overproduction of anti-Tn antibody MLS128 single-chain Fv fragment in *Escherichia coli* cytoplasm using a novel pCold-PDI vector. Protein Expr Purif 82(1):197–204. https://doi.org/10.1016/j.pep.2011.12.010

Tateno H, Nakamura-Tsuruta S, Hirabayashi J (2007) Frontal affinity chromatography: sugar-protein interactions. Nat Protoc 2(10):2529–2537. https://doi.org/10.1038/nprot.2007.357

Vander Meulen KA, Horowitz S, Trievel RC, Butcher SE (2016) Measuring the kinetics of molecular association by isothermal titration calorimetry. Methods Enzymol 567:181–213. https://doi.org/10.1016/bs.mie.2015.08.012

Varki A, Cummings RD, Aebi M, Packer NH, Seeberger PH, Esko JD, Stanley P, Hart G, Darvill A, Kinoshita T, Prestegard JJ, Schnaar RL, Freeze HH, Marth JD, Bertozzi CR, Etzler ME, Frank M, Vliegenthart JF, Lütteke T, Perez S, Bolton E, Rudd P, Paulson J, Kanehisa M, Toukach P, Aoki-Kinoshita KF, Dell A, Narimatsu H, York W, Taniguchi N, Kornfeld S (2015) Symbol nomenclature for graphical representations of glycans. Glycobiology 25(12):1323–1324. https://doi.org/10.1093/glycob/cwv091

Vedadi M, Niesen FH, Allali-Hassani A, Fedorov OY, Finerty PJ Jr, Wasney GA, Yeung R, Arrowsmith C, Ball LJ, Berglund H, Hui R, Marsden BD, Nordlund P, Sundstrom M, Weigelt J, Edwards AM (2006) Chemical screening methods to identify ligands that promote protein stability, protein crystallization, and structure determination. Proc Natl Acad Sci U S A 103(43):15835–15840. https://doi.org/10.1073/pnas.0605224103

Walls AC, Tortorici MA, Frenz B, Snijder J, Li W, Rey FA, DiMaio F, Bosch BJ, Veesler D (2016) Glycan shield and epitope masking of a coronavirus spike protein observed by cryo-electron microscopy. Nat Struct Mol Biol 23(10):899–905. https://doi.org/10.1038/nsmb.3293

Wesener DA, Wangkanont K, McBride R, Song X, Kraft MB, Hodges HL, Zarling LC, Splain RA, Smith DF, Cummings RD, Paulson JC, Forest KT, Kiessling LL (2015) Recognition of microbial glycans by human intelectin-1. Nat Struct Mol Biol 22(8):603–610. https://doi.org/10.1038/nsmb.3053

Chapter 8
Structural Aspects of ER Glycoprotein Quality-Control System Mediated by Glucose Tagging

Tadashi Satoh and Koichi Kato

Abstract *N*-linked oligosaccharides attached to proteins act as tags for glycoprotein quality control, ensuring their appropriate folding and trafficking in cells. Interactions with a variety of intracellular lectins determine glycoprotein fates. Monoglucosylated glycoforms are the hallmarks of incompletely folded glycoproteins in the protein quality-control system, in which glucosidase II and UDP-glucose/glycoprotein glucosyltransferase are, respectively, responsible for glucose trimming and attachment. In this review, we summarize a recently emerging view of the structural basis of the functional mechanisms of these key enzymes as well as substrate *N*-linked oligosaccharides exhibiting flexible structures, as revealed by applying a series of biophysical techniques including small-angle X-ray scattering, X-ray crystallography, high-speed atomic force microscopy, electron microscopy, and computational simulation in conjunction with NMR spectroscopy.

Keywords Endoplasmic reticulum · Glucose · Glucosidase · Glucosyltransferase · *N*-linked oligosaccharide · Protein quality control

8.1 Introduction

Glucose is the most important carbohydrate as an energy source in cells and is stored as polymers including glycogen in animals and starch in plants. It is also a component of the posttranslational modifier tetradecasaccharide ($Glc_3Man_9GlcNAc_2$), which forms a common precursor of asparagine-linked oligosaccharides (*N*-glycans) when attached to newly synthesized proteins (Betenbaugh et al. 2004; Kornfeld and Kornfeld 1985). This tetradecasaccharide has a triantennary structure with three

T. Satoh
Graduate School of Pharmaceutical Sciences, Nagoya City University, Nagoya, Aichi, Japan

K. Kato (✉)
Exploratory Research Center on Life and Living Systems,
National Institutes of Natural Sciences, Okazaki, Aichi, Japan
e-mail: kkatonmr@ims.ac.jp

Y. Yamaguchi, K. Kato (eds.), *Glycobiophysics*, Advances in Experimental Medicine and Biology 1104, https://doi.org/10.1007/978-981-13-2158-0_8

nonreducing terminal branches (termed D1, D2, and D3); the D1 branch is capped with three glucose residues, Glc-α1,2-Glc-α1,3-Glc (Kelleher and Gilmore 2006; Vliegenthart et al. 1983; Fig. 8.1a). *N*-glycans displayed on proteins are involved in various biological processes, including transcriptional regulation, cell-cell communication, bacterial and viral infection, and protein quality control (Stanley et al. 2009). In the quality-control system, a series of *N*-glycan-processing intermediates serve as quality tags for the determination of glycoprotein fates (including folding, transport, or degradation) via interaction with a variety of intracellular lectins (Aebi et al. 2010; D'Alessio et al. 2010; Kamiya et al. 2012; Lederkremer 2009; Satoh et al. 2015; Takeda et al. 2009).

N-glycan processing in the endoplasmic reticulum (ER) is initiated by the elimination of the nonreducing terminal glucose residue on the D1 branch. This unsealing process is accomplished by glucosidase I (GI), which catalyzes cleavage of the Glc-α1,2-Glc glycosidic linkage (Deprez et al. 2005; Grinna and Robbins 1980). GI contains an 80–110 kDa catalytic domain belonging to the glycoside hydrolase (GH) family 63. A crystallographic study demonstrated that the catalytic domain of yeast GI (Cwh41p) consists of a 57 kDa C-terminal $(\alpha/\alpha)_6$ toroid subdomain containing catalytic residues in addition to a 32 kDa N-terminal β-sandwich subdomain of unknown function (Barker and Rose 2013). Docking simulations, together with mutagenesis experiments, suggest that GI enzymatic activity is exerted via a characteristic conformation of the glucotriose substrate, i.e., a bent-back conformation mediated by an intrachain interaction between the first and third glucose residues.

Subsequently, glucosidase II (GII) catalyzes two-step hydrolytic cleavage reactions of the Glc-α1,3-Glc and Glc-α1,3-Man glycosidic linkages, giving rise to $Glc_1Man_9GlcNAc_2$ and $Man_9GlcNAc_2$ glycoforms, respectively (Arendt and Ostergaard 2000; Pelletier et al. 2000; Totani et al. 2008; Trombetta et al. 1996, 2001). In the ER quality-control system, the folding states of newly synthesized proteins are judged based on the presence or absence of a single terminal glucose residue on the D1 branch of *N*-glycan intermediates (Fig. 8.1b; Aebi et al. 2010; D'Alessio et al. 2010; Kamiya et al. 2012; Lederkremer 2009; Satoh et al. 2015; Takeda et al. 2009). During the two-step deglucosylation process, the monoglucosylated glycoform is transiently expressed, serving as a "yet-to-be-folded" tag, which is recognized by the membrane-embedded soluble lectins calnexin (CNX) and calreticulin (CRT), respectively. CNX and CRT constitute chaperone complexes with protein disulfide isomerase family proteins as folding catalysts (ERp57 and ERp29 in mammals) (Jessop et al. 2009; Kozlov et al. 2017; Maattanen et al. 2006; Nakao et al. 2017; Oliver et al. 1997; Park et al. 2005; Sakono et al. 2014). The correct formation of major histocompatibility complex (MHC) class I depends on its interaction with CRT as a component of the huge protein complex known as the peptide-loading complex (PLC). The 5.8-Å resolution cryo-EM structure of human PLC was elucidated very recently, providing insights into the mechanism underlying the workings of this protein machinery (Blees et al. 2017). The MHC-I molecule interacts with multiple subunits of the PLC, i.e., CRT and ERp57, which are essential for the presentation of exogenous antigens on the MHC-I molecule (Hulpke and Tampe 2013). Although cryo-EM data potentially provides insights into the

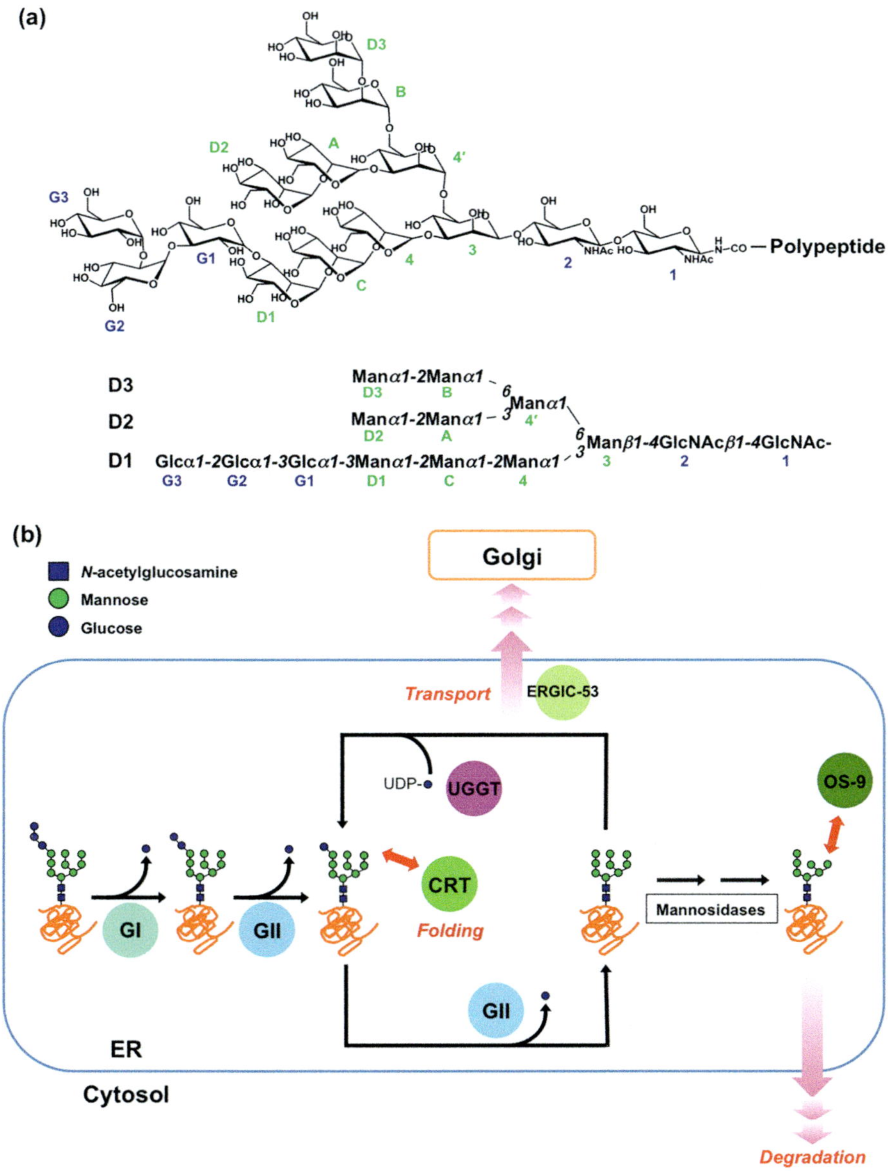

Fig. 8.1 (**a**) Schematic representation of $Glc_3Man_9GlcNAc_2$ with the nomenclature of oligosaccharide residues and branches in accordance with Vliegenthart et al. (1983). (**b**) Schematic cartoon of glycoprotein quality control coupled with *N*-glycan processing and subsequent interactions with a series of intracellular lectins in the ER. GI, glucosidase I; GII, glucosidase II; UGGT, UDP-glucose/glycoprotein glucosyltransferase; CRT, calreticulin; OS-9, osteosarcoma amplified 9; ERGIC-53, ER-Golgi intermediate compartment 53 kDa protein

interactions between MHC-I *N*-glycan and CRT, detailed interpretation is hampered by the limited resolution at present.

After complete deglucosylation of the *N*-glycan by GII, glycoproteins are released from the chaperone complexes for intracellular transport toward the Golgi apparatus, provided that they are correctly folded. In contrast, incompletely folded glycoproteins are recognized by the folding-sensor enzyme UDP-glucose/glycoprotein glucosyltransferase (UGGT), which provides a backup quality-control mechanism by catalyzing re-glucosylation at the D1 branch terminus, regenerating the *glycotope* recognized by CNX and CRT (Caramelo et al. 2003; Ito et al. 2015; Izumi et al. 2012; Taylor et al. 2004; Totani et al. 2009). Despite the accumulation of functional data regarding GII and UGGT over the two decades since they were first described, no structural information about these key enzymes was available until recently.

To elucidate the mechanisms underlying the glycoprotein quality-control system, it is essential to provide detailed structural and dynamic views of the molecules constituting the system by applying a multilateral biophysical approach. Electron microscopy (EM) as well as X-ray crystallography provide 3D structural snapshots of the proteins and their complexes. In particular, the recently advanced cryo-EM can delineate subatomic- to atomic-resolution structures as class-averaged images with their populations through systematic collection of massive numbers of single molecule images in the frozen-hydrated state. On the other hand, small-angle X-ray scattering (SAXS) can provide low-resolution overall structure of dynamic biomacromolecules as an averaged structure in physiological solutions. While the above-mentioned techniques offer only spatial information regarding conformational variability of the biomacromolecules, NMR spectroscopy and high-speed atomic force microscopy (HS-AFM) can provide temporal information on their dynamic behaviors in solutions. Therefore, integration of these methods gives us complementary information for deeper understanding the molecular processes involved in the quality control of glycoproteins.

In recent years, structural and dynamical characterization of glucose-processing enzymes, especially of UGGT, has been performed by utilizing multidisciplinary biophysical techniques including EM, SAXS, X-ray crystallography, and HS-AFM (Calles-Garcia et al. 2017; Caputo et al. 2016; Olson et al. 2013, 2015; Roversi et al. 2017; Satoh et al. 2016a, b, 2017; Zhu et al. 2014). Furthermore, molecular dynamics (MD) simulation combined with NMR spectroscopy has been used to describe the dynamic conformations of a series of high-mannose-type oligosaccharides including the monoglucosylated high-mannose-type oligosaccharide ($Glc_1Man_9GlcNAc_2$), thereby addressing questions about the mechanisms of their interactions with quality-control lectins (Kato and Yamaguchi 2015; Yamaguchi and Kato 2014). This review outlines our current understanding of the molecular mechanisms underlying quality control of glycoproteins mediated by glucosylation based on cutting-edge structural data, especially focusing on GII and UGGT, together with high-mannose-type glycans as their substrate and product.

8.2 Two-Step Glucose Trimming Catalyzed by Glucosidase II Heterodimeric Enzyme

GII consists of a catalytic α-subunit (GIIα) of approximately 110 kDa belonging to the GH family 31 and a 60 kDa multi-domain regulatory β-subunit (GIIβ; Arendt and Ostergaard 2000; Pelletier et al. 2000; Trombetta et al. 1996, 2001). Among the GH31 enzymes, only the GIIα operates with its binding partner GIIβ, which contains an N-terminal GIIα-binding (G2B) domain, a putative coiled-coil segment, and a C-terminal mannose-6-phosphate receptor homology (MRH) domain followed by an ER retention signal (Munro 2001; Quinn et al. 2009; Fig. 8.2a). The MRH domain acts as a lectin that recognizes high-mannose-type oligosaccharides, containing an α1,2-linked mannobiose structure on the D3 branch (Hu et al. 2009)

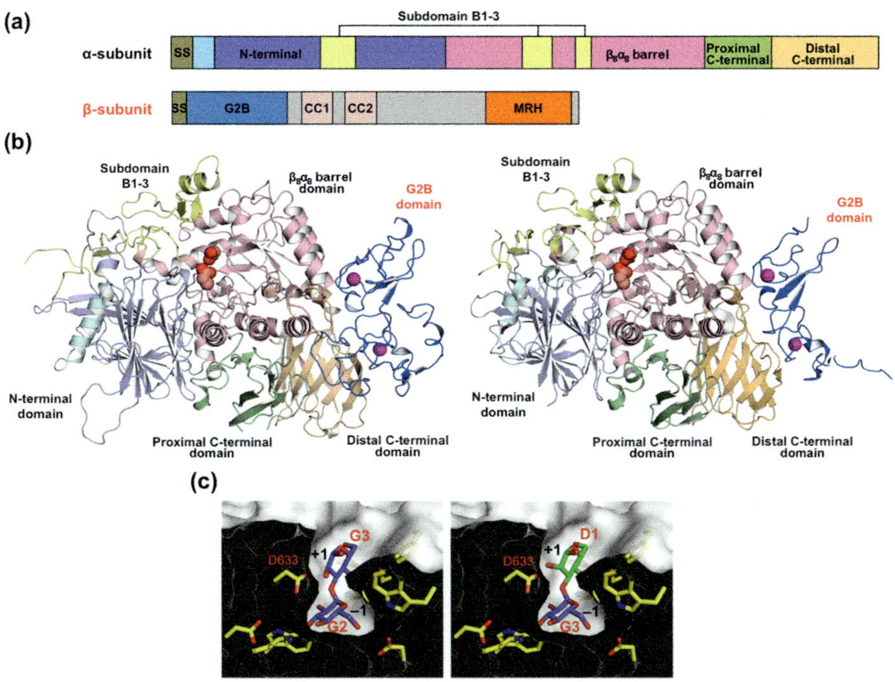

Fig. 8.2 (**a**) Domain structures of α- and β-subunits of glucosidase II. "SS" and "CC" indicate putative signal sequence and coiled-coil segment, respectively. (**b**) Crystal structures of *Chaetomium thermophilum* and murine GIIα complexed with the G2B domain of GIIβ are represented on the left (PDB code: 5JQP) and right (PDB code: 5F0E), respectively. The G2B domain of GIIβ is colored marine blue, and the individual domains of GIIα are colored as follows: N-terminal domain, light-blue; subdomain B1–3, light-yellow; $\beta_8\alpha_8$ barrel domain, light-pink; proximal C-terminal domain, light-green; and distal C-terminal domain, light-orange. These colors correspond to those in Fig. 8.2a. Active-site residues of GIIα and bound Ca^{2+} ions of GIIβ are shown as spheres. (**c**) Substrate-binding pocket of GIIα, represented by sliced surface models: (left) Glc-α1,3-Glc-bound form, (right) Glc-α1,3-Man-bound form. Bound sugar ligands and residues involved in ligand binding are shown as stick models

(Fig. 8.1a), and promotes the enzymatic function of GIIα (D'Alessio and Dahms 2015; Quinn et al. 2009; Stigliano et al. 2009, 2011; Watanabe et al. 2009). Elimination of the α1,2-linked mannobiose from the D3 branch (i.e., ManD3) catalyzed by ER degradation enhancing α-mannosidase-like proteins (EDEMs; yeast homolog Htm1p) is a key step leading to ER-associated degradation mediated by another MRH domain-containing lectin, OS-9 (yeast homolog Yos9p; Hosokawa et al. 2009; Mikami et al. 2010; Ninagawa et al. 2014; Quan et al. 2008; Satoh et al. 2010). Thus, cooperation between the α- and β-subunits of GII is critical to the mechanism underlying the functional extension of this unique GH31 enzyme.

Recently delineated crystal structures of murine and thermophilic fungal GIIs have revealed that GIIα adopts a globular structure comprising four major domains (N-terminal domain, $\beta_8\alpha_8$ barrel domain, proximal C-terminal domain, and distal C-terminal domain) and three subdomains (Caputo et al. 2016; Satoh et al. 2016a, b; Fig. 8.2b), which are similar to other GH31 α-glucosidases (Ernst et al. 2006; Lovering et al. 2005; Ren et al. 2011; Sim et al. 2008, 2010; Tagami et al. 2013). The crystal structures of GIIα were determined in the presence (Caputo et al. 2016; Satoh et al. 2016a; Fig. 8.2b) and absence (Satoh et al. 2016b) of the binding domain of GIIβ, i.e., the G2B domain.

Structural data obtained using iminosugar α-glucosidase inhibitors such as 1-deoxynojirimycin showed that the active-site residues are located in the most highly conserved $\beta_8\alpha_8$ barrel domain (Caputo et al. 2016; Satoh et al. 2016b). It has been proposed that the inhibition of α-glucosidases including GI and GII result in misfolded viral glycoproteins, which impairs the secretion and/or infectivity of several viruses including HIV, dengue virus, and influenza virus, since biosynthesis of most enveloped viruses is dependent on the ER quality-control system of host cells (Sayce et al. 2010; Tyrrell et al. 2017). Therefore, structural information regarding inhibitor-bound GIIα will be important for designing and developing antiviral agents. In addition to inhibitor-bound complexes, crystal structures of fungal GIIα complexed with two different glucosyl ligands containing the scissile bonds of the first and second reactions steps catalyzed by this enzyme were determined using an inactive mutant, D556A; this provided a structural basis for the substrate-binding specificity of two-step deglucosylation (Fig. 8.2c; Satoh et al. 2016b). Importantly, these structural data suggest that the two-step glucose trimming reactions catalyzed in the active-site pocket do not proceed successively on account of the pocket's gourd-shaped bilocular architecture (Fig. 8.2c). After the first glucose trimming reaction, the resultant glycoprotein substrates carrying monoglucosylated glycans must be dissociated from the GII enzyme for the first cleaved glucose product to be released from the deeply buried −1 subsite through the +1 subsite. Presumably, glycoproteins thus have a time window between the first and second step of deglucosylation for chaperone-mediated folding to occur, mediated by the CNX/CRT-containing folding complex that specifically recognizes the terminal monoglucose residue of the *N*-glycan tags. Previous in vitro kinetics data have demonstrated that the first cleavage ($Glc_2Man_9GlcNAc_2 \rightarrow Glc_1Man_9GlcNAc_2$) catalyzed by GII progresses significantly more quickly than the second cleavage ($Glc_1Man_9GlcNAc_2 \rightarrow Man_9GlcNAc_2$) (Totani et al. 2006). Whereas the equatorial

2-OH group of Glc(G3) is not involved in interaction, the axial 2-OH group of Man(D1) forms a hydrogen bond with Asp633 in the Glc(G3)-Man(D1)-bound complex, which may contribute to the delay in the second cleavage reaction.

Crystal structures of GIIα in complex with the G2B domain of GIIβ (Caputo et al. 2016; Satoh et al. 2016a) have also been determined. The G2B domain consists of two Ca^{2+}-binding subdomains with five disulfide bridges, and this has an overall structure similar to that of the low-density lipoprotein receptor class A domain (Fisher et al. 2006). The G2B domain binds the distal C-terminal domain of GIIα mainly through electrostatic interactions involving contiguous double arginine residues in GIIα (Fig. 8.3a). A genetic link has been demonstrated between autosomal-dominant polycystic kidney and liver diseases (ADPKD and ADPLD), which are autosomal recessive disorders, and mutations in *GANAB* and *PRKCSH*, which encode GIIα and GIIβ, respectively (Drenth et al. 2003; Li et al. 2003; Porath et al. 2016). The particular mutations in question affect Arg839 in human GIIα, which is the latter residue of the double arginine motif in the distal C-terminal domain. An R839W mutation in GIIα is likely to impair its association with GIIβ,

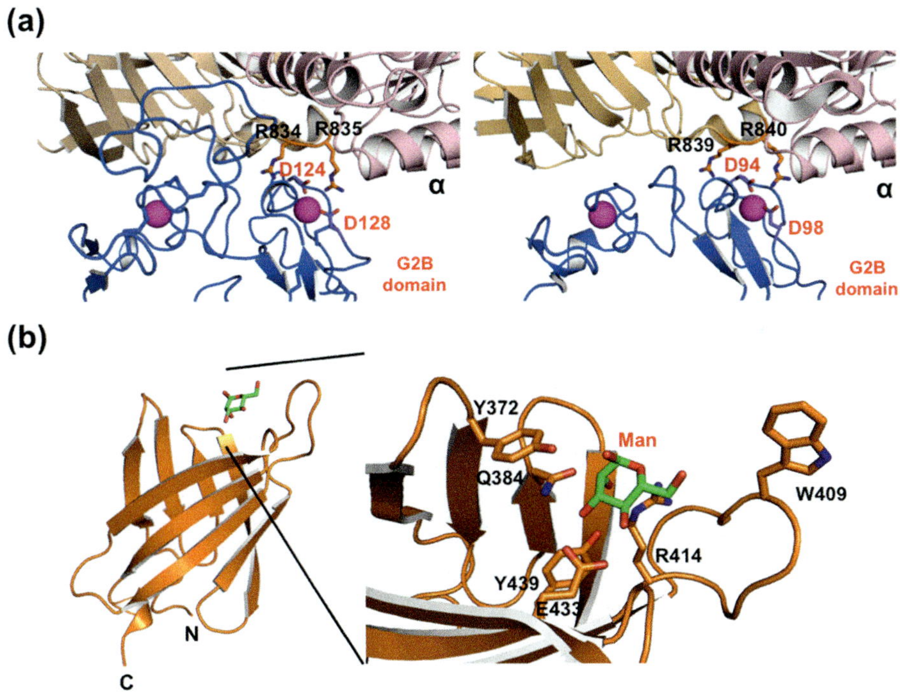

Fig. 8.3 (**a**) Close-up view of the binding interface between the GIIα and G2B domains (*C. thermophilum* [left] and murine [right]). Key residues involved in electrostatic interaction are highlighted. (**b**) Crystal structure of the *Schizosaccharomyces pombe* MRH domain complexed with α-D-mannose (PDB code: 4XQM). The mannose-binding site of the MRH domain is shown on the right. Bound mannose and residues involved in ligand binding are shown as sticks

thereby negatively influencing enzymatic efficiency and subcellular localization of GIIα, since the β-subunit possesses the lectin domain and the ER retention signal.

NMR and X-ray crystallography studies have revealed the 3D structures of the isolated MRH domain of *Schizosaccharomyces pombe* GIIβ, providing a structural explanation for the sugar recognition demonstrated by high-mannose-type glycans (Olson et al. 2013, 2015; Fig. 8.3b). NMR data has identified the binding site of the mannosyl ligands $Man_9GlcNAc_2$ and α1,2-linked mannobiose in the canonical binding pocket of MRH domains in solution, and crystallographic data have revealed that the mannose-binding mode of the GIIβ MRH domain is very similar to those of other MRH domains, including OS-9, which specifically recognizes the Man α1,6-linked D3 branch (Hosokawa et al. 2009; Mikami et al. 2010; Quan et al. 2008; Satoh et al. 2010). Crystal structure also identified tyrosine (Tyr372) and tryptophan (Trp409) as key residues for GII activity, as well as canonical QREY residues conserved among MRH domains (Castonguay et al. 2011; Kim et al. 2009). Previous frontal affinity chromatography data have demonstrated that the MRH domain of GIIβ recognizes the α1,2-linked mannobiose moiety (ManD3-ManC) on the D3 branch (Fig. 8.1a; Hu et al. 2009), while the α-subunit recognizes the D1 branch, resulting in the deglucosylation reaction. Therefore, the crystallographically observed mannose residue in the *S. pombe* GIIβ MRH domain presumably corresponds to the ManD3 residue of high-mannose-type glycans.

Recent SAXS data (Caputo et al. 2016) together with previous ultracentrifugation and limited proteolysis results (Pelletier et al. 2000; Trombetta et al. 2001) have revealed that GIIα/β holoenzyme adopts a highly elongated structure, presumably through the middle non-globular segment that connects the G2B and MRH domains (Fig. 8.2a). These data suggest that the β-subunit has a flexible modular structure in which the MRH lectin domain is tethered to GIIα via its G2B domain with variable spatial arrangements, enabling this enzyme to act on multiple glycosylated proteins at various distances from the tethering point (Satoh et al. 2016a) as suggested by previous in vitro and in vivo studies (Deprez et al. 2005; Olson et al. 2015; Totani et al. 2008). To gain further insight into the working mechanism of GII, structural characterization of intact GII in the presence and absence of substrates using an integrated biophysical approach employing EM and HS-AFM techniques would be necessary.

8.3 Surveillance of Newly Synthesized Glycoproteins by Folding-Sensor Enzyme UGGT

UGGT is a unique soluble glycosyltransferase with a relatively large molecular mass (around 160 kDa), which is thought to be divided into two regions: a 125 kDa N-terminal region and a C-terminal 35 kDa catalytic (CAT) domain belonging to the GT24 family, which is predicted to share structural similarities with the GT8 family (Arnold and Kaufman 2003; D'Alessio et al. 2010; Guerin and

Parodi 2003; Ito et al. 2015). Since it has been shown that the larger, variable N-terminal region is essential for enzymatic activity, it is thought that this region functions as a folding sensor. Higher eukaryotes from green algae to humans share a selenocysteine-containing thioredoxin (Trx)-like oxidoreductase (Labunskyy et al. 2007), Sep15, which binds to UGGT to form a stable complex (Labunskyy et al. 2005), enhancing its glucosyltransferase activity (Takeda et al. 2014). Several studies have demonstrated that UGGT shows enzymatic activity only against incompletely folded glycoproteins (Caramelo et al. 2003; Ito et al. 2015; Izumi et al. 2012; Taylor et al. 2004; Totani et al. 2009). This suggests that this enzyme recognizes both glycan and unstructured polypeptide parts of glycoproteins.

Until recently, no structural information relating to UGGT had been available, possibly due to its instability. To solve this instability problem, various source organisms including *Drosophila melanogaster* (fruit fly; Calles-Garcia et al. 2017), *Penicillium chrysogenum* (mesophilic fungus; Calles-Garcia et al. 2017), *Chaetomium thermophilum* (thermophilic fungus; Roversi et al. 2017; Zhu et al. 2014), and *Thermomyces dupontii* (thermophilic fungus; Satoh et al. 2017) were utilized for structural investigation. Crystallization of UGGTs derived from thermophilic fungi was successfully reported. The first example of UGGT structural crystallization was made on one of three tandem-repeat Trx-like domains, which was predicted to exist in the N-terminal folding-sensor region of *C. thermophilum* UGGT using bioinformatic techniques (Zhu et al. 2014). Thereafter, two groups, including ours, independently delineated the overall structure of the N-terminal folding-sensor region. Roversi et al. reported crystal structures of the full-length *C. thermophilum* UGGT with proteolytic cleavage at the linker connecting the N-terminal region and the CAT domain (Roversi et al. 2017), while we presented the crystal structures of the CAT domain and N-terminal region of *T. dupontii* UGGT separately (Satoh et al. 2017).

These crystallographic studies redefined the previously accepted domain structure of the N-terminal folding-sensor region by identifying an unexpected additional Trx-like domain (designated as Trx4) and two β-sandwich domains with topological structures, unique among Trx-like domain-containing proteins (Fig. 8.4a; Heras et al. 2007; Kozlov et al. 2010a). Where the Trx2, Trx3, and Trx4 domains all showed a canonical Trx-like fold characterized by a three- or four-stranded β-sheet surrounded by six α-helices, the Trx1 domain exhibited a noncanonical Trx-like fold, in which the four-helix subdomain was not within the Trx domain, but instead preceded the N-terminal helix. Although the Trx4 domain adopts a canonical Trx-like fold, it shows unusual topology, which is a result of two nonconsecutive segments. Like the Trx4 domain, strikingly unique domain architecture was also found in the first β-sandwich domain, in which the seven-stranded β-strands consist of three nonconsecutive segments. In contrast to the crystal structures of *C. thermophilum* UGGT, the C-terminal β-sandwich domain comprising seven-stranded β-strands was almost completely disordered in *T. dupontii* UGGT, suggesting that this domain is flexible in nature, although the tertiary structures of the individual domains of the folding-sensor region originating from two fungal species were essentially identical. The overall structure of the folding-sensor region

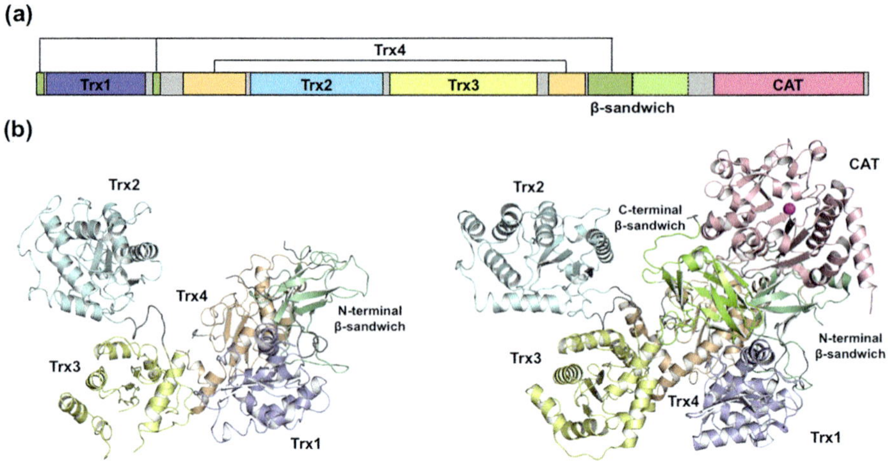

Fig. 8.4 (**a**) Domain structure of UGGT. "Trx" and "CAT" represent thioredoxin-like and catalytic domains, respectively. (**b**) Crystal structures of the folding-sensor region of UGGT from *T. dupontii* (PDB code: 5Y7O) and full-length UGGT from *C. thermophilum* (PDB code: 5MZO) are shown on the left and right, respectively. The individual domains of UGGT are colored as follows: Trx1 domain, light blue; Trx2 domain, pale cyan; Trx3 domain, light yellow; Trx4 domain, light orange; N- and C-terminal β-sandwich domains, light green; and catalytic domain, light pink. These colors correspond to those in Fig. 8.4a

of UGGT exhibits a C-shaped structure containing a large central cavity with hydrophobic patches, suggesting that they may be involved in folding-sensor function (Satoh et al. 2017). In the *C. thermophilum* UGGT, a significant difference was observed in the spatial positioning of the Trx2 domain among the three elucidated crystal structures (Roversi et al. 2017). In the crystal, the Trx2 domain exhibits "open" (PDB code: 5MZO), "intermediate" (5MU1), and "closed" (5N2J) conformations. The crystal structure of the folding-sensor region of *T. dupontii* UGGT (Satoh et al. 2017) was most similar to the open form of the folding-sensor region of *C. thermophilum* UGGT. Intriguingly, double cysteine point mutants that rigidify the UGGT structure upon introduction of extra interdomain disulfide bridges between the Trx2 domain and the N-terminal or the C-terminal β-sandwich domains were shown to impair enzymatic activity, which illustrates the importance of interdomain flexibility in the folding-sensor region (Roversi et al. 2017). Interestingly, in the crystal structures of the *C. thermophilum* UGGT, the CAT domain was shown to be tightly associated with the folding-sensor region through the two β-sandwich domains (Roversi et al. 2017). This point will be discussed later.

The crystal structure of the CAT domain of *T. dupontii* UGGT was determined in its isolated form and in complex with UDP-glucose or UDP in the presence of Ca^{2+} (Satoh et al. 2017), while it was determined in its apo- and Ca^{2+}-bound forms using a full-length construct of *C. thermophilum* UGGT (Fig. 8.5a; Roversi et al. 2017). As with the folding-sensor region, the overall structure of the CAT domains of two fungal species was essentially identical. As expected from previous bioinformatics

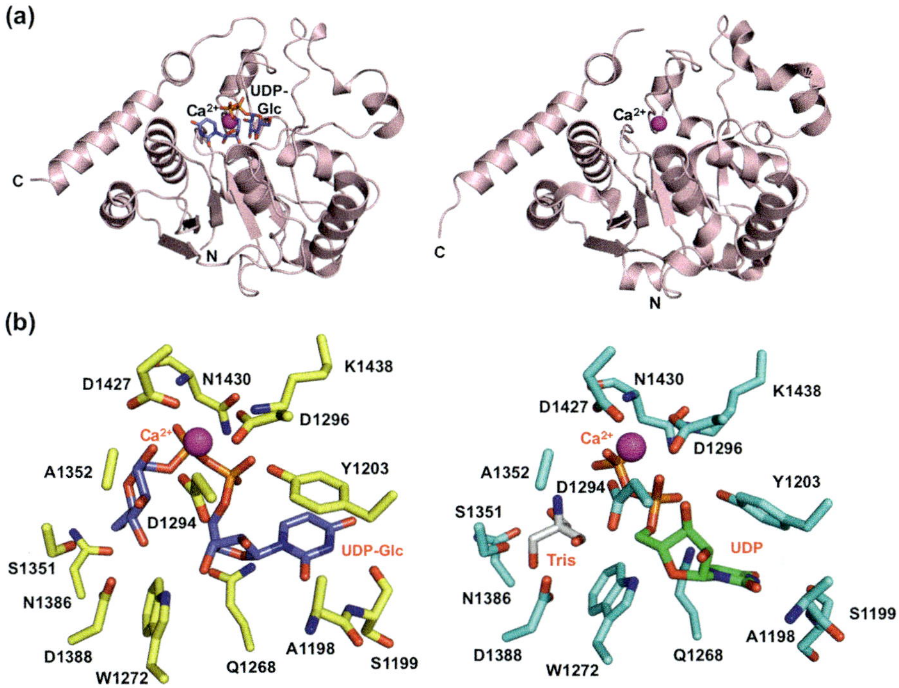

Fig. 8.5 (**a**) Crystal structures of the CAT domain of *T. dupontii* (PDB code: 5H18) and *C. thermophilum* UGGT (PDB code: 5MZO), shown as ribbon models. Bound UDP-Glc and Ca^{2+} are shown as stick and sphere models, respectively. (**b**) The catalytic active site of UGGT in the UDP-Glc- or UDP-bound form is shown on the left and right, respectively. Residues involved in binding to Ca^{2+} and UDP-Glc or UDP are numbered and shown in the stick model. The UDP-Glc and UDP molecules are colored slate and green, respectively

analysis, crystal structures of the CAT domain displayed a typical GT-A fold comprising 9–10 β-strands and 12 α-helices, which share structural similarities with the GT8 family enzyme galactosyltransferase LgtC (Fig. 8.5; Persson et al. 2001). The UDP-glucose and Ca^{2+} ligands were accommodated in an active site containing a DXD motif, in a similar fashion to that of retaining GT8 glycosyltransferase (Fig. 8.5; Persson et al. 2001), implying that the enzymatic mechanisms of UGGT may be evolutionarily conserved. In its UDP/Ca^{2+}-bound form, the glucose-binding site was occupied by a Tris molecule, and two loops containing the active-site residues exhibited significant structural changes upon UDP binding, suggesting its release mechanism of the monoglucosylated product. In all crystal structures (UDP-Glc/Ca^{2+}-, UDP/Ca^{2+}-, Ca^{2+}-bound, and apo forms), surface-exposed hydrophobic patches were observed around the active site, suggesting that the CAT domain may be involved in sensing substrate folding. This hypothesis is supported by the observation that a $Man_9GlcNAc_2$ probe tagged with a hydrophobic moiety selectively binds amino acid residues in the proximity of the hydrophobic patches (Ohara et al. 2015).

In addition to crystallographic data, UGGT enzymes from various species including *D. melanogaster*, *P. chrysogenum* (Calles-Garcia et al. 2017), *C. thermophilum* (Roversi et al. 2017), and *T. dupontii* (Satoh et al. 2017) were structurally characterized by SAXS, negative staining, and cryo-EM techniques. The SAXS and EM data illustrated the monomeric structure of UGGT in solution. The SAXS data regarding *D. melanogaster*, *P. chrysogenum*, and *T. dupontii* UGGTs showed similar anisotropic elongated C-shaped structures (Calles-Garcia et al. 2017; Satoh et al. 2017). In the case of the *T. dupontii* UGGT (Satoh et al. 2017), the crystal structure of the folding-sensor region fitted neatly into the SAXS shape model, while that of the CAT domain did not show a volume density corresponding to the residual space, suggesting that the CAT domain is relatively mobile in solution (Fig. 8.6a). In addition to the CAT domain, a notable bulge was observed around the Trx2 domain, suggesting that it is flexible. The observed conformational heterogeneities of UGGTs were more significant in the EM analysis data, especially in the cryo-EM images (Fig. 8.6b). In the averaged negative-stained EM images, *D. melanogaster*, *P. chrysogenum*, and *T. dupontii* UGGTs exhibited elongated C-shaped structures with a central cavity (Fig. 8.6c), which is consistent with the SAXS-derived structure models (Calles-Garcia et al. 2017; Satoh et al. 2017). Also consistent with the SAXS results, the crystal structure of the folding-sensor region (excepting the Trx2 domain) fitted neatly onto the EM map, whereas that of the CAT domain did not perfectly match the EM map.

The dynamic nature of the *T. dupontii* UGGT was directly observed using HS-AFM techniques (Satoh et al. 2017). By visualizing UGGT using HS-AFM images, it could be seen that the enzyme has one larger lobe and one smaller one, which correspond to the folding-sensor region and the CAT domain, respectively. These two lobes are connected via a flexible linker. The estimated distance between the centers of the two lobes, based on the most frequently observed distances, was 75 Å, which closely matches previous biochemical data that suggest that UGGT can transfer glucose to *N*-glycans at least 40 Å from the unstructured regions of glycoprotein substrates (Taylor et al. 2004). These observations indicate that the *T. dupontii* UGGT exhibits a flexible modular structure in which the CAT domain is tethered to the folding-sensor region with variable spatial arrangements.

In marked contrast, the SAXS data of the *C. thermophilum* UGGT indicated a compact structure, into which the full-length crystal structure including the CAT domain could be fitted, while its cryo-EM data gave no interpretable electron density data regarding the Trx2 domain. This suggests interdomain conformational flexibility through the Trx2 domain but not through the CAT domain (Roversi et al. 2017). Besides species differences, discrepancies in the mobility of the CAT domain (Calles-Garcia et al. 2017; Roversi et al. 2017; Satoh et al. 2017) may be attributed to the presence of an unstructured linker of approximately 40 residues, which is commonly thought to be an intrinsically disordered region in UGGT across species (from fungi to humans). Structural analyses of *D. melanogaster*, *P. chrysogenum*, and *T. dupontii* UGGTs were performed using intact enzymes, whereas a partially degraded enzyme with proteolytic cleavage at the linker was used for the structural characterization of the *C. thermophilum* UGGT. It is possible that the flexible linker

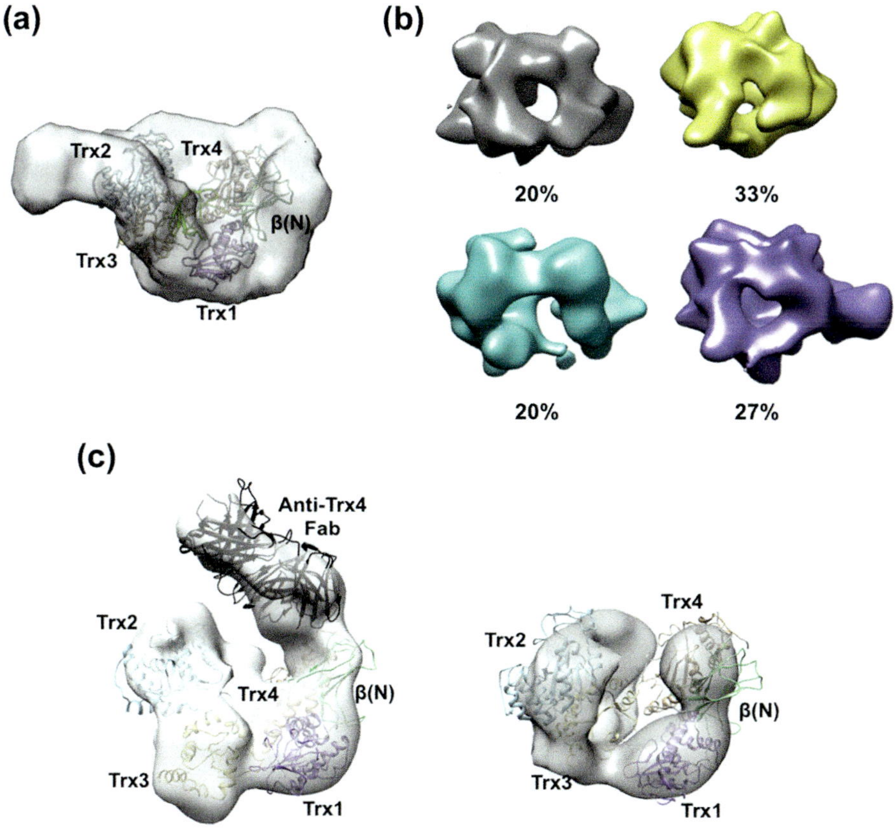

Fig. 8.6 (**a**) SAXS-derived model of full-length *T. dupontii* UGGT. The crystal structure of the folding-sensor region of UGGT from *T. dupontii* (PDB code: 5Y7O) is fitted into the SAXS-derived model. (**b**) 3D classification of the cryo-EM structures of *T. dupontii* UGGT. Each structure showed significant variation in cryo-EM maps but commonly shared a large central cavity. (**c**) Negative-stain EM 3D reconstruction of *T. dupontii* (left) and *Penicillium chrysogenum* (right; EMD code: 8718) UGGTs. The EM data of *T. dupontii* UGGT was obtained as a complex with the Fab of a monoclonal antibody directed against the Trx4 domain. The crystal structure of the folding-sensor region of *T. dupontii* UGGT is fitted into the EM maps

interferes with the interaction between the folding-sensor region and the CAT domain and that this increases the mobility of the CAT domain. To understand the functional importance of structural flexibility between the folding-sensor region and the catalytic domain, further biochemical and biophysical investigation will be necessary.

8.4 Interaction Mechanism Between Monoglucosylated High-Mannose-Type Oligosaccharide and Calreticulin

The key processes in ER glycoprotein quality control involve carbohydrate-protein interactions. Compared with protein molecules (lectins), oligosaccharides have considerable conformational flexibilities in their uncomplexed states. Therefore, carbohydrate-protein interaction is generally associated with the loss of conformational entropy, which is compensated for by the larger enthalpy gain in the interaction system (Burkhalter et al. 2000).

Structural characterization of high-mannose-type oligosaccharides in solution has been performed using theoretical and experimental approaches, i.e., molecular mechanics and dynamics and NMR spectroscopy (Balaji et al. 1994; Kamiya et al. 2013; Petrescu et al. 1997; Wooten et al. 1990; Wormald et al. 2002; Yamaguchi et al. 2013). Recently developed methods to characterize the conformational spaces of oligosaccharides combine replica-exchange MD simulations and paramagnetism-assisted NMR techniques (Kato and Yamaguchi 2015; Yamaguchi and Kato 2014; Zhang et al. 2013; Zhang et al. 2015). To overexpress homogeneous ^{13}C-labeled high-mannose-type oligosaccharides, as is required for NMR analyses, we employed genetically engineered yeast lacking the genes underlying the synthesis of hyperbranched high-mannose-type glycans (Kamiya et al. 2011, 2013, 2014; Yamaguchi et al. 2013). Furthermore, a chemoenzymatic technique employing this genetically engineered yeast, chemical synthesis (UDP-[$^{13}C_6$]glucose), and in vitro glucosylation catalyzed by UGGT to prepare ^{13}C-labeled $Glc_1Man_9GlcNAc_2$ has been established (Suzuki et al. 2017; Zhu et al. 2015).

These techniques enable us to explore the possible conformations of larger, branched high-mannose-type oligosaccharides, including $Man_8GlcNAc_2$ (M8B), $Man_9GlcNAc_2$ (M9), and $Glc_1Man_9GlcNAc_2$ (GM9). The resulting MD trajectories, validated by paramagnetism-assisted NMR data, demonstrate that conformational dynamics are quantitatively very similar between M9 and GM9. This indicates that the attachment of a monoglucose residue on the D1 branch has virtually no conformational impact on the triantennary M9 structure (Fig. 8.7a; Suzuki et al. 2017). This is in marked contrast to M8B, which lacks the terminal mannose on the D2 branch, because this decasaccharide is significantly populated with fold-back conformations (Yamaguchi et al. 2014). NMR-validated MD simulation of GM9 indicates that the spatial positions of nonreducing terminal glucose residues are widely distributed relative to the reducing end in the conformational ensemble (Suzuki et al. 2017), providing versatile opportunities for interactions between carrier glycoproteins and lectin-chaperone complexes, which are thought to recognize not only the terminal glucose residue but also the unstructured polypeptide moiety of the glycoproteins (Oliver et al. 1997). Previous crystallographic data have provided a conformational snapshot of the complex formed between Glc_1Man_3 tetrasaccharides (corresponding to the D1 branch) and the lectin domain of CRT (Fig. 8.7b) in addition to that of the Glc-Man-bound GII complex (Fig. 8.2c; Kozlov et al. 2010b; Satoh et al. 2016b).

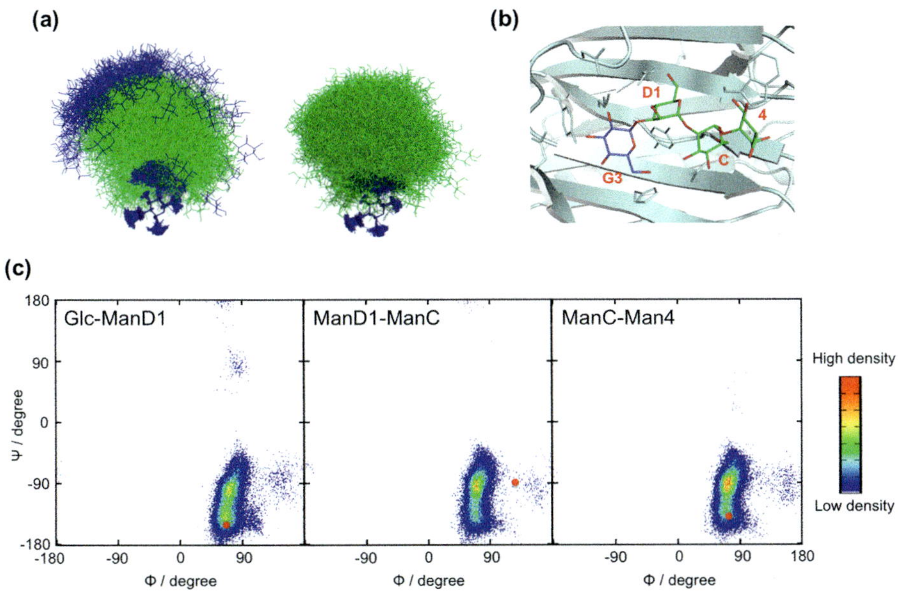

Fig. 8.7 (**a**) Superimposition of each of the 240 conformers of the GM9 and M9 oligosaccharides extracted from the replica-exchange MD simulations validated using NMR data. (**b**). Close-up view of the oligosaccharide-binding site of the lectin domain of mouse CRT (PDB code: 3O0W). (**c**) Density maps of glycosidic linkage torsion angles of the Glc_1Man_3 moiety of the GM9 oligosaccharide obtained from NMR-validated replica-exchange MD simulations. The red dots indicate the glycosidic linkage conformations of the Glc_1Man_3 tetrasaccharide in complex with CRT (PDB code: 3O0W). The Glc-ManD1 glycosidic linkage conformation in complex with GII (PDB code: 5DKZ) was almost identical with that observed in complex with CRT. Figure 8.7c was adapted from Suzuki et al. (2017) with the permission of John Wiley & Sons, Inc.

Intriguingly, the Glc-Man glycosidic linkage conformation observed in a complex was also found within the dynamic conformational ensemble of the D1 branch of the GM9 oligosaccharide, indicating that substrate recognition by GII involves a conformational selection mechanism (Fig. 8.7c). In contrast, the ManD1-ManC glycosidic conformation in its CRT-bound state was apparently isolated from the conformational cluster derived from the uncomplexed state, indicating that oligosaccharide recognition by CRT involves an induced-fit mechanism with an entropic penalty (Fig. 8.7c; Suzuki et al. 2017). It has been speculated that this negatively regulated glycan interaction is important in the quality-control system for selective dual recognition of unfolded glycoproteins carrying GM9 in addition to unstructured polypeptide chains.

8.5 Summary and Perspectives

The glucosylation-mediated ER quality control of glycoproteins is executed via a series of oligosaccharide-recognizing proteins, including glucosidases, glucosyltransferase, and lectins. Their sophisticated biological functions are coupled with internal mobility of the processing enzymes and chaperones composed of multiple domains and/or subunits, along with binding partners. Conformational variabilities in *N*-glycans are also key factors in glycoprotein quality control. Molecular events associated with conformational interconversion have often been characterized based on crystallographic snapshots; however, these should be interpreted with great care, because they can be affected by crystal packing and they provide no temporal information regarding biomacromolecular dynamics. Therefore, for detailed structural characterization, complementary biophysical tools are required. An integrative approach combining SAXS, electron microscopy, X-ray crystallography, and high-speed atomic force microscopy has prompted advancements in our understanding of molecular mechanisms involved in glycoprotein quality control, as exemplified by the glucose-processing enzymes with flexible modular structures, GII and UGGT. The next stage of investigation will address the question of how these molecules interact in such a highly orchestrated manner in the heterogeneous, crowded environment of the ER.

Acknowledgments This work was supported in part by the Okazaki ORION project and Grants-in-Aid for Scientific Research (Grant Numbers JP25121730 to T.S., and JP25102008, JP24249002 to K.K.) from the Ministry of Education, Culture, Sports, Science and Technology, Japan, the PRESTO project (Grant Number JPMJPR13L5 to T.S.) from the Japan Science and Technology Agency, and by Grant-in-Aid for Research in Nagoya City University.

References

Aebi M, Bernasconi R, Clerc S, Molinari M (2010) *N*-glycan structures: recognition and processing in the ER trends. Biochem Sci 35:74–82. https://doi.org/10.1016/j.tibs.2009.10.001

Arendt CW, Ostergaard HL (2000) Two distinct domains of the β -subunit of glucosidase II interact with the catalytic α-subunit. Glycobiology 10:487–492. https://doi.org/10.1093/glycob/10.5.487

Arnold SM, Kaufman RJ (2003) The noncatalytic portion of human UDP-glucose: glycoprotein glucosyltransferase I confers UDP-glucose binding and transferase function to the catalytic domain. J Biol Chem 278:43320–43328. https://doi.org/10.1074/jbc.M305800200

Balaji PV, Qasba PK, Rao VS (1994) Molecular dynamics simulations of high-mannose oligosaccharides. Glycobiology 4:497–515. https://doi.org/10.1093/glycob/4.4.497

Barker MK, Rose DR (2013) Specificity of processing α-glucosidase I is guided by the substrate conformation: crystallographic and in silico studies. J Biol Chem 288:13563–13574. https://doi.org/10.1074/jbc.M113.460436

Betenbaugh MJ, Tomiya N, Narang S, Hsu JT, Lee YC (2004) Biosynthesis of human-type *N*-glycans in heterologous systems. Curr Opin Struct Biol 14:601–606. https://doi.org/10.1016/j.sbi.2004.09.001

Blees A et al (2017) Structure of the human MHC-I peptide-loading complex. Nature 551:525–528. https://doi.org/10.1038/nature24627

Burkhalter NF, Dimick SM, Toone EJ (2000) Protein-carbohydrate interaction: fundamental considerations. In: Ernst B, Hart GW, Sinaÿ P (eds) Carbohydrates in chemistry and biology, part II, vol 2. Wiley-VCH, Weinhein, pp 863–914. https://doi.org/10.1002/9783527618255.ch31

Calles-Garcia D et al (2017) Single-particle electron microscopy structure of UDP-glucose:glycoprotein glucosyltransferase suggests a selectivity mechanism for misfolded proteins. J Biol Chem 292:11499–11507. https://doi.org/10.1074/jbc.M117.789495

Caputo AT et al (2016) Structures of mammalian ER α-glucosidase II capture the binding modes of broad-spectrum iminosugar antivirals. Proc Natl Acad Sci U S A 113(32):E4630–E4638. https://doi.org/10.1073/pnas.1604463113

Caramelo JJ, Castro OA, Alonso LG, De Prat-Gay G, Parodi AJ (2003) UDP-Glc:glycoprotein glucosyltransferase recognizes structured and solvent accessible hydrophobic patches in molten globule-like folding intermediates. Proc Natl Acad Sci U S A 100:86–91. https://doi.org/10.1073/pnas.262661199

Castonguay AC, Olson LJ, Dahms NM (2011) Mannose 6-phosphate receptor homology (MRH) domain-containing lectins in the secretory pathway. Biochim Biophys Acta 1810:815–826. https://doi.org/10.1016/j.bbagen.2011.06.016

D'Alessio C, Caramelo JJ, Parodi AJ (2010) UDP-GlC: glycoprotein glucosyltransferase-glucosidase II, the ying-yang of the ER quality control. Semi Cell Dev Bio 21:491–499. https://doi.org/10.1016/j.semcdb.2009.12.014

D'Alessio C, Dahms NM (2015) Glucosidase II and MRH-domain containing proteins in the secretory pathway. Curr Protein Pept Sci 16:31–48. https://doi.org/10.2174/1389203716666150213160438

Deprez P, Gautschi M, Helenius A (2005) More than one glycan is needed for ER glucosidase II to allow entry of glycoproteins into the calnexin/calreticulin cycle. Mol Cell 19:183–195. https://doi.org/10.1016/j.molcel.2005.05.029

Drenth JP, te Morsche RH, Smink R, Bonifacino JS, Jansen JB (2003) Germline mutations in *PRKCSH* are associated with autosomal dominant polycystic liver disease. Nat Genet 33:345–347. https://doi.org/10.1038/ng1104

Ernst HA, Lo Leggio L, Willemoes M, Leonard G, Blum P, Larsen S (2006) Structure of the *Sulfolobus solfataricus* α-glucosidase: implications for domain conservation and substrate recognition in GH31. J Mol Biol 358:1106–1124. https://doi.org/10.1016/j.jmb.2006.02.056

Fisher C, Beglova N, Blacklow SC (2006) Structure of an LDLR-RAP complex reveals a general mode for ligand recognition by lipoprotein receptors. Mol Cell 22:277–283. https://doi.org/10.1016/j.molcel.2006.02.021

Grinna LS, Robbins PW (1980) Substrate specificities of rat liver microsomal glucosidases which process glycoproteins. J Biol Chem 255:2255–2258

Guerin M, Parodi AJ (2003) The UDP-glucose:glycoprotein glucosyltransferase is organized in at least two tightly bound domains from yeast to mammals. J Biol Chem 278:20540–20546. https://doi.org/10.1074/jbc.M300891200

Heras B, Kurz M, Shouldice SR, Martin JL (2007) The name's bond disulfide bond. Curr Opin Struct Biol 17:691–698. https://doi.org/10.1016/j.sbi.2007.08.009

Hosokawa N, Kamiya Y, Kamiya D, Kato K, Nagata K (2009) Human OS-9, a lectin required for glycoprotein endoplasmic reticulum-associated degradation, recognizes mannose-trimmed *N*-glycans. J Biol Chem 284:17061–17068. https://doi.org/10.1074/jbc.M809725200

Hu D et al (2009) Sugar-binding activity of the MRH domain in the ER α-glucosidase II β subunit is important for efficient glucose trimming. Glycobiology 19:1127–1135. https://doi.org/10.1093/glycob/cwp104

Hulpke S, Tampe R (2013) The MHC I loading complex: a multitasking machinery in adaptive immunity. Trends Biochem Sci 38:412–420. https://doi.org/10.1016/j.tibs.2013.06.003

Ito Y, Takeda Y, Seko A, Izumi M, Kajihara Y (2015) Functional analysis of endoplasmic reticulum glucosyltransferase (UGGT): Synthetic chemistry's initiative in glycobiology. Semin Cell Dev Bio 41:90–98. https://doi.org/10.1016/j.semcdb.2014.11.011

Izumi M et al (2012) Chemical synthesis of intentionally misfolded homogeneous glycoprotein: a unique approach for the study of glycoprotein quality control. J Am Chem Soc 134:7238–7241. https://doi.org/10.1021/ja3013177

Jessop CE, Tavender TJ, Watkins RH, Chambers JE, Bulleid NJ (2009) Substrate specificity of the oxidoreductase ERp57 is determined primarily by its interaction with calnexin and calreticulin. J Biol Chem 284:2194–2202. https://doi.org/10.1074/jbc.M808054200

Kamiya Y, Satoh T, Kato K (2012) Molecular and structural basis for *N*-glycan-dependent determination of glycoprotein fates in cells. Biochim Biophys Acta 1820:1327–1337. https://doi.org/10.1016/j.bbagen.2011.12.017

Kamiya Y, Satoh T, Kato K (2014) Recent advances in glycoprotein production for structural biology: toward tailored design of glycoforms. Curr Opin Struct Biol 26:44–53. https://doi.org/10.1016/j.sbi.2014.03.008

Kamiya Y, Yamamoto S, Chiba Y, Jigami Y, Kato K (2011) Overexpression of a homogeneous oligosaccharide with C labeling by genetically engineered yeast strain. J Biomol NMR 50:397–401. https://doi.org/10.1007/s10858-011-9525-1

Kamiya Y, Yanagi K, Kitajima T, Yamaguchi T, Chiba Y, Kato K (2013) Application of metabolic C labeling in conjunction with high-field nuclear magnetic resonance spectroscopy for comparative conformational analysis of high mannose-type oligosaccharides. Biomol Ther 3:108–123. https://doi.org/10.3390/biom3010108

Kato K, Yamaguchi T (2015) Paramagnetic NMR probes for characterization of the dynamic conformations and interactions of oligosaccharides. Glycoconj J 32:505–513. https://doi.org/10.1007/s10719-015-9599-1

Kelleher DJ, Gilmore R (2006) An evolving view of the eukaryotic oligosaccharyltransferase. Glycobiology 16:47R–62R. https://doi.org/10.1093/glycob/cwj066

Kim JJ, Olson LJ, Dahms NM (2009) Carbohydrate recognition by the mannose-6-phosphate receptors. Curr Opin Struct Biol 19:534–542. https://doi.org/10.1016/j.sbi.2009.09.002

Kornfeld R, Kornfeld S (1985) Assembly of asparagine-linked oligosaccharides. Annu Rev Biochem 54:631–664. https://doi.org/10.1146/annurev.bi.54.070185.003215

Kozlov G, Maattanen P, Thomas DY, Gehring K (2010a) A structural overview of the PDI family of proteins. FEBS J 277:3924–3936. https://doi.org/10.1111/j.1742-4658.2010.07793.x

Kozlov G, Munoz-Escobar J, Castro K, Gehring K (2017) Mapping the ER interactome: The P domains of calnexin and calreticulin as plurivalent adapters for foldases and chaperones. Structure 25:1415–1422 e1413. https://doi.org/10.1016/j.str.2017.07.010

Kozlov G, Pocanschi CL, Rosenauer A, Bastos-Aristizabal S, Gorelik A, Williams DB, Gehring K (2010) Structural basis of carbohydrate recognition by calreticulin. J Biol Chem 285:38612–38620. https://doi.org/10.1074/jbc.M110.168294

Labunskyy VM, Ferguson AD, Fomenko DE, Chelliah Y, Hatfield DL, Gladyshev VN (2005) A novel cysteine-rich domain of Sep15 mediates the interaction with UDP-glucose:glycoprotein glucosyltransferase. J Biol Chem 280:37839–37845. https://doi.org/10.1074/jbc.M508685200

Labunskyy VM, Hatfield DL, Gladyshev VN (2007) The Sep15 protein family: roles in disulfide bond formation and quality control in the endoplasmic reticulum. IUBMB Life 59:1–5. https://doi.org/10.1080/15216540601126694

Lederkremer GZ (2009) Glycoprotein folding, quality control and ER-associated degradation. Curr Opin Struct Biol 19:515–523. https://doi.org/10.1016/j.sbi.2009.06.004

Li A et al (2003) Mutations in *PRKCSH* cause isolated autosomal dominant polycystic liver disease. Am J Hum Genet 72:691–703. https://doi.org/10.1086/368295

Lovering AL, Lee SS, Kim YW, Withers SG, Strynadka NC (2005) Mechanistic and structural analysis of a family 31 α-glycosidase and its glycosyl-enzyme intermediate. J Biol Chem 280:2105–2115. https://doi.org/10.1074/jbc.M410468200

Maattanen P, Kozlov G, Gehring K, Thomas DY (2006) ERp57 and PDI: multifunctional protein disulfide isomerases with similar domain architectures but differing substrate-partner associations. Biochem Cell Biol 84:881–889. https://doi.org/10.1139/o06-186

Mikami K et al (2010) The sugar-binding ability of human OS-9 and its involvement in ER-associated degradation. Glycobiology 20:310–321. https://doi.org/10.1093/glycob/cwp175

Munro S (2001) The MRH domain suggests a shared ancestry for the mannose 6-phosphate receptors and other *N*-glycan-recognising proteins. Curr Biol 11:R499–R501. https://doi.org/10.1016/S0960-9822(01)00302-5

Nakao H, Seko A, Ito Y, Sakono M (2017) PDI family protein ERp29 recognizes P-domain of molecular chaperone calnexin. Biochem Biophys Res Commun 487:763–767. https://doi.org/10.1016/j.bbrc.2017.04.139

Ninagawa S et al (2014) EDEM2 initiates mammalian glycoprotein ERAD by catalyzing the first mannose trimming step. J Cell Biol 206:347–356. https://doi.org/10.1083/jcb.201404075

Ohara K, Takeda Y, Daikoku S, Hachisu M, Seko A, Ito Y (2015) Profiling aglycon-recognizing sites of UDP-glucose:glycoprotein Glucosyltransferase by means of squarate-mediated labeling. Biochemistry 54:4909–4917. https://doi.org/10.1021/acs.biochem.5b00785

Oliver JD, van der Wal FJ, Bulleid NJ, High S (1997) Interaction of the thiol-dependent reductase ERp57 with nascent glycoproteins. Science 275:86–88. https://doi.org/10.1126/science.275.5296.86

Olson LJ et al (2013) Structure of the lectin mannose 6-phosphate receptor homology (MRH) domain of glucosidase II, an enzyme that regulates glycoprotein folding quality control in the endoplasmic reticulum. J Biol Chem 288:16460–16475. https://doi.org/10.1074/jbc.M113.450239

Olson LJ, Orsi R, Peterson FC, Parodi AJ, Kim JJ, D'Alessio C, Dahms NM (2015) Crystal structure and functional analyses of the lectin domain of Glucosidase II: insights into oligomannose recognition. Biochemistry 54:4097–4111. https://doi.org/10.1021/acs.biochem.5b00256

Park S, You KH, Shong M, Goo TW, Yun EY, Kang SW, Kwon OY (2005) Overexpression of ERp29 in the thyrocytes of FRTL-5 cells. Mol Biol Rep 32:7–13. https://doi.org/10.1007/s11033-004-3069-3

Pelletier MF et al (2000) The heterodimeric structure of glucosidase II is required for its activity, solubility, and localization *in vivo*. Glycobiology 10:815–827. https://doi.org/10.1093/glycob/10.8.815

Persson K, Ly HD, Dieckelmann M, Wakarchuk WW, Withers SG, Strynadka NC (2001) Crystal structure of the retaining galactosyltransferase LgtC from *Neisseria meningitidis* in complex with donor and acceptor sugar analogs. Nat Struct Biol 8:166–175. https://doi.org/10.1038/84168

Petrescu AJ, Butters TD, Reinkensmeier G, Petrescu S, Platt FM, Dwek RA, Wormald MR (1997) The solution NMR structure of glucosylated *N*-glycans involved in the early stages of glycoprotein biosynthesis and folding. EMBO J 16:4302–4310. https://doi.org/10.1093/emboj/16.14.4302

Porath B et al (2016) Mutations in *GANAB*, encoding the glucosidase IIα subunit, cause autosomal-dominant polycystic kidney and liver disease. Am J Hum Genet 98:1193–1207. https://doi.org/10.1016/j.ajhg.2016.05.004

Quan EM, Kamiya Y, Kamiya D, Denic V, Weibezahn J, Kato K, Weissman JS (2008) Defining the glycan destruction signal for endoplasmic reticulum-associated degradation. Mol Cell 32:870–877. https://doi.org/10.1016/j.molcel.2008.11.017

Quinn RP, Mahoney SJ, Wilkinson BM, Thornton DJ, Stirling CJ (2009) A novel role for Gtb1p in glucose trimming of *N*-linked glycans. Glycobiology 19:1408–1416. https://doi.org/10.1093/glycob/cwp087

Ren L, Qin X, Cao X, Wang L, Bai F, Bai G, Shen Y (2011) Structural insight into substrate specificity of human intestinal maltase-glucoamylase. Protein Cell 2:827–836. https://doi.org/10.1007/s13238-011-1105-3

Roversi P et al (2017) Interdomain conformational flexibility underpins the activity of UGGT, the eukaryotic glycoprotein secretion checkpoint. Proc Natl Acad Sci U S A 114:8544–8549. https://doi.org/10.1073/pnas.1703682114

Sakono M, Seko A, Takeda Y, Ito Y (2014) PDI family protein ERp29 forms 1:1 complex with lectin chaperone calreticulin. Biochem Biophys Res Commun 452:27–31. https://doi.org/10.1016/j.bbrc.2014.08.041

Satoh T, Chen Y, Hu D, Hanashima S, Yamamoto K, Yamaguchi Y (2010) Structural basis for oligosaccharide recognition of misfolded glycoproteins by OS-9 in ER-associated degradation. Mol Cell 40:905–916. https://doi.org/10.1016/j.molcel.2010.11.017

Satoh T et al (2017) Visualisation of a flexible modular structure of the ER folding-sensor enzyme UGGT. Sci Rep 7:12142. https://doi.org/10.1038/s41598-017-12283-w

Satoh T, Toshimori T, Noda M, Uchiyama S, Kato K (2016a) Interaction mode between catalytic and regulatory subunits in glucosidase II involved in ER glycoprotein quality control. Protein Sci 25:2095–2101. https://doi.org/10.1002/pro.3031

Satoh T, Toshimori T, Yan G, Yamaguchi T, Kato K (2016b) Structural basis for two-step glucose trimming by glucosidase II involved in ER glycoprotein quality control. Sci Rep 6:20575. https://doi.org/10.1038/srep20575

Satoh T, Yamaguchi T, Kato K (2015) Emerging structural insights into glycoprotein quality control coupled with *N*-glycan processing in the endoplasmic reticulum. Molecules 20:2475–2491. https://doi.org/10.3390/molecules20022475

Sayce AC, Miller JL, Zitzmann N (2010) Targeting a host process as an antiviral approach against dengue virus. Trends Microbiol 18:323–330. https://doi.org/10.1016/j.tim.2010.04.003

Sim L, Quezada-Calvillo R, Sterchi EE, Nichols BL, Rose DR (2008) Human intestinal maltase-glucoamylase: crystal structure of the N-terminal catalytic subunit and basis of inhibition and substrate specificity. J Mol Biol 375:782–792. https://doi.org/10.1016/j.jmb.2007.10.069

Sim L, Willemsma C, Mohan S, Naim HY, Pinto BM, Rose DR (2010) Structural basis for substrate selectivity in human maltase-glucoamylase and sucrase-isomaltase N-terminal domains. J Biol Chem 285:17763–17770. https://doi.org/10.1074/jbc.M109.078980

Stanley P, Schachter H, Taniguchi N (2009) N-glycans. In: Varki A, Cummings RD, Esko JD, Freeze HH, Stanley P, Bertozzi CR, Hart GW, Etzler ME (eds) Essentials of Glycobiology. Cold Spring Harbor, Cold Spring Harbor Lab

Stigliano ID, Alculumbre SG, Labriola CA, Parodi AJ, D'Alessio C (2011) Glucosidase II and *N*-glycan mannose content regulate the half-lives of monoglucosylated species in vivo. Mol Biol Cell 22:1810–1823. https://doi.org/10.1091/mbc.E11-01-0019

Stigliano ID, Caramelo JJ, Labriola CA, Parodi AJ, D'Alessio C (2009) Glucosidase II β subunit modulates *N*-glycan trimming in fission yeasts and mammals. Mol Biol Cell 20:3974–3984. https://doi.org/10.1091/mbc.E09-04-0316

Suzuki T et al (2017) Conformational analysis of a high-mannose-type oligosaccharide displaying glucosyl determinant recognised by molecular chaperones using NMR-validated molecular dynamics simulation. Chembiochem 18:396–401. https://doi.org/10.1002/cbic.201600595

Tagami T, Yamashita K, Okuyama M, Mori H, Yao M, Kimura A (2013) Molecular basis for the recognition of long-chain substrates by plant α-glucosidases. J Biol Chem 288:19296–19303. https://doi.org/10.1074/jbc.M113.465211

Takeda Y et al (2014) Both isoforms of human UDP-glucose:glycoprotein glucosyltransferase are enzymatically active. Glycobiology 24:344–350. https://doi.org/10.1093/glycob/cwt163

Takeda Y, Totani K, Matsuo I, Ito Y (2009) Chemical approaches toward understanding glycan-mediated protein quality control. Curr Opin Chem Biol 13:582–591. https://doi.org/10.1016/j.cbpa.2009.09.011

Taylor SC, Ferguson AD, Bergeron JJ, Thomas DY (2004) The ER protein folding sensor UDP-glucose glycoprotein-glucosyltransferase modifies substrates distant to local changes in glycoprotein conformation. Nat Struct Mol Biol 11:128–134. https://doi.org/10.1038/nsmb715

Totani K, Ihara Y, Matsuo I, Ito Y (2006) Substrate specificity analysis of endoplasmic reticulum glucosidase II using synthetic high mannose-type glycans. J Biol Chem 281:31502–31508. https://doi.org/10.1074/jbc.M605457200

Totani K, Ihara Y, Matsuo I, Ito Y (2008) Effects of macromolecular crowding on glycoprotein processing enzymes. J Am Chem Soc 130:2101–2107. https://doi.org/10.1021/ja077570k

Totani K, Ihara Y, Tsujimoto T, Matsuo I, Ito Y (2009) The recognition motif of the glycoprotein-folding sensor enzyme UDP-Glc:glycoprotein glucosyltransferase. Biochemistry 48:2933–2940. https://doi.org/10.1021/bi8020586

Trombetta ES, Fleming KG, Helenius A (2001) Quaternary and domain structure of glycoprotein processing glucosidase II. Biochemistry 40:10717–10722. https://doi.org/10.1021/bi010629u

Trombetta ES, Simons JF, Helenius A (1996) Endoplasmic reticulum glucosidase II is composed of a catalytic subunit, conserved from yeast to mammals, and a tightly bound noncatalytic HDEL-containing subunit. J Biol Chem 271:27509–27516. https://doi.org/10.1074/jbc.271.44.27509

Tyrrell BE, Sayce AC, Warfield KL, Miller JL, Zitzmann N (2017) Iminosugars: Promising therapeutics for influenza infection. Crit Rev Microbiol 43:521–545. https://doi.org/10.1080/1040841X.2016.1242868

Vliegenthart JFG, Dornald L, Halveek HV (1983) High-resolution, H-nuclear magnetic resonance spectroscopy as a tool on the structural analysis of carbohydrates related to glycoproteins. Adv Carbohydr Chem Biochem 41:209–374. https://doi.org/10.1016/S0065-2318(08)60059-1

Watanabe T, Totani K, Matsuo I, Maruyama J, Kitamoto K, Ito Y (2009) Genetic analysis of glucosidase II β -subunit in trimming of high-mannosetype glycans. Glycobiology 19:834–840. https://doi.org/10.1093/glycob/cwp061

Wooten EW, Bazzo R, Edge CJ, Zamze S, Dwek RA, Rademacher TW (1990) Primary sequence dependence of conformation in oligomannose oligosaccharides. Eur Biophys J 18:139–148. https://doi.org/10.1007/BF02427373

Wormald MR, Petrescu AJ, Pao YL, Glithero A, Elliott T, Dwek RA (2002) Conformational studies of oligosaccharides and glycopeptides: complementarity of NMR, X-ray crystallography, and molecular modelling. Chem Rev 102:371–386. https://doi.org/10.1021/cr990368i

Yamaguchi T, Kamiya Y, Choo Y-M, Yamamoto S, Kato K (2013) Terminal spin labeling of a high-mannose-type oligosaccharide for quantitative NMR analysis of its dynamic conformation. Chem Lett 42:544–546. https://doi.org/10.1246/cl.130040

Yamaguchi T, Kato K (2014) Paramagnetism-assisted nuclear magnetic resonance analysis of dynamic conformations and interactions of oligosaccharides glycoscience. In: Taniguchi N, Endo T, Hart GW, Seeberger P, Wong C-H (eds) Biology and medicine. Springer (Japan) in press, https://doi.org/10.1007/978-4-431-54836-2_101-1

Yamaguchi T, Sakae Y, Zhang Y, Yamamoto S, Okamoto Y, Kato K (2014) Exploration of conformational spaces of high-mannose-type oligosaccharides by an NMR-validated simulation. Angewandte Chemie 53:10941–10944. https://doi.org/10.1002/anie.201406145

Zhang Y, Yamaguchi T, Kato K (2013) New NMR tools for characterizing the dynamic conformations and interactions of oligosaccharides. Chem Lett 42:1455–1462. https://doi.org/10.1246/cl.130789

Zhang Y et al (2015) Conformational dynamics of oligosaccharides characterized by paramagnetism-assisted NMR spectroscopy in conjunction with molecular dynamics simulation. Adv Exp Med Biol 842:217–230. https://doi.org/10.1007/978-3-319-11280-0_14

Zhu T, Satoh T, Kato K (2014) Structural insight into substrate recognition by the endoplasmic reticulum folding-sensor enzyme: crystal structure of third thioredoxin-like domain of UDP-glucose:glycoprotein glucosyltransferase. Sci Rep 4:7322. https://doi.org/10.1038/srep07322

Zhu T, Yamaguchi T, Satoh T, Kato K (2015) A hybrid strategy for the preparation of ^{13}C-Labeled high-mannose-type oligosaccharides with terminal Glucosylation for NMR study. Chem Lett 44:1744–1746. https://doi.org/10.1246/cl.150898

Chapter 9
Structural Basis of Protein Asn-Glycosylation by Oligosaccharyltransferases

Daisuke Kohda

Abstract Glycosylation of asparagine residues is a ubiquitous protein modification. This N-glycosylation is essential in Eukaryotes, but principally nonessential in Prokaryotes (Archaea and Eubacteria), although it facilitates their survival and pathogenicity. In many reviews, Archaea have received far less attention than Eubacteria, but this review will cover the N-glycosylation in the three domains of life. The oligosaccharide chain is preassembled on a lipid-phospho carrier to form a donor substrate, lipid-linked oligosaccharide (LLO). The en bloc transfer of an oligosaccharide from LLO to selected Asn residues in the Asn-X-Ser/Thr (X≠Pro) sequons in a polypeptide chain is catalyzed by a membrane-bound enzyme, oligosaccharyltransferase (OST). Over the last 10 years, the three-dimensional structures of the catalytic subunits of the Stt3/AglB/PglB proteins, with an acceptor peptide and a donor LLO, have been determined by X-ray crystallography, and recently the complex structures with other subunits have been determined by cryo-electron microscopy. Structural comparisons within the same species and across the different domains of life yielded a unified view of the structures and functions of OSTs. A catalytic structure in the TM region accounts for the amide bond twisting, which increases the reactivity of the side-chain nitrogen atom of the acceptor Asn residue in the sequon. The Ser/Thr-binding pocket in the C-terminal domain explains the requirement for hydroxy amino acid residues in the sequon. As expected, the two functional structures are formed by the involvement of short amino acid motifs conserved across the three domains of life.

Keywords Amide-bond twisting · Co-translational modification · Free oligosaccharide · Lipid-linked oligosaccharide · Membrane protein · N-glycosylation · Oligosaccharyltransferase · Sequon

---

D. Kohda (✉)
Division of Structural Biology, Medical Institute of Bioregulation, Kyushu University, Fukuoka, Japan
e-mail: kohda@bioreg.kyushu-u.ac.jp

Y. Yamaguchi, K. Kato (eds.), *Glycobiophysics*, Advances in Experimental Medicine and Biology 1104, https://doi.org/10.1007/978-981-13-2158-0_9

Abbreviations

AFM	Atomic force microscopy
Ara4N	4-amino-4-deoxy-L-arabinose
AglB	Archaeal glycosylation B
CmanT	C-Mannosyltransferase
CDG	Congenital disorders of glycosylation
EL	External loop
EM	Electron microscopy
ER	Endoplasmic reticulum
fOS	Free oligosaccharide
FNG	Free N-glycan
HGT	Horizontal gene transfer
LLO	Lipid-linked oligosaccharide
NGT	Cytosolic N-glycosyltransferase
OST	Oligosaccharyltransferase
OTase	Oligosaccharyltransferase
PglB	Protein glycosylation B
RTC	Ribosome translocon complex
STT3	Staurosporine and temperature sensitivity 3
TM	Transmembrane
Und-P	Undecaprenyl phosphate

9.1 Introduction

Glycosylation of asparagine residues is one of the most ubiquitous protein modifications (Aebi 2013; Breitling and Aebi 2013). Strictly speaking, "N-glycosylation" refers to the sugar attachment on a nitrogen atom, but in this chapter the term N-glycosylation is used in the sense of the glycosylation on Asn residues, according to the standard practice. Fortunately, this is not a problem because the one-letter code of Asn is N. The N-glycosylation is a highly coordinated and complex process that involves many glycotransferases and glycosidases (Fig. 9.1). The entire process can be divided into three steps: (1) biosynthesis of lipid-linked oligosaccharides (LLO), (2) oligosaccharyl transfer from LLO to Asn residues in proteins, and (3) processing of oligosaccharide structures on proteins.

Within the complex N-glycosylation process, the covalent bond formation between an oligosaccharide chain and a protein is the defining event. The enzyme that catalyzes this reaction is a membrane-bound enzyme, oligosaccharyltransferase (OST or OTase) (Kelleher and Gilmore 2006; Knauer and Lehle 1999b; Mohorko et al. 2011). LLO serves as the sugar donor for OST (Hartley and Imperiali 2012). In the biosynthesis of LLO, monosaccharides are added one at a time by a specific glycotransferase from a nucleotide-monosaccharide or a dolichol-monophosphate-monosaccharide to the growing oligosaccharide chain on a lipid-phospho carrier. In

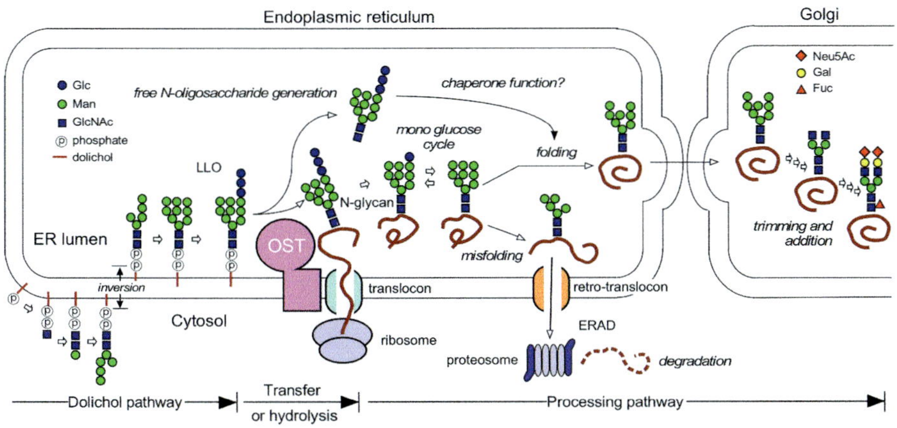

Fig. 9.1 A schematic overview of the glycosylation reaction on asparagine residues in proteins. The structure of the LLO is a higher eukaryotic (i.e., animals, plants, insects, and fungi) version, which consists of 14 monosaccharide residues, 2 phosphate groups, and a dolichol chain. On the cytoplasmic face of ER, dolichol phosphate receives GlcNAc-phosphate to generate Dol-PP-GlcNAc. After 1 GlcNAc residue and 5 mannose residues are sequentially added from UDP-GlcNAc and GDP-Man, respectively, the LLO is inverted to the luminal side of ER by as-yet-unidentified "flippase." Then, additional 4 mannose residues are added from Dol-P-Man and 3 glucose residues from Dol-P-Glc, to complete the 14-residue N-glycan structure. OST catalyzes the transfer of the oligosaccharide chain from LLO to asparagine residues in proteins. OST also hydrolyzes LLO to release the N-oligosaccharide into the luminal space of the ER. The released N-oligosaccharide might function as a chaperone of protein folding. ER Glucosidases remove the three terminal glucose residues. The mono-glucose cycle, in which removal and re-addition of the first glucose residue take place, assists the folding of glycoproteins and determines their fate, either translocation to the Golgi apparatus for secretion or into the cytoplasm for degradation by proteasome. ER mannosidase removes some mannose residues during the selection process. In the Golgi, further removal of mannose residues and addition of fucose, GlcNAc, galactose, and sialic acid residues generate a complex N-glycan structure

contrast, the oligosaccharide in LLO is transferred en bloc to Asn residues in a polypeptide chain, by the action of OST. The acceptor Asn, in principle, resides in the consensus sequence, Asn-X-Thr or Asn-X-Ser, where X can be any amino acid residue except Pro. The N-glycosylation consensus, Asn-X-Ser/Thr, is referred to as the "sequon," and the N-oligosaccharide chain attached to a protein is referred to as the "N-glycan."

The attachment of oligosaccharides to Asn residues is essential for the structures and functions of secretory and membrane glycoproteins (Varki et al. 2017). At the molecular level, the N-glycans affect various physicochemical properties of glycoproteins, including solubility, folding, oligomerization, and aggregation. In terms of the biological consequences, the N-glycans play key roles in protein quality control and degradation, as well as sorting and transport of proteins inside cells (Fig. 9.1). The N-glycans are also indispensable in cell-cell and host-pathogen interactions on cell surfaces and the clearance of glycoproteins from the bloodstream. In these processes, the N-glycans function as tags that are recognized by other proteins.

N-glycosylation is essential in eukaryotic organisms. In human beings, genetic defects in N-glycosylation (also O-glycosylation and lipid glycosylation) result in rare genetic metabolic diseases, collectively referred to as congenital disorders of glycosylation (CDG) (Hennet and Cabalzar 2015; Jaeken and Peanne 2017). CDG patients show extremely broad spectrum of clinical symptoms. More than 100 types of CDG have been identified according to the large number of responsible genes. In contrast, N-glycosylation is principally nonessential in prokaryotic organisms. A possible exception is a group of Archaea, referred to as Crenarchaeota. Defects of N-glycosylation in the crenachaeon *Sulfolobus acidocaldarius* are reportedly lethal (Meyer and Albers 2014). Even in other prokaryotes in which N-glycosylation is not essential, N-glycosylation improved their survival. For the pathogenic *Campylobacter*, N-glycosylation is necessary for efficient attachment to and invasion of the host human cells (Szymanski and Wren 2005).

9.2 Free N-oligosaccharide Generation by OST

In addition to the vital roles of N-glycans on glycoproteins, the free forms of the N-oligosaccharides (fOS or FNG) in the endoplasmic reticulum (ER) in eukaryotic cells and in the periplasmic space of prokaryotic cells have recently attracted attention (Chantret and Moore 2008; Dwivedi et al. 2013). OST has a special hydrolytic activity of LLO to release the free oligosaccharides into the lumen of the ER or the extracellular and periplasmic spaces and is considered to be the main producer of the free forms of N-oligosaccharides (Harada et al. 2013) (Fig. 9.1). Presumably, this hydrolytic activity is not a simple byproduct of the oligosaccharyl transfer reaction, but rather a second enzymatic activity acquired in evolution driven by necessity, considering the large amounts of free N-oligosaccharides accumulated in cells (Dwivedi et al. 2013).

What is the biological function of the free oligosaccharides? Free N-oligosaccharides act on several proteins as chaperones in vitro (Kimura et al. 1998). High-mannose-type N-oligosaccharides, generated by OST, were more effective than complex-type N-oligosaccharides (Jitsuhara et al. 2002). The molecular basis is considered to be the hydrophobic surface provided by the oligomannose portion, which strengthens the oligosaccharide-protein interactions. The free N-oligosaccharides could assist with the protein folding of secretory (glyco)proteins in the ER and extracellular and cytoplasmic spaces. Further experiments are required to clarify the biological functions of the free N-oligosaccharides inside cells. The TREX1 protein reportedly interacted with the human OST complex and suppressed the free oligosaccharide generation (Hasan et al. 2015). TREX1 is an ER-membrane bound DNase. Analyses of the DNase-independent function of TREX1 in immune cells revealed that the C-terminal truncation or phosphorylation in the linker segment between the DNase domain and the TM region of TREX1 disrupted the interaction with the human OST complex, thus enhancing the free oligosaccharide production (Kucej et al. 2017). Given the fact that TREX1 mutations are associated with autoimmune and autoinflammatory diseases, studies of the TREX1 protein are potentially interesting, but further studies are necessary to prove the convincing association with OST.

9.3 Biological Significance of the OST Enzyme

The OST enzyme has special biological significance. First, OST is a unique type of glycotransferase. Most glycotransferases transfer a monosaccharide to acceptor molecules, but OST transfers an oligosaccharide chain en bloc. Secondly, OST can activate the nitrogen atom in the side chain of an Asn residue, to form a covalent bond with an anomeric carbon of a monosaccharide residue. The nitrogen atom in a planar carboxyamide group is only weakly reactive, due to the delocalization of an electron on the nitrogen atom to the carbonyl oxygen atom (Liu and Szostak 2017). Thus, the molecular basis of the amide nitrogen activation is very interesting, from the viewpoints of chemistry and applications. Thirdly, the oligosaccharyl transfer reaction is occasionally coupled with the transmembrane passage of a nascent polypeptide chain synthesized by a ribosome. Accordingly, in a strict sense, the protein modification catalyzed by OST is a *co-translational* reaction, although the N-glycosylation is generally treated as a representative of *posttranslational* protein modifications. In addition to OST, very few enzymes, including signal peptidases (Nyathi et al. 2013) and acetyltransferases (Gautschi et al. 2003), work in a co-translational mode. The signal peptidase cleaves the N-terminal signal peptide, and the acetyltransferase transfers an acetyl group from acetyl coenzyme A to the α-amino group of the N-terminal residue. The recognition modes and catalytic mechanisms of these co-translational enzymes must be quite special, as they perform efficient searches of the target sites in nascent polypeptide chains and compete with the folding process.

9.4 Evolutional View of N-glycosylation

Figure 9.2 shows a phylogenetic tree of life, from the viewpoint of the N-oligosaccharyl transfer reaction. N-glycosylation is widespread not only in Eukarya, but also in virtually all Archaea (Abu-Qarn et al. 2008). N-glycosylation also occurs in some species belonging to the δ- and ε-proteobacterial groups in Eubacteria (Mills et al. 2016). The genus *Campylobacter* is a clinically important example, since it causes food poisoning. The eubacterial N-glycosylation system presumably originated from horizontal gene transfer (HGT). The HGT probably occurred independently and multiple times from Archaea (or common ancestors of Eukarya and Archaea). This speculation is based on the fact that some deep-sea-dwelling organisms, which are distantly separated from the genus *Campylobacter*, also have the N-glycosylation system (Mills et al. 2016).

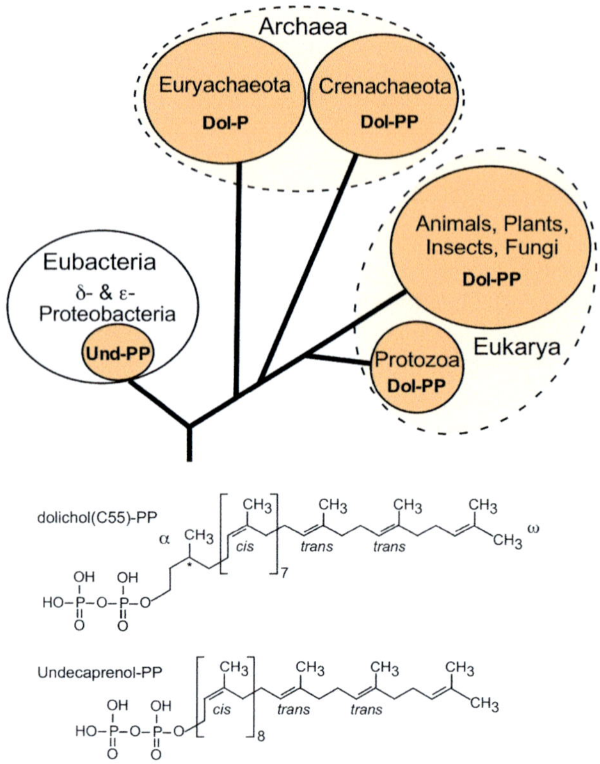

Fig. 9.2 Evolutionary distribution of N-glycosylation. N-glycosylation occurs in all Eukarya and Archaea and subsets of δ- and ε-Proteobacteria in Eubacteria. The chemical structure of the lipid-phospho carrier of N-oligosaccharides is different in each sub-domain of life. Und, undecaprenol; Dol, dolichol; P, monophosphate; PP, diphosphate. The C3 carbon (denoted by the asterisk) of the α-isoprene unit in dolichol has an S configuration

Archaea are frequently used as model organisms in structural biology, because many archaeal proteins have simpler subunit compositions than their eukaryotic counterparts, as in the case of the archaeal OST (Table 9.1). Moreover, proteins from thermophilic archaea are naturally stable and suitable for structural and biophysical studies. In fact, some hyperthermophilic archaeal cells grow at temperatures around 100 °C. Archaeal strains are now easily obtainable from various bio-resource banks. Most Archaea grow anaerobically, but the culturing of selected archaeal species is very easy in conventional laboratories. However, among the many reviews and research papers on N-glycosylation, Archaea have received far less attention than Eubacteria. This review comprehensively addresses N-glycosylation in the three domains of life.

Table 9.1 Biochemical properties of N-glycosylation in the three domains of life

Domain of life	Sub-domain	OST/ catalytic subunits	Donor[a]			Acceptor[b]
Eukarya	Higher Eukaryote (animal, plant, insect, fungus)	Multi-subunit	Dolichol-PP-OS			N-!P-[ST]
		STT3	C70–C110, $\sigma = 1$	Di-phos	14 res	
	Lower Eukaryote (Protozoa)	Single subunit	Dolichol-PP-OS			
		STT3	C55–65 $\sigma = 1$	Di-phos	11 res	
Archaea	Crenarchaeota	Single subunit	Dolichol-PP-OS			
		AglB	C30–C50 $\sigma = 3–5$	Di-phos	6–10 res	
	Euryarchaeota	Single subunit	Dolichol-PP-OS			
		AglB	C55–C60 $\sigma = 1–5$	Mono-phos	3–7 res	
Eubacteria	Limited distribution in the δ- and ε-Proteobacteria (including *Campylobacter*, *Helicobacter*, and deep-sea vent Proteobacteria)	Single subunit	Polyprenol-PP-OS			[DE]-!P-N-!P-[ST]
		PglB	C55 $\sigma = 0$	Di-phos	7 res	

[a]C*number* denotes the length of the dolichol/polyprenol in the five-carbon isoprene unit. For example, C55 denotes a dolichol/polyprenol with 11 isoprene units. σ represents the number of saturations of the double bonds in the dolichol/polyprenol chain. By definition, the dolichol has one saturated unit at the α-terminus (on the side with the OH group), in addition to the other saturated units, which are probably close to the opposite ω-terminus

[b]The acidic residue at the -2 position is not absolutely necessary in some eubacterial species. !P denotes any amino acid residue except for Pro

9.5 Catalytic Subunit of the OST Enzyme

The catalytic subunit of the OST enzyme is a polypeptide chain, referred to as Stt3 in Eukarya, AglB in Archaea, and PglB in Eubacteria (Table 9.1). Even though they have different names, they share a common evolutionary origin. Each genome usually encodes multiple paralogous genes of the catalytic subunit. One important exception is yeast (*Saccharomyces cerevisiae*), in which the genome encodes a single Stt3 gene. The other subunits in the multi-subunit OST complex of higher Eukaryotes, including yeast, can modulate the catalytic function of the Stt3 subunit or add new functions (Table 9.2) (Kelleher and Gilmore 2006). In the case of the single-subunit OSTs (lower eukaryotic Protozoa, Archaea, and Eubacteria), the

Table 9.2 Various properties of OSTs in the three domains of life

Domain	Eukarya			Archaea		Eubacteria
Phylum	Metazoa	Fungi	Euglenozoa	Crenarchaeota	Euryarchaeota	ε-Proteobacteria
Species	*Homo sapiens*	*Saccharomyces cerevisiae*	*Trypanosoma brucei*	*Pyrobaculum calidifontis*	*Archaeoglobus fulgidus*	*Campylobacter jejuni*
Number of subunits	8	8	1	1	1	1
Catalytic subunit	Stt3A, Stt3B	Stt3	Stt3A[a], Stt3B, Stt3C, Stt3D	AglB	AglB-L, AglB-S1, AglB-S2	PglB[b]
Subunit composition	See Table 9.3	See Table 9.3	Single	Single	Single	Single
Amino acid motifs	2 DXDs, SVSE, WWDYG, DK	2 DXDs, SVSE, WWDYG, DK	2 DXDs, SVSE, WWDYG, DK	2 DXDs, TIXE, WWDYG, DK	2 DXDs, TIXE, WWDYG, DK*i* [c]	2 DXDs, TIXE, WWDYG, MI
Structure of the catalytic subunit	LLO-bound forms complexed with a ribosome (6FTG)[d], (6FTI, 6FTJ)[e]	Apo forms of the OST complex (6EZN, 6C26)[e]	Not determined	Not determined	Apo form (3WAK)[f], sulfate ion bound (3WAJ)[f], sequon bound (5GMY)[f]	Sequon bound (3RCE)[f], sequon-LLO-ternary form (5OGL)[f]
Structures of other subunits	Tusc3 (4M8G, 4M90, 4M91, 4M92)[f], Ost4(2LAT)[g]	Ost6(3G7Y, 3G9B, 3GA4)[f], Ost4(1RKL)[g]	–	–	–	–

[a]A closely related species, *Trypanosoma cruzi*, possesses a single Stt3
[b]Some *Helicobacter* species possess two PglB homologs
[c]DK*i* is a variant of the DK motif with an insertion (Matsumoto et al. 2012). The former name is DM (Maita et al. 2010)
[d]CryoEM subtomogram averaging
[e]CryoEM single particle analysis
[f]X-ray crystal structure
[g]NMR structure

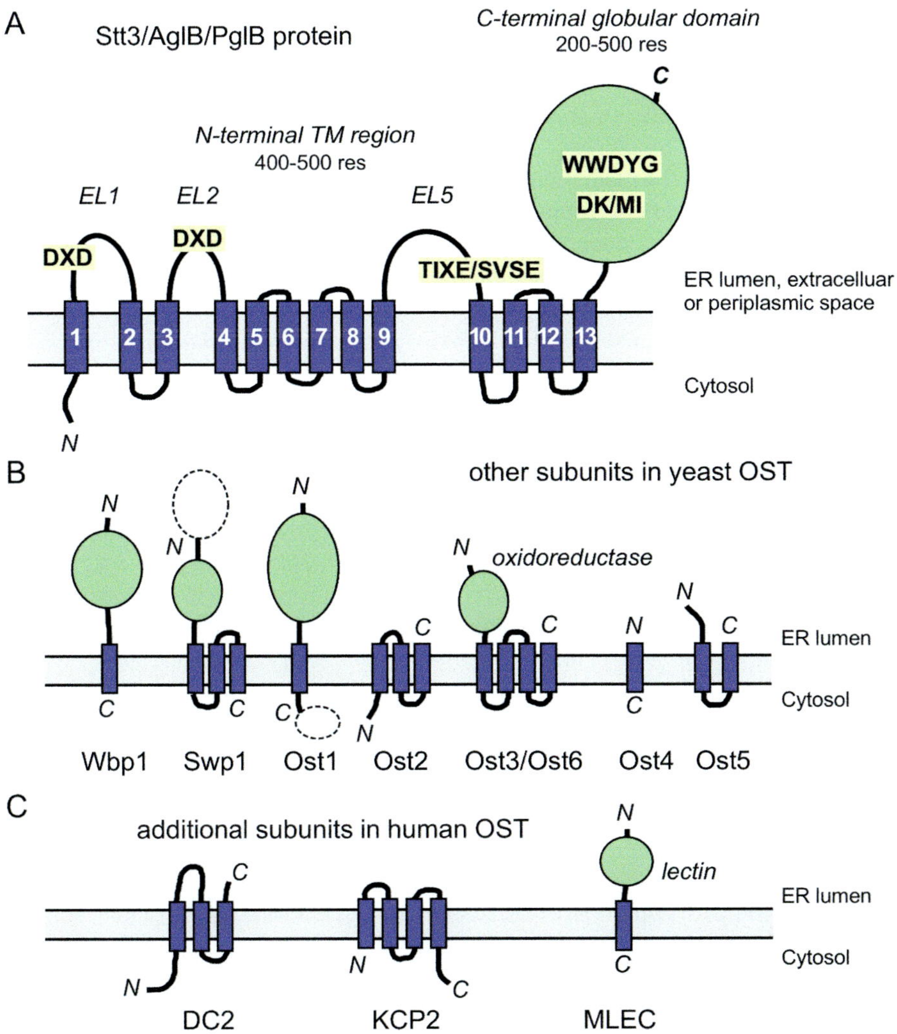

Fig. 9.3 Domain structures of the subunits of OST. (**a**) Stt3/AglB/PglB. EL1, EL2, and EL5 are the external loops. Short conserved amino acid motifs are located in these loop regions and in the C-terminal globular domain. (**b**) Other obligatory subunits in yeast OST. Extra domains present in human OST are depicted as dashed ovals. (**c**) Additional subunits in human OST

variations in the homologous Stt3/AglB/PglB proteins alter the oligosaccharyl transfer functions, as substitutes for the non-Stt3 subunits. For example, the amino acid variations in the single-subunit OSTs confer different preferences for the amino acid sequences around a sequon and for the oligosaccharide structures in the LLO (Izquierdo et al. 2009; Taguchi et al. 2016).

The Stt3/AglB/PglB proteins consist of 600–1,000 amino acid residues. The N-terminal half (400–500 residues) of the primary sequence is a transmembrane region, with 13 transmembrane (TM) α-helices, whereas the C-terminal half (200–500 residues) of the primary sequence forms a soluble, globular domain (Fig. 9.3a).

The overall sequence conservation is very low across the three domains of life, but a few conserved short amino acid motifs have been identified. The N-terminal TM region contains two diacidic DXD motifs on the first and second external loops (EL1 and EL2) and an SVSE/TIXE motif on the fifth external loop (EL5). The C-terminal globular domain contains a well-conserved five-residue motif, WWDYG, and a DK/MI motif. The latter amino acid motif was identified by the 3D structure-based sequence alignment (Maita et al. 2010).

The recombinant production of a higher eukaryotic Stt3 protein has not been reported. This is probably because the higher eukaryotic OST enzyme is a multi-subunit protein complex. In contrast, single-subunit OSTs (i.e., the stand-alone catalytic subunits) can be expressed in active forms. A lower eukaryotic Stt3 from *Trypanosoma brucei* was produced in insect cells (Ramirez et al. 2017), and archaeal AglBs and eubacterial PglBs were produced in the membrane fractions of *E. coli* cells (Jaffee and Imperiali 2013; Maita et al. 2010; Matsumoto et al. 2013b). The multi-subunit OST complex from yeast can be purified to homogeneity with an affinity tag, which was previously incorporated into a subunit (Chavan et al. 2006; Harada et al. 2013). An archaeal OST, AglB, was also isolated from cultured archaeal cells using specific polyclonal antibodies (Igura et al. 2008). This native AglB protein is suitable for studies of protein modifications. In fact, the N-glycosylation of yeast and mammalian Stt3s and *Pyrobaculum calidifontis* AglB were experimentally confirmed (Braunger et al. 2018; Fujinami et al. 2017; Li et al. 2005).

9.6 Non-catalytic Subunits of the OST Enzyme

Yeast OST has been studied as a model of the eukaryotic multi-subunit OST complex. The yeast OST consists of eight membrane-protein subunits in an equimolar ratio (Stt3, Wbp1, Swp1, Ost1, Ost2, Ost3 or Ost6, Ost4, Ost5) (Table 9.3 and Fig. 9.3b). Ost3 and Ost6 are homologous to each other and interchangeable to form two different complexes, containing either Ost3 or Ost6 (Spirig et al. 2005). The five subunits (Stt3, Wbp1, Swp1, Ost1, Ost2) are essential, and their deletions from the genome are lethal (Kelleher and Gilmore 2006). The other three subunits (Ost3 or Ost6, Ost4, Ost5) are nonessential, but their deletions affect the state of protein N-glycosylation in yeast cells. The Ost3 and Ost6 double knock out led to severe underglycosylation of proteins (Knauer and Lehle 1999a).

The Ost3 and Ost6 proteins have a thioredoxin fold in the N-terminal soluble domain. The oxidoreductase activity is considered to suppress the disulfide bond formation during the oligosaccharyl transfer reaction (Schulz et al. 2009). Ost4 is a small, 33-residue protein consisting of one membrane-spanning TM helix. Ost4 is required for the incorporation of Ost3 or Ost6 into the yeast OST complex (Spirig et al. 2005). The functions of the other subunits remain to be determined. Interestingly, recent cryoEM structures of the yeast OST complex suggested that the N-terminal soluble domain of Ost1 might bind to a glycosylated polypeptide segment, to prevent it from sliding back to the catalytic site of Stt3 (Bai et al. 2018).

Table 9.3 Subunit composition of the OST complex

<table>
<tr><th colspan="2">Homo sapiens (human)[a]</th><th>Saccharomyces cerevisiae (yeast)</th><th>Function</th></tr>
<tr><td>STT3A complex</td><td>STT3B complex</td><td>Stt3 complex</td><td></td></tr>
<tr><td colspan="3">Obligatory subunit</td><td></td></tr>
<tr><td>STT3A</td><td>STT3B</td><td>Stt3[b]</td><td>Catalytic subunit</td></tr>
<tr><td colspan="2">DDOST|OST48</td><td>Wbp1</td><td>?</td></tr>
<tr><td colspan="2">RPN1|ribophorin I[c]</td><td>Ost1</td><td>Extra cytosolic domain of RPN1 that binds to ribosome</td></tr>
<tr><td colspan="2">RPN2|ribophorin II[d]</td><td>Swp1</td><td>?</td></tr>
<tr><td colspan="2">DAD1</td><td>Ost2</td><td>?</td></tr>
<tr><td colspan="2">OST4</td><td>Ost4</td><td>Single TM protein that binds to Stt3</td></tr>
<tr><td colspan="2">TMEM258</td><td>Ost5</td><td>?</td></tr>
<tr><td>–</td><td>TUSC3|N33 or MAGT1|IAP[e]</td><td>Ost3 or Ost6[e]</td><td>Oxidoreductase activity</td></tr>
<tr><td>DC2|OSTC</td><td>–</td><td></td><td>mediates the interaction with the translocon channel,</td></tr>
<tr><td>KCP2|KRTCAP2</td><td>–</td><td></td><td>probably as described above</td></tr>
<tr><td colspan="3">Accessory subunit</td><td></td></tr>
<tr><td colspan="2">Malectin|MLEC</td><td></td><td>Lectin activity, direct interaction with RPN1</td></tr>
<tr><td colspan="2">TREX1[f]</td><td></td><td>modulation of the fOS generation</td></tr>
</table>

[a]"|" denotes two alias names
[b]Yeast Stt3 is related to human STT3B
[c]RPN1 has an extra C-terminal domain that is not present in Ost1 (Fig. 9.3b)
[d]RPN2 has a large extra N-terminal domain that is not present in Swp1 (Fig. 9.3b)
[e]Two subunits are mutually exclusive
[f]Further studies are necessary to prove that TREX1 is an accessory subunit of OST

Mammalian genomes encode two Stt3 paralogs, Stt3A and Stt3B, leading to two OST complexes in human cells (Cherepanova et al. 2016). The human OST complexes also possess subunits homologous to their yeast counterparts as obligatory subunits (Table 9.3). The Stt3A OST complex primarily acts in a co-translational mode to transfer an oligosaccharide to a sequon as soon as it emerges into the lumen of the ER, while the Stt3B OST complex modifies a sequon that is skipped by the Stt3A complex (Shrimal et al. 2015). It should be noted, however, that there is no sharp separation of the functions between the Stt3A and Stt3B OST complexes. The functional difference can be explained by the subunits corresponding to the yeast Ost3 and Ost6 (Fig. 9.3b). The Stt3A OST complex contains the DC2 and KCP2 subunits (Shrimal et al. 2017a), whereas the Stt3B OST complex contains the TUSC3 or MAGT1 subunit (Cherepanova et al. 2014). The TUSC3 and MAGT1 proteins are Ost3 and Ost6 homologs, respectively, and the two proteins have the N-terminal oxidoreductase domain in addition to the C-terminal TM region (Mohorko et al. 2014). The DC2 and KCP2 proteins lack the oxidoreductase domain

(Fig. 9.3c) and can mediate the interactions with the protein translocation channel. Among the other obligatory subunits, RPN2|ribophorin II ("|" denotes two alias names of one protein) is related to yeast Swp1 but has acquired an extra amino acid sequence corresponding to the N-terminal 350-residue domain during evolution (Fig. 9.3b). RPN1|ribophorin I is related to yeast Ost1 and has an extra amino acid sequence corresponding to the C-terminal 150-residue domain (Fig. 9.3b). The human OST complexes contain the accessory subunit, Malectin|MLEC (Fig. 9.3c). Malectin|MLEC is a lectin that binds to the Glc-α-1,3-Glc disaccharide in the $Glc_2Man_9GlcNAc_2$ trimming intermediate of protein-bound N-glycans and interacts with RPN1|ribophorin I (Cherepanova et al. 2016).

The subunit composition of the plant OST complex from *Arabidopsis thaliana* was reported (Jeong et al. 2018). There are two homologous genes for not only the Stt3 proteins, but also the homologs of human RPN1, DAD1, TUSC3, OST4, and MAGT1. The HAP6 and DGL1 proteins are single counterparts of human RPN2 and DDOST, respectively. The counterpart of human TMEM258 has not been identified yet.

9.7 The Sequon as the N-Oligosaccharide Acceptor

The N-glycosylation sequon is common across the three domains of life. The consensus sequence is Asn-X-Ser or Asn-X-Thr, where the middle X position can be any amino acid residue except Pro (Gavel and von Heijne 1990). The residue next to the C-terminus of the hydroxy amino acid residue (the +3 position) is also preferably a non-Pro residue, but the exclusion of Pro is not strict (Ben-Dor et al. 2004; Petrescu et al. 2004). The glycosylation of the Asn residue in an atypical sequon, Asn-X-Cys, has occasionally been reported (Sato et al. 2000). It is reasonable to consider that the SH group of a Cys residue can substitute for the role of the OH group of the Ser/Thr residue in the sequon. Statistical studies of glycosylated sequons revealed that Thr was preferable to Ser (Ben-Dor et al. 2004; Petrescu et al. 2004), and many experiments confirmed the preference in in vitro oligosaccharyl transfer (Chen et al. 2007; Igura and Kohda 2011a). Eubacteria use an extended version of the five-residue sequon, Asp-X-Asn-X-Ser/Thr or Glu-X-Asn-X-Ser/Thr, where X represents a non-Pro residue (Kowarik et al. 2006b). The statistics of glycosylated eubacterial sequons showed that Asp is preferable to Glu at the -2 position. However, the presence of an acidic residue at the -2 position is not absolutely required, as sequons lacking the acidic residue at the -2 position were glycosylated in some eubacterial species (Ollis et al. 2015).

Although they are rare, the N-glycan modifications of the Asn residues in atypical sequences, such as Asn-Gly-X, Asn-X-Val, and Ser/Thr-X-Asn, and the modification of the Gln residue in a Gln-Gly-Thr sequence were found in glycoproteome analyses (Valliere-Douglass et al. 2010; Zielinska et al. 2010). Highly sensitive mass spectrometry enabled the detection of such rare N-glycosylation events. In in vitro assays, the use of high concentrations of acceptor peptides and the PglB protein (Gerber et al. 2013; Lizak et al. 2013) and the cross-linking of acceptor peptides to AglB (Matsumoto et al. 2017) generated the glycosylated Asn and Gln residues in atypical sequons. Therefore, in contrast to the descriptions in many textbooks that the Asn-X-Ser/Thr sequon (X ≠ Pro) is necessary for protein N-glycosylation, neither an Asn residue at the first position nor a hydroxy amino acid at the third position in the sequon is absolutely required for the N-oligosaccharyl transfer reaction. This fact is important for the elucidation of the catalytic mechanism of the N-oligosaccharyl transfer reaction (see below).

It is interesting to discuss the amino acid bias at the X position of the sequon. Statistical analyses of many glycosylated sites in glycoproteins revealed little preference for a particular amino acid at the X position, except for the strict Pro exclusion (Abu-Qarn and Eichler 2007; Ben-Dor et al. 2004; Petrescu et al. 2004). When a certain protein was selected as an acceptor protein in vivo, the analysis of the N-glycosylation of a particular site in the acceptor protein showed some tendency of amino acid preference at the X position (Chen et al. 2007; Kasturi et al. 1997). These observations suggested that the amino acid sequence variation in the flanking region of a sequon affects the amino acid preference at the X position. In other words, to observe the effects of the amino acid variation at the X position, the amino acid sequence within and around the sequon must be fixed. In fact, in in vitro OST assays with a short peptide as an acceptor substrate, strong amino acid preference was observed at the X position. For example, the *Pyrococcus furiosus* and *Archaeoglobus fulgidus* AglBs preferentially glycosylated sequons with Val and Glu at the X position, respectively (Igura and Kohda 2011a; Matsumoto et al. 2017), and *Campylobacter jejuni* PglB preferred Ala, Ser, Arg, Lys, and Val at the X position (Chen et al. 2007). Interestingly, the in vitro bias can be modulated by amino acid mutations in the *P. furiosus* AglB protein, to some extent (Igura and Kohda 2011b), and the tethering of an acceptor peptide via a disulfide bond almost completely suppressed the preference at the X position (Matsumoto et al. 2017). These results indicated that the OST determines the bias at the X position in glycosylated sequons. Finally, the Pro exclusion at the X position remains enigmatic, even though the crystal structures of the AglB/PglB-sequon complexes were determined. The structural basis of the Pro exclusion at the X position is a challenge to address in the future.

9.8 Site Occupancy

OST determines the site occupancy of the N-glycan attachment in a sequon. Incomplete oligosaccharide transfer results in the macro-heterogeneity of glycoproteins (i.e., incomplete N-glycan attachment), which is particularly bothersome in pharmaceutical glycoprotein production, as well as micro-heterogeneity (i.e., variations in N-glycan structures). In this context, studies of the OST enzyme will be useful for the production of homogenous glycoproteins, using engineered cultured cells. When focusing on one particular sequon in a protein, the occurrence of

N-glycan attachment is almost 100 % or 0 % under the usual in vivo conditions. Partial N-glycan attachments may occur when a glycoprotein is overproduced in cultured cells, probably due to the shortage of the LLO supply. Statistics revealed that about one-third of the sequons are not glycosylated (Abu-Qarn and Eichler 2007; Ben-Dor et al. 2004; Petrescu et al. 2004). Thus, the sequon is necessary, but not sufficient, for the N-oligosaccharyl transfer reaction. The factors that affect the site occupancy remain unknown. The site occupancy rule of N-glycosylation should be clarified by the conformations of the polypeptide chains in the states bound to the Stt3/AglB/PglB protein.

9.9 Co- vs. Posttranslational Modes

The OST-catalyzed reaction is quite unique, with respect to the substrate recognition. In the co-translational mode, the active site must recognize a sequon quickly in a rapidly growing polypeptide chain and quickly release the glycosylated sequon from the active site. In the multi-subunit OSTs, other subunit proteins may facilitate the recognition and release processes. As a good example, the oxidoreductase domains of TUSC3/MAGT1 and Ost3/Ost6 were suggested to capture a polypeptide chain, before the polypeptide bound to the active site of Stt3 (Bai et al. 2018; Wild et al. 2018). It would be interesting to examine whether the oxidoreductase activity or simply the binding ability to the substrate polypeptide chains is necessary for the lookout role.

The TM region of DC2 (and probably KCP2) bridges the gap between Stt3A and the Sec61 translocon channel in the mammalian OST complex (Braunger et al. 2018; Shrimal et al. 2017a). In other words, the presence or absence of DC2 (and KCP2) determines the reaction mode of the N-oligosaccharyl transfer. As stated above, in human cells, the Stt3B OST complex transfers N-glycans in a posttranslational manner to the sequons located close to the C-termini of polypeptides (Shrimal et al. 2013). This is explained by the fact that the Stt3B OST complex does not contain the DC2 subunit. The yeast OST does not contain DC2, either, suggesting that N-glycosylation in yeast cells is principally posttranslational.

The single-subunit OSTs should work in a posttranslational mode. In fact, the *C. jejuni* OST, consisting of a single PglB protein, modified a sequon in flexible segments of folded proteins in a posttranslational manner (Kowarik et al. 2006a). Alternatively, the single-subunit OSTs may work in a co-translational mode, in the presence of as-yet-unidentified protein subunits. In fact, the N-oligosaccharyl transfer by PglB occurred before complete folding and might be coupled to the translocation process to the periplasm across the membrane (Silverman and Imperiali 2016). The mode of archaeal OSTs consisting of a single AglB protein has not been examined yet. The limited number of glycoproteins in species belonging to the phylum Euryarchaeota suggests that euryachaeal AglBs work in a posttranslational mode. In contrast, considering the wide variety of glycoproteins in species belonging to the phylum Crenarchaeota (Fujinami et al. 2017; Palmieri et al. 2013), the co-translational mode of crenarchaeal AglBs is not impossible from necessity of efficient glycosylation in harsh environments.

9.10 LLO as the N-Oligosaccharide Donor

The chemical structures of the glycan donor, LLO, are diverse across the three domains of life in two ways: the variation of the chemical structure of the oligosaccharide portion and the variation of the chemical structure of the lipid-phospho carrier (Table 9.1). In Eukarya, the canonical 14-residue oligosaccharide structure, $Glc_3Man_9GlcNAc_2$, is widely conserved. Lower eukaryotes use a shorter version of the 14-residue structure, lacking the terminal three glucose residues (Samuelson et al. 2005). In contrast, Archaea and Eubacteria use completely different sets of oligosaccharide structures from species to species (Jarrell et al. 2014). Previously, the archaeal oligosaccharides were considered to be shorter and to lack notable branching structures, but recent studies have revealed the rich world of archaeal oligosaccharide structures. In particular, the N-oligosaccharide of *P. calidifontis* is a biantennary, high-mannose type and is the most similar to the eukaryotic counterpart, among all archaeal and eubacterial N-glycans reported to date (Fujinami et al. 2017). Structural differences exist in the anomeric configuration between Man and GlcNAc, and the chemical modifications in the chitobiose structure, GlcNAc-β-1,4-GlcNAc. These differences make *P. calidifontis* LLO unexchangeable with yeast LLO in the oligosaccharyl transfer reaction. Another interesting example is the N-oligosaccharide of *A. fulgidus*, which exclusively consists of hexose residues (Fujinami et al. 2015). Finally, how many monosaccharide residues are necessary for recognition by the OST enzyme? The in vitro assays with a synthetic LLO analog indicated that the minimum numbers of monosaccharide residues are one for *Campylobacter* PglB (Ishiwata et al. 2015) and two for *Methanococcus voltae* AglB (Larkin et al. 2013) and *Trypanosoma* Stt3 (Ramirez et al. 2017).

As for the lipid-phospho part, Eukarya and Archaea use dolichol, but Eubacteria use polyprenol (Hartley and Imperiali 2012) (Table 9.1). One isoprene unit, $-CH_2-CH=C(CH_3)-CH_2-$, consists of five carbon atoms and contains one double bond. Polyprenol is a long-chain primary alcohol, consisting of many isoprene units, and dolichol is a special type of polyprenol with one saturated isoprene unit at the α-terminus (i.e., the side with the OH group) (Fig. 9.2). The numbers of isoprene units and the degrees of double-bond saturation are variable among the three domains of life. Higher Eukaryotes use long 14–22-unit dolichol chains, with no saturation other than the α-isoprene unit. Lower Eukaryotes (Protozoa) use shorter dolichols with 11–13 units. Archaea use shorter dolichol chains with 6–12 units, whereas Eubacteria use 11-unit undecaprenol chains. The dolichol chains of Archaea contain zero to four saturated isoprene units besides the α-isoprene unit, which are presumably close to the opposite ω-terminus (Kuntz et al. 1997). Regarding the number of phosphate groups in the lipid-phospho carrier of LLO, the two phosphate-type LLOs are widely used. Exceptionally, a subset of Archaea uses the one phosphate-type LLO (Chang et al. 2015; Larkin et al. 2013; Taguchi et al. 2016). It was proposed that the monophosphate-type LLO is used in the phylum Euryarchaeota, and the diphosphate-type LLO is used in the phylum Crenarchaeota (Taguchi et al. 2016). The distribution of LLOs with different numbers of phosphate groups is consistent with the closer relationship of the phylum Crenarchaeota to Eukarya than that of the phylum Euryarchaeota, in the phylogenetic tree of life (Fig. 9.2).

New phyla in the domain Archaea have been identified, based on the information obtained from the DNA sequences of environmental samples and biological samples (Zaremba-Niedzwiedzka et al. 2017). Unfortunately, these new Archaea have not yet been cultured in laboratories. Since the purification of LLO from cultured cells is necessary, due to the stringent specificity of AglB for its cognate LLO, the biochemical studies of N-glycosylation in the new Archaea are currently impossible. In contrast, studies of the proteins from these Archaea are feasible, by using synthetic DNAs encoding the deduced protein sequences for the production of the recombinant proteins in heterologous host cells.

9.11 Crystal Structures of Archaeal AglB and Eubacterial PglB

As often found in structural studies of membrane proteins, the crystal structures of the C-terminal soluble, globular domains of the AglB and PglB proteins were first determined for six archaeal AglB proteins from four archaeal species (Igura et al. 2008; Matsumoto et al. 2012, 2013a; Nyirenda et al. 2013) and one eubacterial PglB protein from *C. jejuni* (Maita et al. 2010). Although they are enzymatically inactive, the detailed comparison of the structures revealed a conserved short motif, named the DK/MI motif (Maita et al. 2010), and the unusual plasticity of the region containing the WWDYG motif in the C-terminal globular domain (Nyirenda et al. 2013).

The crystal structure of the full-length PglB protein from *Campylobacter lari* was determined to a resolution of 3.4 Å (Lizak et al. 2011), and subsequently that of the full-length AglB protein from *A. fulgidus* was solved at the same resolution (Matsumoto et al. 2013b), both in the presence of detergents (Fig. 9.4a). The structures of the C-terminal globular domains were used as the template structures for the molecular replacement in the structure determinations. The PglB structure was a sequon peptide substrate-bound form, whereas the AglB structure was an apo form. To obtain the AglB structure in the peptide-bound form, an acceptor peptide was tethered to the protein through an engineered disulfide bond (Matsumoto et al. 2017). The position of the anchoring Cys residue in the AglB protein was selected by reference to the PglB structure. The Cys-containing mutant AglB protein was cross-linked to a peptide containing a sequon and a Cys residue. The crystal structure of the cross-linked AglB-sequon complex was determined to a resolution of 3.5 Å (Matsumoto et al. 2017). The addition of LLO to the cross-linked complex

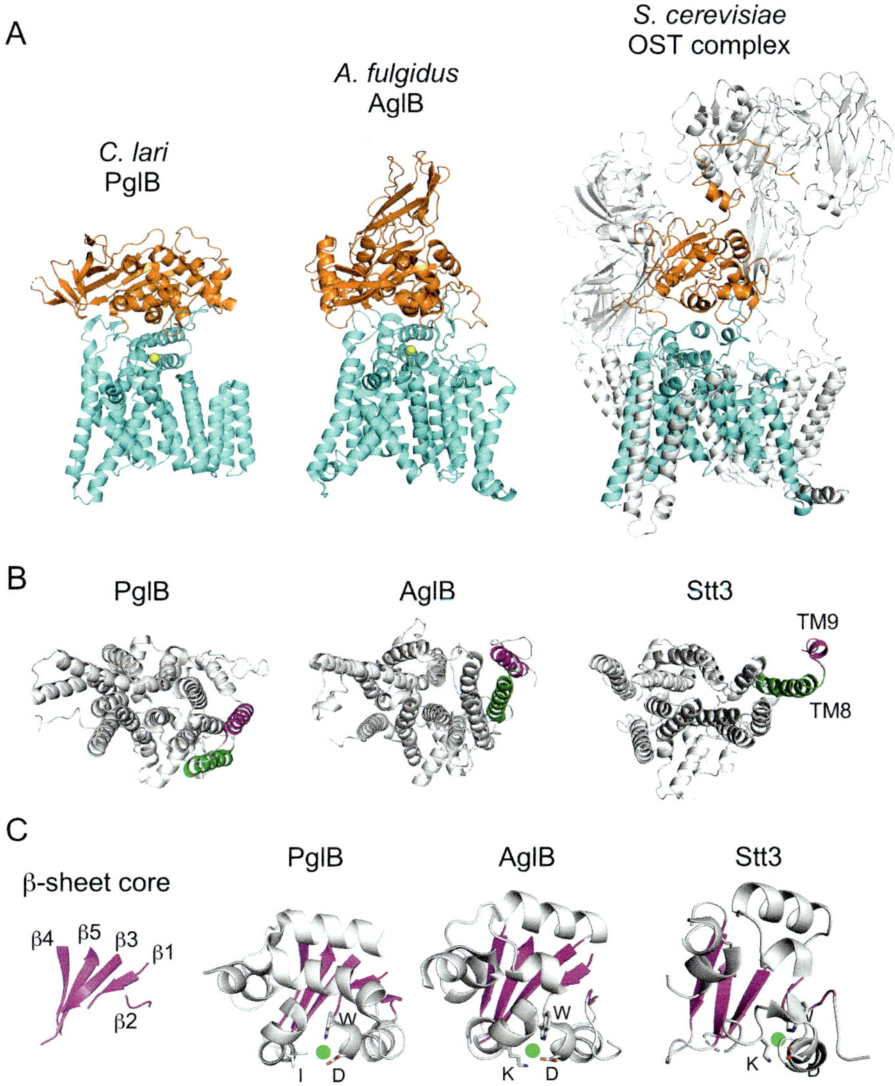

Fig. 9.4 Structures of the catalytic subunit of OST. (**a**) Crystal structures of *C. lari* PglB (PDB entry 3RCE), *A. fulgidus* AglB (3WAK), and EM structure of *S. cerevisiae* (yeast) Stt3 (6EZN). The last structure is the complex structure with seven other subunits. The small spheres in the PglB and AglB structures are the divalent metal ions bound to the active site. (**b**) The 13 TM helices, viewed from the cytoplasmic side. The TM helix 8 and helix 9 are labeled. (**c**) The CC (C-terminal core) structures conserved in the Stt3/AglB/PglB proteins. The leftmost panel shows the frame structure consisting of five β-strands. The dots show the position of the Ser/Thr-binding pocket formed by the first Trp and Asp in the WWDYG motif and the Lys/Ile residue in the DK/MI motif. The CC structural unit in PglB consists of 100 residues (D458-G482, S493-E513+V561-Q578, E675-L710), that in AglB consists of 104 residues (Y545-Q571, A584-V624, S671-S699+Y770-R776), and that in Stt3 consists of 94 residues (K512-N536, G547-F573+D582-M590, E622-K635+L663-Q681)

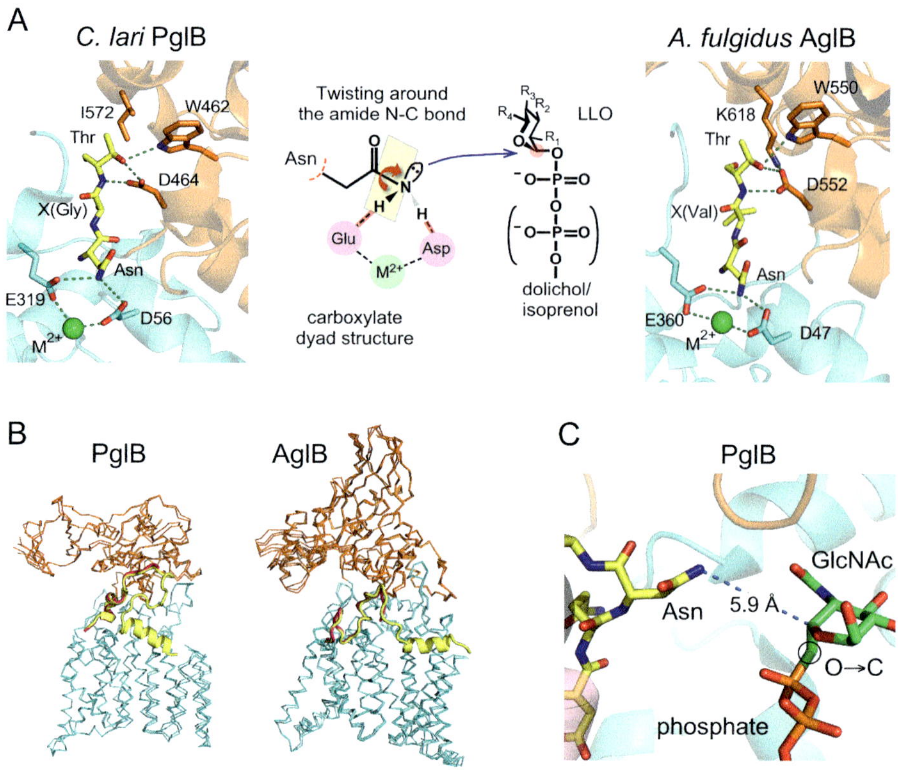

Fig. 9.5 Twisted amide mechanism. (**a**) Close-up views of the sequon peptide binding site of *C. lari* PglB (PDB entry 3RCE) and *A. fulgidus* AglB (5GMY). The hypothetical activation mechanism of the nitrogen atom in the side-chain carboxyamide group of the acceptor Asn is illustrated in the central panel. (**b**) Variable ordered and disordered states of the EL5 loop. The PglB EL5 (G282-I322) was fully ordered (yellow) in the ternary complex (5OGL) but partially ordered (red) in the peptide-bound state (3RCE). The AglB EL5 (I334-T373) was fully ordered (yellow) in the apo form (3WAK), partially ordered (red) in the product-releasing form (3WAJ), and fully disordered (not shown) in the peptide-bound form (5GMY). If we assume that the sequon binding and LLO binding accompany the conformational transition from the disordered state to the ordered state in the crystal structures, then the fully ordered state in the apo form of AglB (3WAK) is inconsistent with this assumption, for some unaccountable reason. (**c**) Close-up view of the sequon peptide and a nonhydrolyzable LLO analog in the ternary complex of PglB (5OGL). The distance between the side-chain nitrogen atom of Asn and the C1 carbon of the reducing-end monosaccharide residue is 5.9 Å

resulted in the oligosaccharyl transfer from LLO to the peptide tethered on the enzyme. The production of a glycopeptide within the same molecule indicated that the cross-linked complex was catalytically competent and provided a rational basis for the determined structure.

Close-up views of the sequon-bound sites are shown in Fig. 9.5a. In the C-terminal globular domains, the first Trp and Asp residues of the WWDYG motif and the Lys and Ile residues in the DK/MI motif form the Ser/Thr-binding pocket. The hydroxy group of the Ser/Thr residue in the sequon is recognized by the carboxy group of the Asp residue in the WWDYG motif. The methyl group of the Thr residue in the sequon contacts the alkyl chain portion of the Lys residue in the DK motif in the AglB structure and the Ile residue in the MI motif in the PglB structure. The interactions with the methyl group are absent for the Ser residue. The extra contacts involving the methyl group explain the structural basis for the preference of a Thr-containing sequon over a Ser-containing sequon.

The functional differences between the DK-based Ser/Thr-binding pocket and the MI-based Ser/Thr-binding pocket remain an open question. They may be linked to the differences in the dynamical properties of the pocket. The Ser/Thr-binding pocket catches and releases the Ser and Thr residues in the sequon during the catalytic cycle. A working hypothesis is that the DK-based Ser/Thr-binding pocket is more dynamic than the MI-based Ser/Thr-pocket, which makes the Stt3 and AlgB proteins more suitable for the co-translational mode of the N-oligosaccharyl transfer reaction than the PglB protein.

9.12 Electron Microscopy (EM) Structures of the Yeast OST Complex

Single particle analyses of the yeast OST complex, isolated from the membrane fractions of yeast cells, were reported using negative-stained images (Chavan et al. 2006) and amorphously frozen images (Li et al. 2008). The resolutions of these 3D EM maps were low (25 Å and 12 Å), since these studies were performed before the development of direct electron detectors. Recently, two new single particle analysis structures of the yeast OST complex have been reported (Bai et al. 2018; Wild et al. 2018). Owing to the latest sensitive detectors, the resolutions of the 3D EM maps reached 3.3–3.5 Å, which were sufficient to place atomic models into the EM maps without reference to any predetermined high-resolution crystal structures (Fig. 9.4a). The atomic models include an ordered N-glycan structure attached to the subunit proteins and several ordered phospholipids between the subunits. The ordered N-glycan moiety is attached to a conserved Asn residue in the yeast Stt3 (residue 539). The ordered N-glycan may contribute to the recognition of the N-oligosaccharide portion of the LLO molecule by forming a wide cavity with Wbp1 and Swp1 in the yeast OST complex. The eight subunits of yeast OST can be divided into three subcomplexes: subcomplex 1 contains Ost1 and Ost5; subcomplex 2 contains Stt3, Ost3, and Ost4; and subcomplex 3 contains Ost2, Wbp1, and Swp. The ordered phospholipids are located at the membrane-embedded interfaces between these subcomplexes. For purification from yeast cells, an affinity tag sequence was fused to the C-terminus of the Ost4 subunit in the yeast genome. After solubilization in the presence of detergents (digitonin or a mixture of *n*-dodecyl-β-D-maltopyranoside (DDM) and cholesteryl hemisuccinate (CHS)), the OST complex was affinity purified by binding to tag-specific monoclonal antibodies. In one case, the OST complex was then reconstituted into a nanodisc. In spite of the differences in the preparative methods, the two structures are almost identical to each other. The two structures can be superimposed with a root-mean-square displacement of 1.7 Å over 1,963 aligned Cα atoms.

Another single particle analysis has been reported, even though the resolution was limited to medium resolutions (4.2–4.7 Å) (Braunger et al. 2018). Ribosome translocon complex (RTC) was prepared from canine rough ER membranes in the presence of digitonin. The atomic models of Stt3A, DC2, and the C-terminal half of RPN1 in the canine OST complex were constructed as part of a ribosome associated with a translocation channel. The DC2 protein mediates the interaction between the Stt3A subunit and the Sec61 translocon channel and occupies the position corresponding to Ost3 in the yeast OST complex. The C-terminal cytosolic domain of RPN1 interacts with the ribosome. The DC2 and the extra C-terminal domain of RPN1, which are absent in the yeast OST complex (Table 9.3 and Fig. 9.3), provide the structural basis of the co-translational oligosaccharyl transfer mode of the mammalian Stt3A OST complex.

Although these EM structures allow many interesting speculations, the information about the catalytic and substrate recognition mechanisms is limited, due to the absence of substrates in the EM maps. For example, a divalent ion, which is essential for the catalytic activity, is absent in these structures. The high-resolution EM analysis with a sequon-containing (poly)peptide and (soluble analogs of) LLO will provide fruitful information in the future.

9.13 Conserved Protein Architectures Among the Stt3, AglB, and PglB Proteins

The amino acid sequence conservation among the Stt3/AglB/PglB proteins is very low (< 20 %) across the three domains of life, but common protein architectures can still be recognized. The N-terminal transmembrane region consists of 13 transmembrane helices with the same topology (Fig. 9.4a). The positions of the TM helix 8 and TM helix 9 are similar between Stt3 and AglB but different in PglB (Fig. 9.4b). The conformational changes of the EL5 loop (Fig. 9.5b), which continues directly to the C-terminus of the TM helix 9 in the primary sequence, are considered to be critical for the LLO binding. The dynamic behaviors of the two TM helices and the EL5 loop are key elements for understanding the LLO recognition and the catalytic mechanism (Lizak et al. 2014; Shrimal et al. 2017b).

The C-terminal globular domain contains a common 100-residue structural unit, referred to as the C-terminal common unit (CC) in the comparison of AglB and PglB (Maita et al. 2010). Visual inspection revealed that the same CC structure is conserved in Stt3 (Fig. 9.4c). The framework of the CC unit is a five-stranded β-sheet structure. The WWDYG motif is located at the N-terminus of the α-helix that connects the first and second β-strands. The DK/MI motif is located on the helical structure (probably in a 3_{10}-helix conformation) after the third β-strand. The significance of the DK/MI motif has been questioned in the past, since the chemical properties of Lys and Ile are very different. The identification of the conserved CC structure in Stt3 clearly demonstrated that the Lys and Ile residues in the motif occupy the identical positions and contribute to the formation of the Ser/Thr-binding pocket (Fig. 9.4c). In summary, the "13TM + CC structure" is well conserved in all of the Stt3/AglB/PglB proteins.

9.14 Common Features Shared with Other Glycotransferases

Some glycosyltransferase enzymes utilize polyprenol/dolichol-phosphate-linked monosaccharides as the sugar donor. Among them, protein C-mannosyltransferase (CmanT) is particularly interesting. CmanT transfers a mannose residue from dolichol-monophosphate-Man to a Trp residue in the WXXW motif in selected proteins, to form a C-C bond. The DPY-19 protein was identified as a CmanT in *Caenorhabditis elegans* (Buettner et al. 2013). The discovery was guided by the topological homology of the TM helices to those of Stt3/AglB/PglB. However, the evolutionary relationship between CmanT and Stt3/AglB/PglB is currently enigmatic but will be clarified in the future when the three-dimensional structure of CmanT is solved.

Another interesting case is the ArnT glycosyltransferase, which catalyzes the transfer of 4-amino-4-deoxy-L-arabinose (Ara4N) from undecaprenol-monophosphate-Ara4N to a phosphate group in the lipid A moiety of the lipopolysaccharide. The crystal structures of ArnT from *Cupriavidus metallidurans* were determined in the apo form and in the complex with undecaprenyl-phosphate (Und-P) to resolutions of 2.8 Å and 3.2 Å, respectively (Petrou et al. 2016). Strikingly, the structure of ArnT is reminiscent of the Stt3/AglB/PglB structure. The ArnT protein consists of the N-terminal TM region, containing 13 helices, and the smaller C-terminal globular domain. The topology of 8 TM helices (1-7+13) among the 13 TM helices is identical, and a flexible periplasmic loop 4 (PL4), which connects TM helix 7 and helix 8, was disordered in the apo form and ordered in the Und-P bound form, similar to the EL5 loop in the Stt3/AglB/PglB protein. The five-stranded β-sheet structure in the CC unit of Stt3/AglB/PglB (Fig. 9.4c) also exists in the C-terminal domain of ArnT. An interesting feature, which was not mentioned in the previous paper (Petrou et al. 2016), is that Glu306 is located on the N-terminus of the α-helix that connects the first and second β-strands in the five-stranded β-sheet structure, and it exactly corresponds to the Asp residue in the WWDYG motif of the Stt3/AglB/PglB protein. In fact, the Glu residue in ArnT is perfectly conserved among the 25 members of the ArnT family and was suggested to be involved in the Ara4N binding (Petrou et al. 2016).

9.15 Twisted Mechanism for the Activation of the Amide Nitrogen

In the past, the Asn-X-Ser/Thr segment in the sequon was thought to adopt a turn-like structure in the bound state, to bring the side chain of the acceptor Asn residue closer to the third residue, thus letting the hydroxy group of the third residue function as a catalytic group in the oligosaccharyl transfer reaction (Knauer and Lehle 1999b). Recent in vitro experiments, however, have cast doubt on the possibility of the substrate-assisted catalysis. In fact, a sequon without the hydroxy or sulfhydryl group of the third amino acid residue was active as an acceptor (Zielinska et al. 2010). Unexpectedly, no alternate chemical groups were found around the side chain of the acceptor Asn residue in the AglB/PglB-sequon complexes (Lizak et al. 2011; Matsumoto et al. 2013b). In the field of organic chemistry, the N-C bond twisting of an inherently planar carboxyamide group is a general concept to increase the nucleophilic reactivity of the inert amide nitrogen. NMR analyses suggested that amide bond twisting underlies the catalytic mechanism of protein splicing (Romanelli et al. 2004). Lizak et al. proposed that amide bond twisting is involved in the mechanism of the N-oligosaccharyl transfer reaction (Lizak et al. 2011, 2013). In the N-terminal transmembrane region, the carboxy group of the second Asp residue in the first DXD motif on the EL1 loop and the carboxy group of the Glu residue in the TIXE/SVSE motif in the EL5 loop constitute a catalytic entity with a bound divalent metal ion, presumably a magnesium ion (Fig. 9.5a). Acceptor peptide binding to the active site is aided by the interaction of the Ser/Thr residue in the sequon with the Ser/Thr-binding pocket in the C-terminal globular domain. Through the bipartite interactions with the two conserved carboxy groups, the side chain of the Asn residue in the sequon becomes part of a catalytic entity, termed the "carboxylate dyad structure," and results in the twisting of the N-C bond in the side-chain carboxyamide group of the Asn residue (Fig. 9.5a). The distortion of the planar carboxyamide group frees the lone electron pair on the nitrogen atom for a nucleophilic attack on the C1 anomeric carbon of the reducing-end monosaccharide residue in LLO. This hypothetical mechanism is referred to as the "twisted amide mechanism." In addition, the involvement of the N-acetylamino group on the C-2 carbon (at the R_1 position in Fig. 9.5a) of the reducing-end monosaccharide was suggested (Lizak et al. 2011). Note that some archaeal LLOs lack the N-acetylamino group. The *A. fulgidus* LLO, consisting of only hexose residues, is a good example (Fujinami et al. 2015). Since these special LLOs are the monophosphate type, the details of the catalytic mechanism may differ from those of the Stt3/AglB/PglB enzymes, which use the diphosphate-type LLOs.

The amide bond twisting could occur in the crystals of the AglB/PglB-sequon complexes, but the resolution of the electron density maps was not sufficient to discuss the formation of the twisted amide (Lizak et al. 2011; Matsumoto et al. 2017). Although a theoretical study with a molecular dynamics simulation suggested the generation of the twisted amide in the absence of LLO (Pedebos et al.

2015), the stabilization by the binding of LLO may be necessary for the twisted amide formation. The ternary complex structure of *C. lari* PglB with a peptide and LLO has been reported, to a resolution of 2.7 Å (Napiorkowska et al. 2017). The LLO used for crystallization was a chemically synthesized, soluble analog with a single GlcNAc residue and a short C_{20} polyprenol chain. The oxygen atom directly attached to the C1 carbon of the GlcNAc residue was replaced by a carbon atom, to create a nonhydrolyzable LLO analog. In the complex structure, the distance between the amide nitrogen of the acceptor Asn residue and the C1 carbon of the GlcNAc residue is 5.9 Å, which is too far for the nucleophilic attack to occur (Fig. 9.5c). Thus, unfortunately, the crystal structure did not provide any experimental evidence for the twisted amide formation. The replacement of the oxygen atom by a carbon atom in the nonhydrolyzable analog probably had an unwanted effect. Thus, further studies are necessary to prove the hypothetical twisted amide mechanism of the OST enzyme.

NMR spectroscopy is a promising method for the analysis of the amide bond twisting. It was used to detect twisted amides in low-molecular-weight compounds (Yamada 1996) and a 200-residue soluble protein, the *Mycobacterium xenopi* DNA gyrase A intein (Romanelli et al. 2004). Although the application of NMR to high-molecular-weight membrane proteins is challenging, the high-field shifts of the chemical shift values of the amide ^{1}H and ^{15}N nuclei and the decreased ^{1}J coupling constant of the N-C bond in the side-chain carboxyamide group of the Asn residue could directly prove the existence of the twisted carboxyamide group.

The cytoplasmic N-glycosyltransferase (NGT) is a functional homolog of the OST (Naegeli et al. 2014). The NGT enzyme catalyzes the transfer of a monosaccharide (mainly Glc) to the side chain of Asn in polypeptides from nucleotide-activated monosaccharides (UDP-Glc, GDP-Glc). Only a few genomes in the γ-Proteobacteria class encode this special enzyme, including *Haemophilus influenzae* and *Actinobacillus pleuropneumoniae*. Since the same consensus sequence, Asn-X-Ser/Thr (X≠Pro), is preferably N-glycosylated, the catalytic mechanism of NGT is interesting. The crystal structures of *A. pleuropneumoniae* NGT (631-residue soluble protein) in the apo form and the UDP-Glc-bound form revealed that a catalytic mechanism based on amide bond twisting was unlikely, because no structure similar to the carboxylate dyad was present (Naegeli et al. 2014). In fact, in terms of the amino acid homology, NGT is more similar to the eukaryotic O-GlcNAc transferase.

9.16 Dynamic Nature of the OST Enzyme

The structures of two forms of the *C. lari* PglB protein, a sequon-bound form and a ternary complex form, have been determined. For the *A. fulgidus* AglB protein, three different forms are now available, an apo form, a sequon-bound form, and a product-releasing form. The product-releasing form (2.5 Å resolution) is a sulfate ion-bound structure, and the sulfate ion was supposed to mimic the phosphate group of the monophosphate-type LLO (Matsumoto et al. 2013b; Petrou et al. 2016). The structures of two apo forms of yeast Stt3 in the multi-subunit OST complex have been solved. The comparison of these catalytic subunit structures in different states suggested that the conformation of the EL5 loop (30–50 residues) is quite variable (Fig. 9.5b). Interestingly, the N-terminal half and the C-terminal half of the EL5 loop adopt ordered and disordered states in the electron maps, independently of each other (Lizak et al. 2014). The ordered state of the N-terminal half of EL5 contains a short α-helix that interacts with the LLO. In the ordered state of the C-terminal half of EL5, the Glu residue in the TIXE/SVSE motif participates in the formation of the carboxylate dyad structure, to interact with the acceptor Asn residue in the sequon. In summary, the conformations of the N-terminal half and C-terminal half of EL5 are directly related to the formation and disruption of the LLO binding site and the sequon binding site, respectively. It is quite possible that the conformational changes of the N-terminal and C-terminal halves of EL5 are highly coordinated for the formation and disruption of the Ser/Thr-binding pocket during the catalytic cycle. These dynamic natures are considered to be necessary to facilitate the efficient scanning of a nascent or folded polypeptide chain, to find glycosylation sequons.

9.17 Concluding Remarks

The first crystal structure of the catalytic subunit, Stt3/AglB/PglB, of an oligosaccharyltransferase was the C-terminal globular domain of *P. furiosus* AglB, in 2008. Over the past decade, the full-length structures of the AglB and PglB proteins were determined by X-ray crystallography, and those of the yeast and mammalian OST complexes were determined to atomic and near-atomic resolutions by cryoEM single particle analyses. Those structures include apo forms and various complex states with acceptor peptides containing a sequon and/or LLOs. The detailed structural comparisons of the various states within the same species and across the different domains of life (Tables 9.1, 9.2 and 9.3) yielded a unified view of the structure of the OST enzyme. OST has a conserved structure consisting of 13 TM helices and the CC structural unit (Fig. 9.4). The carboxylate dyad structure, consisting of the acceptor Asn, two conserved acidic residues, and a divalent metal ion, formed on the luminal side of the TM region, accounts for the amide bond twisting mechanism to increase the nucleophilicity of the amide nitrogen atom of the acceptor Asn residue (Fig. 9.5a). The Ser/Thr-binding pocket in the CC unit in the C-terminal domain explains the requirement of the hydroxy amino acid residues in the N-glycosylation sequon (Figs. 9.4c and 9.5a). Not surprisingly, the two functional structures are formed by short conserved amino acid motifs, the DXD, TIXE/SVSE, WWDYG, and DK/MI motifs (Fig. 9.3a). Note that the hydroxy group is dispensable for the oligosaccharyl transfer reaction under special conditions, because it serves as a simple marker for the N-glycosylation sequon.

Since the information obtained from crystallography and cryoEM analyses is inevitably static, other techniques, such as NMR and AFM and single-molecule observations using fluorescent labels, are desired in the next research stage. The dynamical features of the EL5 loop, the TM helices 8 and 9, and the Ser/Thr-binding pocket are valuable concepts in the promotion of the next-generation research of the structures and functions of OST enzymes.

References

Abu-Qarn M, Eichler J (2007) An analysis of amino acid sequences surrounding archaeal glycoprotein sequons. Archaea 2:73–81

Abu-Qarn M, Eichler J, Sharon N (2008) Not just for Eukarya anymore: protein glycosylation in Bacteria and Archaea. Curr Opin Struct Biol 18:544–550. https://doi.org/10.1016/j.sbi.2008.06.010

Aebi M (2013) N-linked protein glycosylation in the ER. Biochim Biophys Acta 1833:2430–2437. https://doi.org/10.1016/j.bbamcr.2013.04.001

Bai L, Wang T, Zhao G, Kovach A, Li H (2018) The atomic structure of a eukaryotic oligosaccharyltransferase complex. Nature. https://doi.org/10.1038/nature25755

Ben-Dor S, Esterman N, Rubin E, Sharon N (2004) Biases and complex patterns in the residues flanking protein N-glycosylation sites. Glycobiology 14:95–101. https://doi.org/10.1093/glycob/cwh004

Braunger K et al (2018) Structural basis for coupling protein transport and N-glycosylation at the mammalian endoplasmic reticulum. Science. https://doi.org/10.1126/science.aar7899

Breitling J, Aebi M (2013) N-linked protein glycosylation in the endoplasmic reticulum. Cold Spring Harb Perspect Biol 5:a013359. https://doi.org/10.1101/cshperspect.a013359

Buettner FF, Ashikov A, Tiemann B, Lehle L, Bakker H (2013) C. elegans DPY-19 is a C-mannosyltransferase glycosylating thrombospondin repeats. Mol Cell 50:295–302. https://doi.org/10.1016/j.molcel.2013.03.003

Chang MM, Imperiali B, Eichler J, Guan Z (2015) N-Linked Glycans Are Assembled on Highly Reduced Dolichol Phosphate Carriers in the Hyperthermophilic Archaea Pyrococcus furiosus. PLoS One 10:e0130482. https://doi.org/10.1371/journal.pone.0130482

Chantret I, Moore SE (2008) Free oligosaccharide regulation during mammalian protein N-glycosylation. Glycobiology 18:210–224. https://doi.org/10.1093/glycob/cwn003

Chavan M, Chen Z, Li G, Schindelin H, Lennarz WJ, Li H (2006) Dimeric organization of the yeast oligosaccharyl transferase complex. Proc Natl Acad Sci U S A 103:8947–8952. https://doi.org/10.1073/pnas.0603262103

Chen MM, Glover KJ, Imperiali B (2007) From peptide to protein: comparative analysis of the substrate specificity of N-linked glycosylation in C. jejuni. Biochemistry 46:5579–5585. https://doi.org/10.1021/bi602633n

Cherepanova NA, Shrimal S, Gilmore R (2014) Oxidoreductase activity is necessary for N-glycosylation of cysteine-proximal acceptor sites in glycoproteins. J Cell Biol 206:525–539. https://doi.org/10.1083/jcb.201404083

Cherepanova N, Shrimal S, Gilmore R (2016) N-linked glycosylation and homeostasis of the endoplasmic reticulum. Curr Opin Cell Biol 41:57–65. https://doi.org/10.1016/j.ceb.2016.03.021

Dwivedi R, Nothaft H, Reiz B, Whittal RM, Szymanski CM (2013) Generation of free oligosaccharides from bacterial protein N-linked glycosylation systems. Biopolymers 99:772–783. https://doi.org/10.1002/bip.22296

Fujinami D, Nyirenda J, Matsumoto S, Kohda D (2015) Structural elucidation of an asparagine-linked oligosaccharide from the hyperthermophilic archaeon, Archaeoglobus fulgidus. Carbohydr Res 413:55–62. https://doi.org/10.1016/j.carres.2015.05.010

Fujinami D, Taguchi Y, Kohda D (2017) Asn-linked oligosaccharide chain of a crenarchaeon, Pyrobaculum calidifontis, is reminiscent of the eukaryotic high-mannose-type glycan. Glycobiology. https://doi.org/10.1093/glycob/cwx044

Gautschi M et al (2003) The yeast N(alpha)-acetyltransferase NatA is quantitatively anchored to the ribosome and interacts with nascent polypeptides. Mol Cell Biol 23:7403–7414

Gavel Y, von Heijne G (1990) Sequence differences between glycosylated and non-glycosylated Asn-X-Thr/Ser acceptor sites: implications for protein engineering. Protein Eng 3:433–442

Gerber S et al (2013) Mechanism of bacterial oligosaccharyltransferase: in vitro quantification of sequon binding and catalysis. J Biol Chem 288:8849–8861. https://doi.org/10.1074/jbc.M112.445940

Harada Y, Buser R, Ngwa EM, Hirayama H, Aebi M, Suzuki T (2013) Eukaryotic oligosaccharyltransferase generates free oligosaccharides during N-glycosylation. J Biol Chem 288:32673–32684. https://doi.org/10.1074/jbc.M113.486985

Hartley MD, Imperiali B (2012) At the membrane frontier: a prospectus on the remarkable evolutionary conservation of polyprenols and polyprenyl-phosphates. Arch Biochem Biophys 517:83–97. https://doi.org/10.1016/j.abb.2011.10.018

Hasan M et al (2015) Cytosolic nuclease TREX1 regulates oligosaccharyltransferase activity independent of nuclease activity to suppress immune activation. Immunity 43:463–474. https://doi.org/10.1016/j.immuni.2015.07.022

Hennet T, Cabalzar J (2015) Congenital disorders of glycosylation: a concise chart of glycocalyx dysfunction. Trends Biochem Sci 40:377–384. https://doi.org/10.1016/j.tibs.2015.03.002

Igura M, Kohda D (2011a) Quantitative assessment of the preferences for the amino acid residues flanking archaeal N-linked glycosylation sites. Glycobiology 21:575–583. https://doi.org/10.1093/glycob/cwq196

Igura M, Kohda D (2011b) Selective control of oligosaccharide transfer efficiency for the N-glycosylation sequon by a point mutation in oligosaccharyltransferase. J Biol Chem 286:13255–13260. https://doi.org/10.1074/jbc.M110.213900

Igura M, Maita N, Kamishikiryo J, Yamada M, Obita T, Maenaka K, Kohda D (2008) Structure-guided identification of a new catalytic motif of oligosaccharyltransferase. EMBO J 27:234–243. https://doi.org/10.1038/sj.emboj.7601940

Ishiwata A, Taguchi Y, Lee YJ, Watanabe T, Kohda D, Ito Y (2015) N-Glycosylation with synthetic undecaprenyl pyrophosphate-linked oligosaccharide to oligopeptides by PglB oligosaccharyltransferase from Campylobacter jejuni. Chembiochem 16:731–737. https://doi.org/10.1002/cbic.201402658

Izquierdo L et al (2009) Distinct donor and acceptor specificities of Trypanosoma brucei oligosaccharyltransferases. EMBO J 28:2650–2661. https://doi.org/10.1038/emboj.2009.203

Jaeken J, Peanne R (2017) What is new in CDG? J Inherit Metab Dis 40:569–586. https://doi.org/10.1007/s10545-017-0050-6

Jaffee MB, Imperiali B (2013) Optimized protocol for expression and purification of membrane-bound PglB, a bacterial oligosaccharyl transferase. Protein Expr Purif 89:241–250. https://doi.org/10.1016/j.pep.2013.04.001

Jarrell KF, Ding Y, Meyer BH, Albers SV, Kaminski L, Eichler J (2014) N-linked glycosylation in Archaea: a structural, functional, and genetic analysis. Microbiol Mol Biol Rev 78:304–341. https://doi.org/10.1128/MMBR.00052-13

Jeong IS et al (2018) Purification and characterization of Arabidopsis thaliana oligosaccharyltransferase complexes from the native host: a protein super-expression system for structural studies. Plant J. https://doi.org/10.1111/tpj.13847

Jitsuhara Y, Toyoda T, Itai T, Yamaguchi H (2002) Chaperone-like functions of high-mannose type and complex-type N-glycans and their molecular basis. J Biochem 132:803–811

Kasturi L, Chen H, Shakin-Eshleman SH (1997) Regulation of N-linked core glycosylation: use of a site-directed mutagenesis approach to identify Asn-Xaa-Ser/Thr sequons that are poor oligosaccharide acceptors. Biochem J 323(Pt 2):415–419

Kelleher DJ, Gilmore R (2006) An evolving view of the eukaryotic oligosaccharyltransferase. Glycobiology 16:47R–62R. https://doi.org/10.1093/glycob/cwj066

Kimura N, Uchida M, Nishimura S, Yamaguchi H (1998) Promotion of polypeptide folding by interactions with Asn-Glycans. J Biochem 124:857–862

Knauer R, Lehle L (1999a) The oligosaccharyltransferase complex from Saccharomyces cerevisiae. Isolation of the OST6 gene, its synthetic interaction with OST3, and analysis of the native complex. J Biol Chem 274:17249–17256

Knauer R, Lehle L (1999b) The oligosaccharyltransferase complex from yeast. Biochim Biophys Acta 1426:259–273

Kowarik M et al (2006a) N-linked glycosylation of folded proteins by the bacterial oligosaccharyltransferase. Science 314:1148–1150. https://doi.org/10.1126/science.1134351

Kowarik M et al (2006b) Definition of the bacterial N-glycosylation site consensus sequence. EMBO J 25:1957–1966. https://doi.org/10.1038/sj.emboj.7601087

Kucej M, Fermaintt CS, Yang K, Irizarry-Caro RA, Yan N (2017) Mitotic Phosphorylation of TREX1 C Terminus Disrupts TREX1 Regulation of the Oligosaccharyltransferase Complex. Cell Rep 18:2600–2607. https://doi.org/10.1016/j.celrep.2017.02.051

Kuntz C, Sonnenbichler J, Sonnenbichler I, Sumper M, Zeitler R (1997) Isolation and characterization of dolichol-linked oligosaccharides from Haloferax volcanii. Glycobiology 7:897–904

Larkin A, Chang MM, Whitworth GE, Imperiali B (2013) Biochemical evidence for an alternate pathway in N-linked glycoprotein biosynthesis. Nat Chem Biol 9:367–373. https://doi.org/10.1038/nchembio.1249

Li G, Yan Q, Nita-Lazar A, Haltiwanger RS, Lennarz WJ (2005) Studies on the N-glycosylation of the subunits of oligosaccharyl transferase in Saccharomyces cerevisiae. J Biol Chem 280:1864–1871. https://doi.org/10.1074/jbc.M410969200

Li H, Chavan M, Schindelin H, Lennarz WJ, Li H (2008) Structure of the oligosaccharyl transferase complex at 12 A resolution Structure 16:432-440 doi:https://doi.org/10.1016/j.str.2007.12.013

Liu C, Szostak M (2017) Twisted amides: from obscurity to broadly useful transition-metal-catalyzed reactions by N-C amide bond activation. Chemistry 23:7157–7173. https://doi.org/10.1002/chem.201605012

Lizak C et al (2013) Unexpected reactivity and mechanism of carboxamide activation in bacterial N-linked protein glycosylation. Nat Commun 4:2627. https://doi.org/10.1038/ncomms3627

Lizak C, Gerber S, Numao S, Aebi M, Locher KP (2011) X-ray structure of a bacterial oligosaccharyltransferase. Nature 474:350–355. https://doi.org/10.1038/nature10151

Lizak C et al (2014) A catalytically essential motif in external loop 5 of the bacterial oligosaccharyltransferase PglB. J Biol Chem 289:735–746. https://doi.org/10.1074/jbc.M113.524751

Maita N, Nyirenda J, Igura M, Kamishikiryo J, Kohda D (2010) Comparative structural biology of eubacterial and archaeal oligosaccharyltransferases. J Biol Chem 285:4941–4950. https://doi.org/10.1074/jbc.M109.081752

Matsumoto S et al (2012) Crystal structure of the C-terminal globular domain of oligosaccharyltransferase from Archaeoglobus fulgidus at 1.75 A resolution. Biochemistry 51:4157–4166. https://doi.org/10.1021/bi300076u

Matsumoto S, Shimada A, Kohda D (2013a) Crystal structure of the C-terminal globular domain of the third paralog of the Archaeoglobus fulgidus oligosaccharyltransferases. BMC Struct Biol 13:11. https://doi.org/10.1186/1472-6807-13-11

Matsumoto S, Shimada A, Nyirenda J, Igura M, Kawano Y, Kohda D (2013b) Crystal structures of an archaeal oligosaccharyltransferase provide insights into the catalytic cycle of N-linked protein glycosylation. Proc Natl Acad Sci U S A 110:17868–17873. https://doi.org/10.1073/pnas.1309777110

Matsumoto S, Taguchi Y, Shimada A, Igura M, Kohda D (2017) Tethering an N-glycosylation sequon-containing peptide creates a catalytically competent oligosaccharyltransferase complex. Biochemistry 56:602–611. https://doi.org/10.1021/acs.biochem.6b01089
Meyer BH, Albers SV (2014) AglB, catalyzing the oligosaccharyl transferase step of the archaeal N-glycosylation process, is essential in the thermoacidophilic crenarchaeon Sulfolobus acidocaldarius. Microbiologyopen 3:531–543. https://doi.org/10.1002/mbo3.185
Mills DC, Jervis AJ, Abouelhadid S, Yates LE, Cuccui J, Linton D, Wren BW (2016) Functional analysis of N-linking oligosaccharyl transferase enzymes encoded by deep-sea vent proteobacteria Glycobiology 26:398-409 doi:https://doi.org/10.1093/glycob/cwv111
Mohorko E, Glockshuber R, Aebi M (2011) Oligosaccharyltransferase: the central enzyme of N-linked protein glycosylation. J Inherit Metab Dis 34:869–878. https://doi.org/10.1007/s10545-011-9337-1
Mohorko E, Owen RL, Malojcic G, Brozzo MS, Aebi M, Glockshuber R (2014) Structural basis of substrate specificity of human oligosaccharyl transferase subunit N33/Tusc3 and its role in regulating protein N-glycosylation. Structure 22:590–601. https://doi.org/10.1016/j.str.2014.02.013
Naegeli A et al (2014) Substrate specificity of cytoplasmic N-glycosyltransferase. J Biol Chem 289:24521–24532. https://doi.org/10.1074/jbc.M114.579326
Napiorkowska M, Boilevin J, Sovdat T, Darbre T, Reymond JL, Aebi M, Locher KP (2017) Molecular basis of lipid-linked oligosaccharide recognition and processing by bacterial oligosaccharyltransferase. Nat Struct Mol Biol 24:1100–1106. https://doi.org/10.1038/nsmb.3491
Nyathi Y, Wilkinson BM, Pool MR (2013) Co-translational targeting and translocation of proteins to the endoplasmic reticulum. Biochim Biophys Acta 1833:2392–2402. https://doi.org/10.1016/j.bbamcr.2013.02.021
Nyirenda J, Matsumoto S, Saitoh T, Maita N, Noda NN, Inagaki F, Kohda D (2013) Crystallographic and NMR evidence for flexibility in oligosaccharyltransferases and its catalytic significance. Structure 21:32–41. https://doi.org/10.1016/j.str.2012.10.011
Ollis AA et al (2015) Substitute sweeteners: diverse bacterial oligosaccharyltransferases with unique N-glycosylation site preferences. Sci Rep 5:15237. https://doi.org/10.1038/srep15237
Palmieri G, Balestrieri M, Peter-Katalinic J, Pohlentz G, Rossi M, Fiume I, Pocsfalvi G (2013) Surface-exposed glycoproteins of hyperthermophilic Sulfolobus solfataricus P2 show a common N-glycosylation profile. J Proteome Res 12:2779–2790. https://doi.org/10.1021/pr400123z
Pedebos C, Arantes PR, Giesel GM, Verli H (2015) In silico Investigation of the PglB active site reveals transient catalytic states and octahedral metal ion coordination. Glycobiology 25:1183–1195. https://doi.org/10.1093/glycob/cwv053
Petrescu AJ, Milac AL, Petrescu SM, Dwek RA, Wormald MR (2004) Statistical analysis of the protein environment of N-glycosylation sites: implications for occupancy, structure, and folding. Glycobiology 14:103–114. https://doi.org/10.1093/glycob/cwh008
Petrou VI et al (2016) Structures of aminoarabinose transferase ArnT suggest a molecular basis for lipid A glycosylation. Science 351:608–612. https://doi.org/10.1126/science.aad1172
Ramirez AS et al (2017) Characterization of the single-subunit oligosaccharyltransferase STT3A from Trypanosoma brucei using synthetic peptides and lipid-linked oligosaccharide analogs. Glycobiology 27:525–535. https://doi.org/10.1093/glycob/cwx017
Romanelli A, Shekhtman A, Cowburn D, Muir TW (2004) Semisynthesis of a segmental isotopically labeled protein splicing precursor: NMR evidence for an unusual peptide bond at the N-extein-intein junction. Proc Natl Acad Sci U S A 101:6397–6402. https://doi.org/10.1073/pnas.0306616101
Samuelson J, Banerjee S, Magnelli P, Cui J, Kelleher DJ, Gilmore R, Robbins PW (2005) The diversity of dolichol-linked precursors to Asn-linked glycans likely results from secondary loss of sets of glycosyltransferases. Proc Natl Acad Sci U S A 102:1548–1553. https://doi.org/10.1073/pnas.0409460102

Sato C, Kim JH, Abe Y, Saito K, Yokoyama S, Kohda D (2000) Characterization of the N-oligosaccharides attached to the atypical Asn-X-Cys sequence of recombinant human epidermal growth factor receptor. J Biochem 127:65–72
Schulz BL et al (2009) Oxidoreductase activity of oligosaccharyltransferase subunits Ost3p and Ost6p defines site-specific glycosylation efficiency. Proc Natl Acad Sci U S A 106:11061–11066. https://doi.org/10.1073/pnas.0812515106
Shrimal S, Trueman SF, Gilmore R (2013) Extreme C-terminal sites are posttranslocationally glycosylated by the STT3B isoform of the OST. J Cell Biol 201:81–95. https://doi.org/10.1083/jcb.201301031
Shrimal S, Cherepanova NA, Gilmore R (2015) Cotranslational and posttranslocational N-glycosylation of proteins in the endoplasmic reticulum. Semin Cell Dev Biol 41:71–78. https://doi.org/10.1016/j.semcdb.2014.11.005
Shrimal S, Cherepanova NA, Gilmore R (2017a) DC2 and KCP2 mediate the interaction between the oligosaccharyltransferase and the ER translocon. J Cell Biol 216:3625–3638. https://doi.org/10.1083/jcb.201702159
Shrimal S, Cherepanova NA, Gilmore R (2017b) One flexible loop in OST lassos both substrates. Nat Struct Mol Biol 24:1009–1010. https://doi.org/10.1038/nsmb.3508
Silverman JM, Imperiali B (2016) Bacterial N-glycosylation efficiency is dependent on the structural context of target sequons. J Biol Chem 291:22001–22010. https://doi.org/10.1074/jbc.M116.747121
Spirig U, Bodmer D, Wacker M, Burda P, Aebi M (2005) The 3.4-kDa Ost4 protein is required for the assembly of two distinct oligosaccharyltransferase complexes in yeast. Glycobiology 15:1396–1406. https://doi.org/10.1093/glycob/cwj025
Szymanski CM, Wren BW (2005) Protein glycosylation in bacterial mucosal pathogens. Nat Rev Microbiol 3:225–237. https://doi.org/10.1038/nrmicro1100
Taguchi Y, Fujinami D, Kohda D (2016) Comparative analysis of archaeal lipid-linked oligosaccharides that serve as oligosaccharide donors for Asn glycosylation. J Biol Chem 291:11042–11054. https://doi.org/10.1074/jbc.M115.713156
Valliere-Douglass JF, Eakin CM, Wallace A, Ketchem RR, Wang W, Treuheit MJ, Balland A (2010) Glutamine-linked and non-consensus asparagine-linked oligosaccharides present in human recombinant antibodies define novel protein glycosylation motifs. J Biol Chem 285:16012–16022. https://doi.org/10.1074/jbc.M109.096412
Varki A et al (2017) Essentials of glycobiology, 3rd edn. Cold Spring Harbor Laboratory Press, Cold Spring Harbor, New York
Wild R, Kowal J, Eyring J, Ngwa EM, Aebi M, Locher KP (2018) Structure of the yeast oligosaccharyltransferase complex gives insight into eukaryotic N-glycosylation. Science 359:545–550. https://doi.org/10.1126/science.aar5140
Yamada S (1996) Effects of C(O)-N bond rotation on the 13C, 15N, and 17O NMR chemical shifts, and infrared carbonyl absorption in a series of twisted amides. J Org Chem 61:941–946
Zaremba-Niedzwiedzka K et al (2017) Asgard archaea illuminate the origin of eukaryotic cellular complexity. Nature 541:353–358. https://doi.org/10.1038/nature21031
Zielinska DF, Gnad F, Wisniewski JR, Mann M (2010) Precision mapping of an in vivo N-glycoproteome reveals rigid topological and sequence constraints. Cell 141:897–907. https://doi.org/10.1016/j.cell.2010.04.012

Chapter 10
Visualization of Functional Structure and Kinetic Dynamics of Cellulases

Akihiko Nakamura and Ryota Iino

Abstract Cellulose is the most abundant carbohydrate on earth and hydrolyzed by cellulases in nature. During catalysis, cellulase transfers protons to and from the oxygen atoms of the glycosidic bond and a water molecule. Since cellulose is an insoluble polymer, some kinds of cellulases, with high activity toward crystalline cellulose, move on the crystal surface with continuous hydrolysis of the molecular chain. In addition, binding and dissociation on/from the crystal surface are also important elementary steps of the reaction cycle. Recently, these interesting features of cellulases can be directly analyzed, due to the development of visualization techniques. In this chapter, we introduce (1) visualization of the protonation state of the catalytic residue by neutron crystallography, (2) visualization of processive movement on the crystal surface by high-speed atomic force microscopy, and (3) visualization of binding and dissociation events by single-molecule fluorescence microscopy.

Keywords Cellulase · cellulose · Processivity · Molecular motor · Single-molecule analysis · Neutron crystallography · Proton pathway

---

A. Nakamura (✉)
Okazaki Institute for Integrative Bioscience, National Institutes of Natural Sciences, Aichi, Japan

Department of Functional Molecular Science, School of Physical Sciences, Kanagawa, Japan
e-mail: aki-naka@ims.ac.jp

R. Iino
Okazaki Institute for Integrative Bioscience, National Institutes of Natural Sciences, Aichi, Japan

Department of Functional Molecular Science, School of Physical Sciences, Kanagawa, Japan

Institute for Molecular Science, National Institutes of Natural Sciences, Aichi, Japan

Y. Yamaguchi, K. Kato (eds.), *Glycobiophysics*, Advances in Experimental Medicine and Biology 1104, https://doi.org/10.1007/978-981-13-2158-0_10

10.1 Introduction

Cellulase hydrolyzes cellulose, the main component of plant cell walls, to mono- or oligosaccharides. Cellulase is mainly produced by fungi and bacteria that live in soil (Štursová et al. 2012), coexist with animals (Julliand et al. 1999), or are pathogens of plants (Wood 1960). Cellulases are classified as members of the glycoside hydrolase (GH) family based on amino acid sequences and three-dimensional structures in the Carbohydrate-Active enZymes (CAZy) database (Lombard et al. 2014). At this time (Feb., 2018), 149 GH families exist, and cellulases are classified in 12 families (GH5, 6, 7, 8, 9, 12, 44, 45, 48, 51, 74, and 124). Traditional classifications of them, based on activity, are cellobiohydrolases and endoglucanases (EC 3.2.1.4). The former is additionally separated into nonreducing-end cellobiohydrolases (EC 3.2.1.91) and reducing-end cellobiohydrolases (EC 3.2.1.176), depending on different preferences for the ends of the molecular chain (Fig. 10.1). Cellobiohydrolases mainly produce cellobiose (a β-1,4-linked glucose dimer) from chain ends, and endoglucanase cuts the middle of chains in amorphous regions. Many cellulases consist of a catalytic domain (CD) and a cellulose-binding domain (CBD). These two domains of fungal cellulases are connected by an intrinsically disordered region modified with sugars, and those of bacterial cellulase (Leschine 1995) are connected by linker domain (i.e., fibronectin type III-like domain). Although, cellulases without CBD also exist. In this chapter, we focus on fungal cellulase.

Two reaction mechanisms have been proposed for cellulases depending on the anomeric structure of the products: one is the retaining mechanism, and the other is the inverting mechanism (McCarter and Stephen Withers 1994). In the former case, the catalytic nucleophile residue attacks the C1 carbon of glucose, and the catalytic acid/base residue protonates the oxygen of the glycosidic bond. After making a glycosyl-enzyme intermediate, the catalytic acid/base residue receives a proton from a water molecule, and the glycosidic bond of the intermediate is hydrolyzed.

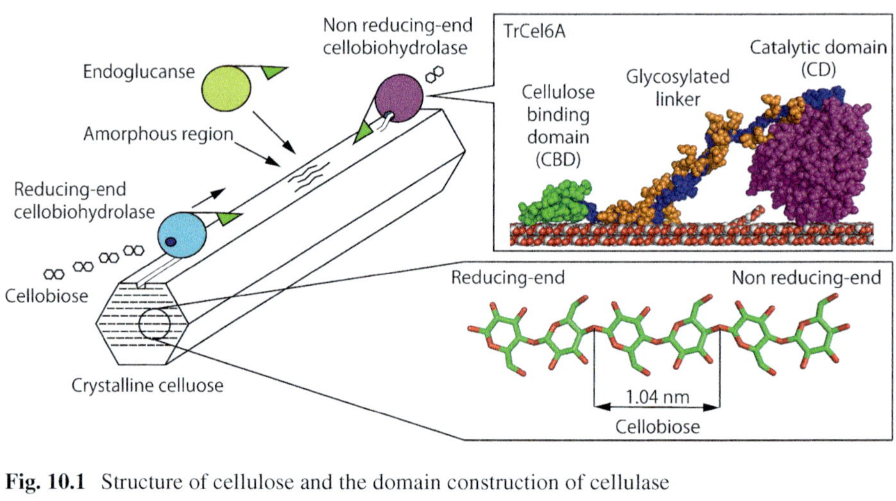

Fig. 10.1 Structure of cellulose and the domain construction of cellulase

On the other hand, inverting cellulases hydrolyze cellulose at once with protonation of the oxygen of the glycosidic bond and activation of water by the catalytic acid and base residues, respectively. Both mechanisms basically need two acidic residues (aspartate or glutamate) and catalyze proton exchanges between cellulose and water molecules. Therefore, visualization of the protonated states of the residues around the substrate-binding site is useful to determine the catalytic residues and mechanisms.

The most unique feature of cellulase is the heterogeneous reaction at the liquid-solid interface. Cellulose is an insoluble polymer of β-1,4-linked glucose, and the molecular chains make stabile crystal structures. During the reaction, cellulase needs to bind on the surface of cellulose, hydrolyze the glycosidic bond, and dissociate from the surface. In the case of crystalline cellulose hydrolysis, not all enzyme-binding events are productive, because molecular chains are tightly packed and hydrolyzable parts are limited (Jalak et al. 2012). The constants (e.g., turnover and dissociation constant) obtained by biochemical assays are averages of all states. Therefore, single-molecule analysis is needed to determine the binding rate constant (k_{on}), dissociation rate constant (k_{off}), and turnover of "truly" productive molecules. Actually, the movement of cellobiohydrolases on crystalline cellulose that had been expected from biochemical results was directly visualized by high-speed atomic force microscopy (HS-AFM), as described below (Igarashi et al. 2009).

In this chapter, we will introduce visualization of (1) protonation states of the catalytic residue by neutron and X-ray crystallography, (2) processive movement by HS-AFM, and (3) binding and dissociation on/from cellulose crystals by single-molecule fluorescence microscopy. These visualization methods should be useful not only for the characterization of cellulases but also for other carbohydrate active enzymes.

10.2 Visualization of Protonated States by X-Ray and Neutron Crystallography

The hydrogen atom is the most abundant atom in an enzyme molecule. Hydrogen atom is an important element of hydrogen bonding that is indispensable for stabilization of higher-order enzyme structures and interactions between enzymes and substrates. It is also well known that proton exchange (hydrogen nucleus, generally also called a hydrogen ion or proton) occurs between substrate or solvent and amino acids that have polar groups. Structural analysis, including hydrogen atoms and protons, is extremely important for understanding the structure and reaction mechanism of enzymes.

10.2.1 Advantages and Challenges of Neutron Crystallography

In terms of three-dimensional structure determination of an enzyme, the most common method is X-ray crystallography, and over 100,000 structures have been registered in the Protein Data Bank (PDB), to date. In X-ray crystallography, the electron density of the atoms constituting the enzyme molecule are observed using X-ray as a probe. Carbon, oxygen, and nitrogen atoms that have a large atomic number and a large number of electrons are easily observed, compared to hydrogen atoms. Ultrahigh (sub-angstroms) resolution is needed for observing hydrogen atom, hydride, or proton with coordinate bond (Ogata et al. 2015). On the other hand, neutron crystallography uses neutron as a probe, and neutrons diffracted by nuclei are observed. Therefore, the visibilities of atoms are completely different from X-ray crystallography, as shown in Fig. 10.2 (Varley 1992; McMaster et al. 1969). For example, sulfur atoms, easily observed by X-ray crystallography, show weak scattering intensity in neutron crystallography. Additionally, oxygen and nitrogen atoms are easily distinguished from each other by neutron crystallography because nitrogen atoms clearly show stronger scattering intensity. The most interesting feature of neutron crystallography is the differences of scattering intensities between isotopes. Due to this property, a deuterium atom (D) shows a scattering intensity equivalent to that of carbon atoms, whereas a hydrogen atom (H) shows an apparent negative scattering intensity, because the phase of the scattered neutron is reversed. If the hydrogen atoms of an enzyme are fully exchanged to deuterium atoms, the protonation states of the targeted residues can be easily visualized. During the diffraction measurement, the enzyme crystal will not be damaged or reduced even at room temperature. The energy of the neutron beam of 81.8 meV (7.9 kJ/mol) for a wavelength of 1Å is not enough to cleave the covalent bonds of amino acids. This is the other advantage of neutron crystallography.

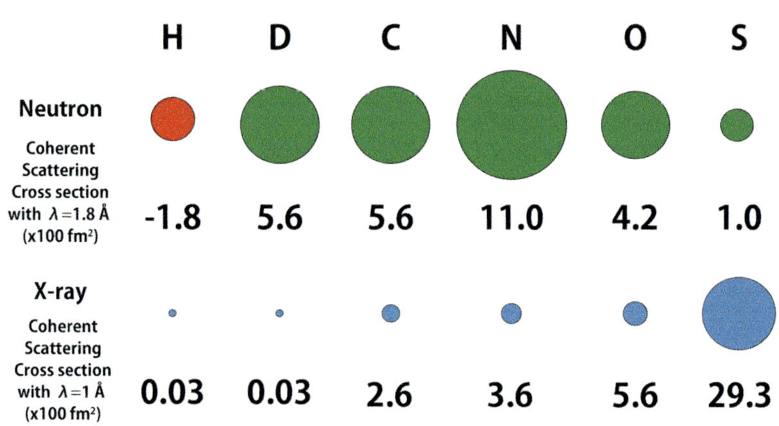

Fig. 10.2 Difference in scattering efficiency among atoms in neutron and X-ray crystallography

There are two types of neutron sources for neutron crystallography. One is a nuclear reactor, which generates a mixed wavelength neutron beam, and the desired wavelength of the neutron beam is separated by a monochromator. This type of sources is available in Japan (Japan Research Reactor No.3), France (High-Flux Reactor), Germany (Forschungs-Neutronenquelle Heinz Maier-Leibnitz), and the USA (High Flux Isotope Reactor). The other is a short-pulsed nuclear spallation neutron source that generates a neutron from nuclear fission of a metal target (such as tungsten or liquid mercury) by pulsed proton beams from a synchrotron or linear accelerator. The neutron wavelength diffracted by the crystal can be selected using a synchronized chopper, because the timing of neutron generation is known. These sources are now available in Japan (Materials and Life Science Experimental Facility; MLF), the United Kingdom (ISIS spallation neutron source), and the USA (Spallation Neutron Source and Protein Crystallography Station). Additionally, a European Spallation Source is under construction in Sweden. Especially for the MLF BL02 iBIX of the Japan Proton Accelerator Research Complex (J-PARC), detectors synchronized with the neutron generator and chopper are used to analyze diffraction spots by a time-resolved Laue method. This system showed one of the highest efficiencies of measurement in the world (Kusaka et al. 2013).

Even with the continuous development of neutron crystallography, the biggest problem is the weak brightness of the neutron source. Because of this disadvantage, usually more than 1 mm^3 of protein crystal and a half-month of measurement are needed for diffraction analysis. The probability of big crystal formation can be increased with detailed analysis of the crystallization condition (Nakamura et al. 2013a), but crystal preparation is still the bottleneck of neutron crystallography. Further improvement of facilities and crystallization techniques are necessary to make neutron crystallography more generalized.

10.2.2 An Example of Protonation Analysis of Cellulase

Some cellulases do not have a pair of acidic residues at the expected catalytic site from substrate-bound structures. For example, a GH45 inverting cellulase from a basidiomycete *Phanerochaete chrysosporium* (PcCel45A) has an aspartic acid residue as the catalytic acid and produces products from cellulose, but an acidic residue was not found at a suitable position for the catalytic base residue (Nakamura et al. 2015). The asparagine 92 residue was identified as a possible catalytic base from structural comparison with the most studied GH45 cellulase from the ascomycete *Humicola insolens* (HiCel45A) (Fig. 10.3a). In a biochemical assay, the PcCel45A N92D mutant showed drastically lower activity than the WT, indicating that Asn92 is acting as the catalytic residue (Fig. 10.3b). These results mean that PcCel45A has a different mechanism from other known cellulases. Normally, however, asparagine cannot receive a proton at a moderate pH, because protonation of nitrogen and oxygen atoms at a side chain amide need pH -7 or pH 0, respectively. Therefore, neutron crystallography was applied to determine the protonation states of Asn92 and

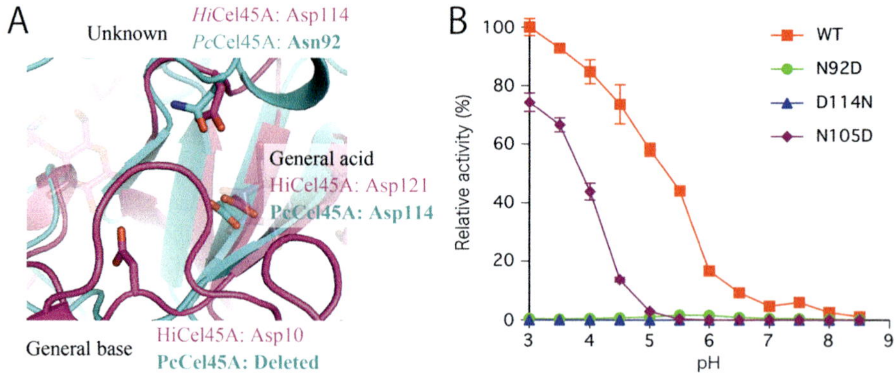

Fig. 10.3 (**a**) Superposition of the active sites of PcCel45A and HiCel45A (**b**) pH dependencies of the activity of PcCel45A WT, D114A, N92D, and N105D

the reaction mechanism of PcCel45A. Neutron diffraction from the 6 mm^3 crystal was measured for 13.5 days at the iBIX in the MLF of J-PARC with 300 kW accelerator power. As a result, it was revealed that the Asn92 residue is not in the usual "amide form," but in the "imidic acid form," that is, the tautomer of amide (Fig. 10.4). However, under normal conditions, the imidic acid form is unstable, and the equilibrium is biased to the amide form. Then, further analysis of the hydrogen bond around the catalytic residue revealed that the oxygen of the carbonyl of the Asn92 side chain and the nitrogen of the amide of the main chain are connected by the positive scattering power map of neutron crystallography. Since no scattering map was observed there with X-ray crystallography, it was expected that deuterium ion was shared between the two atoms. Next, the hydrogen bonds between the two catalytic residues (Asn92 and Asp114) were also analyzed in detail, and it was confirmed that the two catalytic residues are connected by a series of the positive scattering power map of neutron (Fig. 10.4). To verify these results, the activity profile of PcCel45A N105D was analyzed, because Asn105 does not directly participate in substrate recognition or catalysis itself, but it looks to be relaying protons between the catalytic residues. As expected, the N105D mutant retained 80% activity, compared to wild type at pH 3.0, whereas at pH 5.5, the activity of the N105D mutant was completely lost (Fig. 10.3b). This is likely because the proton relay pathway is disconnected, due to the deprotonation of the aspartic acid side chain (p*K*a 3.9) at a lower pH than the wild type. From these results, it was shown that PcCel45A uses the imidic acid form of Asn92 as the catalytic base and that protonation of the glycosidic bond and activation of the water molecule is coupled to the proton pathway. In this model, the imidic acid form of Asn92 can be regenerated via reverse proton relays by protonation of Asp114; thus it can be explained why the optimal pH of the reaction is acidic.

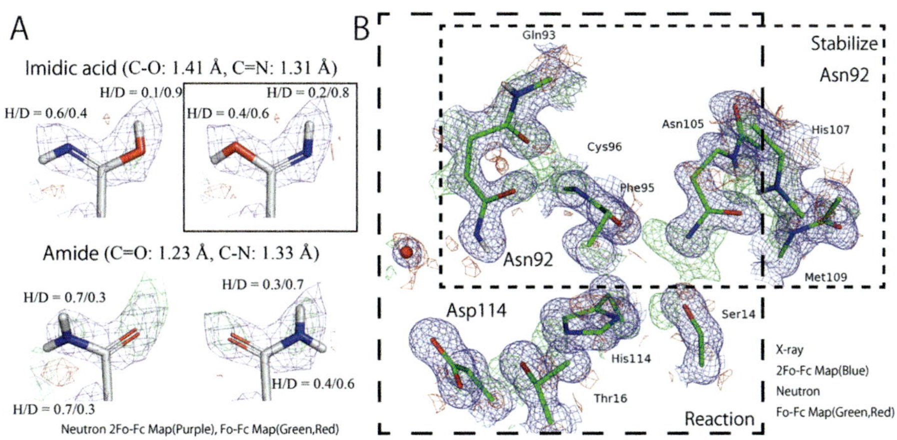

Fig. 10.4 (**a**) Imidic acid form of Asn92 in PcCel45A WT. (**b**) Neutron scattering power map between Asn92 and Asp114

10.3 Visualization of Processive Movement of Cellulase by HS-AFM

High-speed atomic force microscopy (HS-AFM) is the cutting-edge technique in single-molecule observation. The most important point of this system is that enzymes can be visualized without labeling. Higher frame rates and lower tapping force than conventional AFM were achieved using small cantilevers and high-speed scanners (Ando 2013). Actually, unidirectional movement of myosin V on actin filaments (Kodera et al. 2010) and chitinase on the crystal chitin surface (Igarashi et al. 2014) was visualized. In addition, the structural change of bacteriorhodopsin with photo activation (Shibata et al. 2010) and cooperative movements of the ring of a rotating motor F_1-ATPase ($\alpha_3\beta_3$ subunits), without the rotator (γ-subunit), were observed (Uchihashi et al. 2011). Recently, the distance distribution between two domains, connected by an intrinsically disordered region, was also analyzed (Kodera et al. 2015). These results mean that even the structural dynamics of enzymes can be visualized by HS-AFM. Please read reviews by developers about the detailed working mechanism of HS-AFM (Ando 2017).

One of the most exciting findings in the cellulase research is the direct observation of the processive movement of GH7 cellobiohydrolase from the ascomycete *Trichoderma reesei* (TrCel7A) using HS-AFM (Igarashi et al. 2009), like shown in Fig. 10.5. As described as above, cellobiohydrolases have been thought of as processive enzymes, because they specifically produce cellobioses, that is, the repeating unit of the cellulose molecular chain. From the structural comparison between cellobiohydrolase and endoglucanase in the same GH family, it was shown that only cellobiohydrolases have a tunnellike structure to keep the substrate-bound state

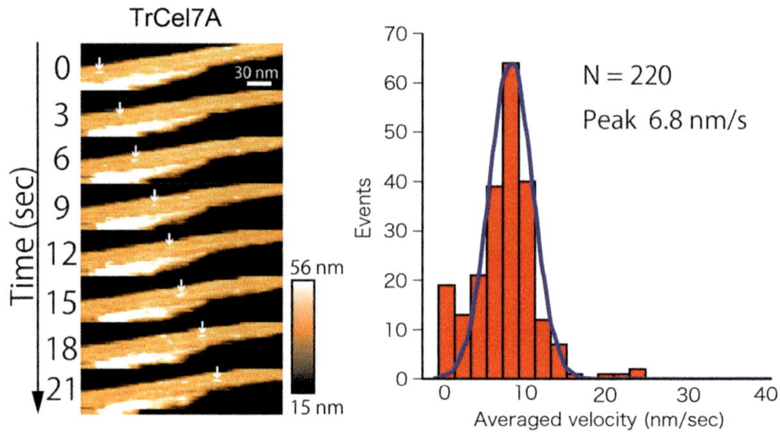

Fig. 10.5 Processive movement of TrCel7A observed by HS-AFM

(Davies and Henrissat 1995). Furthermore, sharpened edges of cellulose crystals, after degradation by cellobiohydrolases, were shown by electron microscopy (Imai et al. 1998). Processivity of cellulases have been discussed for a long time with these indirect evidences. Visualization of processive movement by HS-AFM provided a new direction of analysis for cellulase characterization, like moving velocity (meaning k_{cat} of a productive molecule) and moving length (indicating processivity).

10.3.1 Analysis of Velocity and Processivity by HS-AFM

For the observation of cellulase movement, cellulose crystals should be stabilized on the surface of a graphite disc or hydrophobized mica, stuck on a glass stage. Fortunately, cellulose crystals are bound on their surface by just putting a drop of the cellulose crystal suspension, because a cellulose crystal has hydrophobic planes on its surface. Usually, 2 μl of 20 μM enzyme are added into 80 μl of observation buffer in the liquid chamber. In the case of a weak-binding mutant, like a separated CD, about ten times higher concentration of enzyme solution is required to find moving molecules. The highest frame rate reported previously is 5 fps. Higher frame rates will provide better time resolution for moving time analysis and processivity estimation, but they also make it difficult to track moving molecules across a long distance, due to the limitation of the field of view. Tracking of moving molecules is achieved by peak estimation with a 2D-Gaussian distribution fitting to the molecule image of height per frame. Moving distance and time are calculated from the difference of peak positions and number of frames between the start and end frames of movement. From these values, velocities of each molecule are determined, and the moving velocity of the enzyme can be estimated from the distributions of these velocities of observed molecules. In single-molecule analysis, a velocity of an observed molecule does not always represent the typical character of the target enzyme, because the values of the velocities of molecules follow a Gaussian distribution, if the enzyme has a single mode of movement. Therefore, velocities of a high enough number of molecules should be analyzed to discuss the overall enzyme properties. The analysis of moving time follows the same manner. Distribution of moving times of observed molecules is fitted with an exponential decay function, and moving time (or dissociation constant) can be determined. Moving distance can be calculated using the values of velocity and moving time, and processivity can be calculated when the step size of cellobiohydrolases is assumed as 1 nm from the size of cellobiose (1.04 nm).

Not many kinds of cellobiohydrolases were observed with HS-AFM. Values of velocity and processivity have only been compared among TrCel7A and two isozymes of GH7 cellobiohydrolases from *P. chrysosporium* (PcCel7C and PcCel7D) (Nakamura et al. 2014). All of them hydrolyze cellulose from reducing ends. At first, the processive movement of TrCel7A was compared, across the wild-type enzyme, only the CD, an inactive mutant (E212Q) and a less-active mutant to crystalline cellulose (W40A) (Igarashi et al. 2009). Processive movement of the TrCel7A CD indicated that the CBD and linker region are not necessary for processive movement. On the other hand, TrCel7A E212Q just bound to cellulose and did not move. From these results, it was clearly shown that TrCel7A moves on cellulose with hydrolysis of the chain. Additionally, processive movement of TrCel7A W40A was also not found, although Trp40 is only located at the entrance of the tunnel structure and far from the catalytic site. The tryptophan residue at the entrance of the tunnel is highly conserved in GH7 cellobiohydrolases (Nakamura et al. 2013b). In a biochemical assay, TrCel7A W40A showed lower activity per bound enzyme to crystalline cellulose, but higher specific activity toward amorphous cellulose. The role of Trp40 in TrCel7A was identified from the results of molecular dynamics simulations. When the glucose ring of the reducing end of the cellulose chain was placed on Trp40, wild-type TrCel7A can take the molecular chain into the tunnellike structure. In contrast, W40A could not hold the reducing end at the entrance of the tunnel and failed to take in the chain. However, when the chain end was placed in the tunnel, there was no difference between the wild type and W40A, indicating that Trp40 is just important for initial chain catching at the entrance of the tunnel. Trp40 is important only for crystalline cellulose degradation, because the number and length of available chain ends might be limited on the surface of crystalline cellulose, compared to disordered amorphous cellulose.

Compared to cellobiohydrolases from *Ascomycetes* like TrCel7A, cellobiohydrolases from *Basidiomycetes* have more open tunnel structures, due to the deletion of loops (Nakamura et al. 2014). These structural features are expected to cause the difference in processivities, but there has been no clear evidence. In a direct comparison of the processive movement with HS-AFM, the processivity of TrCel7A was 34.0, while the processivities of PcCel7D and PcCel7C were 29.4 and 28.8, respectively (Table 10.1). TrCel7A showed higher processivity than PcCel7D and PcCel7C, as expected. Their moving times showed a similar trend, and the values

Table 10.1 Moving velocities, moving time, and processivity of GH7 cellobiohydrolases

Enzyme	Velocity (nm/s)	Moving time (s)	Processivity
TrCel7A	6.8 ± 3.5	3.5 ± 0.2	34.0
PcCel7D	9.4 ± 3.7	2.2 ± 0.1	29.4
PcCel7C	14.7 ± 9.1	1.4 ± 0.0	28.8

From Nakamura et al. (2014)

were 5.0 s (TrCel7A), 3.1 s (PcCel7D), and 2.0 s (PcCel7C). Interestingly, the order of moving velocities was opposite, and TrCel7A showed the slowest velocity (6.8 nm/s) in the three. The velocity of PcCel7C (14.7 nm/s) was higher than that of PcCel7D (9.4 nm/s). This trade-off relationship between velocity and processivity, or moving time, indicates that a high affinity to the cellulose chain causes slow-moving actions. These results show that HS-AFM is a powerful tool to analyze the relationship between structure and function of processive cellobiohydrolases.

One of the GH6 cellobiohydrolases from *T. reesei* (TrCel6A) that hydrolyzes cellulose from nonreducing ends was also observed by HS-AFM (Igarashi et al. 2011), but clear movement of TrCel6A was not observed, even though TrCel6A showed crystals with sharpened edges in electron microscopy analysis (Chanzy and Henrissat 1985). Some GH6 enzymes from cellulose-degrading bacterial species showed processive motion on cellulose crystals, but this has not yet been published. Further analysis is needed to clarify the structural and functional differences of moving and nonmoving GH6 cellobiohydrolases in HS-AFM observation.

10.4 Visualization of Binding and Dissociation by Single-Molecule Fluorescence Microscopy

Single-molecule fluorescence microscopy is a more conventional method than HS-AFM, but it is still a very useful technique (Iino et al. 2017). To visualize target molecules, samples must be labeled with fluorescence probes. One of the easiest labeling strategies is to make a fusion enzyme with a fluorescent protein (e.g., the green fluorescence protein, GFP). The labeling ratio can be very high, if the linker peptide between the protein and GFP is not cleaved by proteases. But for cellulase analysis, the size of GFP is almost same as the size of the CD. Thus, it has the possibility to inhibit the binding. The other method of fluorescence labeling is conjugation with small dyes. Many kinds of combinations of dyes and functional groups are commercially available. Cy3-maleimide is used for labeling of cellulase, in that a free cysteine residue is introduced far from the expected cellulose-binding surface. Cy3 itself does not bind to cellulose; thus, labeling will not affect the binding properties of cellulase. Because cellulases work outside the cell, an unexpected reaction with an intrinsic free cysteine rarely happens. Therefore, conjugation with a maleimide group is easier for specific labeling than using isothiocyanate and

N-hydroxysuccinimide ester groups that react with amide groups. After labeling with Cy3, the unreacted dye should be completely removed with size exclusion chromatography or ultrafiltration to calculate the labeling ratio of the enzyme. This labeling ratio is needed to normalize the observed k_{on} value. Cellulose crystals were stabilized on glass for observation (usually, cover glass with a thickness of 0.12–0.17 mm is used) by spin coating. The glass is treated with 10 M KOH overnight and carefully washed with ultrapure water to remove fluorescence by debris. If a His6-tag is used for enzyme purification, it should be removed by protease treatment before observation. Histidine residues have positive charges at acidic pH and cause non-specific binding to the glass surface.

Total internal reflection fluorescence microscopy was used to observe the fluorescence from single dye molecules. Illumination by total internal reflected light can excite the molecules just near the glass surface (~200 nm). Therefore, background intensity from unbound molecules is kept low, and only bound molecules on the crystal surface are observed. Observation conditions for k_{on} and k_{off} analysis are 5 fps with 0.14 μW/μm^2 of laser power (532 nm) to visualize the fast dissociation events and long binding events without photobleaching. The photobleaching time of Cy3 in 50 mM sodium acetate buffer, pH 5.0, at 25 °C is 18.8 ± 0.8 s under these observation conditions. For observation of single molecules, 25 to 250 pM of 20 μl enzyme solution is dropped on the glass, and cellulose fibrils are stained by 10 nM labeled enzyme. A movie of binding and dissociation is overlaid with the stained cellulose image using ImageJ software, and the number of bound molecules on each crystal and the binding times of each molecule are counted during the observation time. Lengths of cellulose fibrils are calculated from the length and pixel size in the stained image. The value of k_{on} is defined by the number of bound molecules per concentration of enzyme (M), standard length of the fibril (μm), and analyzed time (s). The value of k_{on} for the target enzyme is estimated from the peak of a Gaussian fit to the distribution of k_{on} for each fibril. The value of k_{off} is obtained by fitting with exponential decays to the distribution of the binding time of the molecules. Movies for the analysis of moving velocities and processivity are recorded at 0.5 fps with 0.28 μW/μm^2 laser power to improve the position precision. The values of precision on the x- and y-axis are 6 to 8 nm, and the photobleaching time is 15.0 ± 0.7 s. Moving distances and times of molecules that move more than three frames along with cellulose are analyzed. Velocity, moving time, and processivity are analyzed with the same method used for HS-AFM observation.

To understand the complete reaction cycle of cellulase, not only the velocity of processive movement (k_{tr}) but also the rate constants of binding (k_{on}) and dissociation (k_{off}) should be analyzed. In HS-AFM observation, the total number of binding molecules is difficult to count, due to the limited field of view and non-specific binding of enzymes to graphite or the hydrophobic, coated mica surface. TrCel7A and TrCel6A have been observed by single-molecule fluorescence microscopy (Shibafuji et al. 2014; Nakamura et al. 2016). The results of the kinetic constant analysis are summarized in the following subsections.

10.4.1 Analysis of the Binding and Dissociation Rate Constants of Cellulase

Distributions of binding constants of both TrCel7A and TrCel6A showed multiple peaks (Fig. 10.6). Peak values are roughly quantized in both cases, due to the bundled forms of fibrils. By HS-AFM observation of the cellulose crystals, bundles of crystals were observed. The width of single crystals of cellulose I_α and III_I that are the different shapes of crystals were 24 nm and 21 nm, respectively. Thus, bundles of cellulose were not resolved by fluorescence microscopy (special resolution is ~ $\lambda/2$), and k_{on} values might show quantized values. First peaks of k_{on} (corresponding to the single crystals) for cellulose I_α of TrCel7A and TrCel6A were $7.3 \times 10^8 \pm 0.3 \times 10^8$ $M^{-1}\mu m^{-1}s^{-1}$ and $7.5 \times 10^8 \pm 0.3 \times 10^8$ $M^{-1}\mu m^{-1}s^{-1}$, respectively. These two

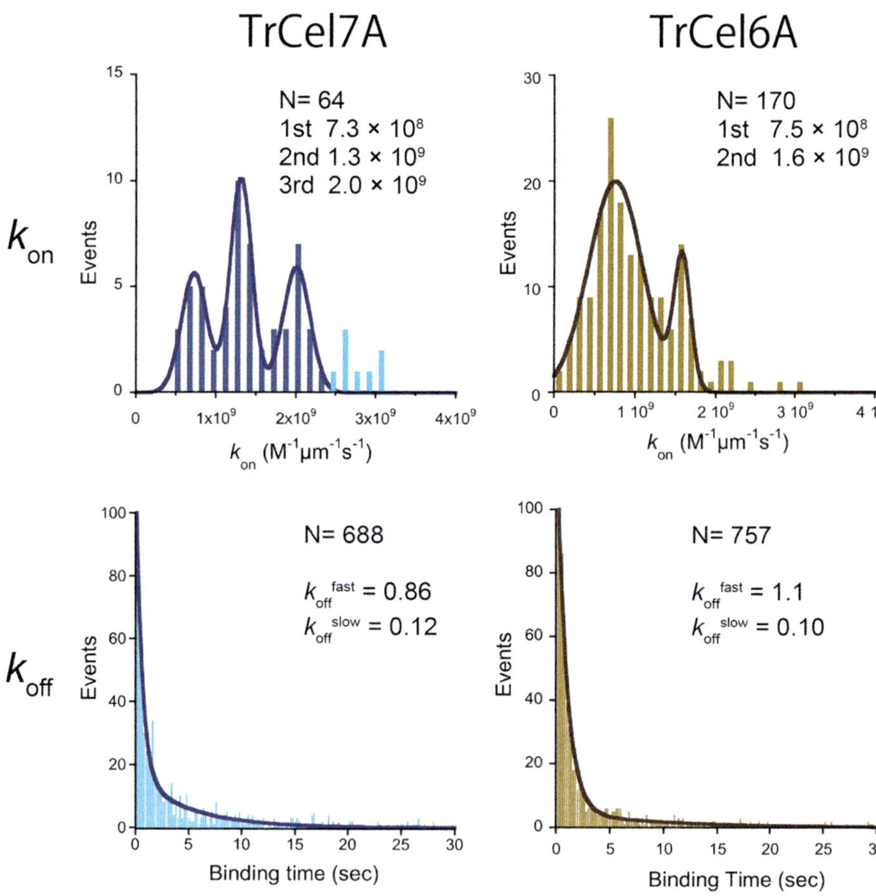

Fig. 10.6 k_{on} and k_{off} of TrCel7A and TrCel6A for cellulose I_α measured by single-molecule fluorescence microscopy

enzymes both consist of the CBD, a linker, and the CD. These similar values of k_{on} represent the similarities of the domain structures. When the k_{on} values of the CD, CBD-linker, and CBD of TrCel6A are compared with the value for the whole structure of TrCel6A, the values of the CD ($5.2 \times 10^7 \pm 0.4 \times 10^7$ $M^{-1}\mu m^{-1}s^{-1}$) and the CBD ($5.6 \times 10^7 \pm 0.2 \times 10^7$ $M^{-1}\mu m^{-1}s^{-1}$) were less than one tenth of the whole enzyme. Although that of the CBD-linker was $3.9 \times 10^8 \pm 0.05 \times 10^8$ $M^{-1}\mu m^{-1}s^{-1}$, that is just a half of the full-length TrCel6A. These results mean that the linker region is very important for the initial interaction with the cellulose surface. The linker region of cellobiohydrolases from fungi is comprised of serine- and threonine-rich sequences and highly O-glycosylated. The role of glycosylation has been thought to be for protection from proteolysis. Recently, the interaction between sugars of the linker and the cellulose crystal has been proposed by molecular dynamics simulations (Payne et al. 2013). The results of the k_{on} analysis by single-molecule fluorescence microscopy are highly consistent with this hypothesis.

Distributions of the binding time for TrCel7A and TrCel6A on cellulose I_α were fitted with a double exponential decay (Fig. 10.6). The values of k_{off} and their ratios of TrCel7A are 0.12 ± 0.01/s (48%) and 0.86 ± 0.03 /s (58%). Those of TrCel6A are 0.10 ± 0.02/s (30%) and 1.1 ± 0.03/s (70%). These results mean that they have at least two binding modes, with different binding energies for the cellulose surface. There are many possibilities for multiple binding modes. Expected reasons are, for example, binding with the CBD and the CD (nonproductive and productive), binding to the crystal region or amorphous region, and differences in the binding surfaces (hydrophobic and hydrophilic). The most probable one is differences at the binding surface. For observation, highly crystalline cellulose, purified from green algae, was used. The crystallinity of cellulose was more than 99%, indicating that binding to the amorphous region is rare. Additionally, only the CBD showed two binding modes with similar k_{off} values and ratios to the whole protein, meaning the double domain structure is not the main reason.

10.4.2 Analysis of Velocity and Processivity of Cellulase by Single-Molecule Fluorescence Microscopy

TrCel6A had been thought of as a processive cellulase from biochemical assays and electron microscopy observation, but processive movement was not found in HS-AFM observation. Therefore, the processive movement of TrCel6A was monitored with improved position precision. As a result, molecules moving along with cellulose fibrils were found, and the distribution of velocities was fitted by two Gaussian functions (peaks at 8.8 ± 5.5 nm/s and 34.9 ± 16.3 nm/s), as shown in Fig. 10.7. To determine the velocity of processive movement, an inactive mutant of TrCel6A (D221A) was also observed in the same way. The distribution of velocities of the inactive mutant showed a single peak at 35.5 ± 17.4 nm/s. Moving times of fast-moving TrCel6A and inactive mutant are almost less than 10 s, but that of

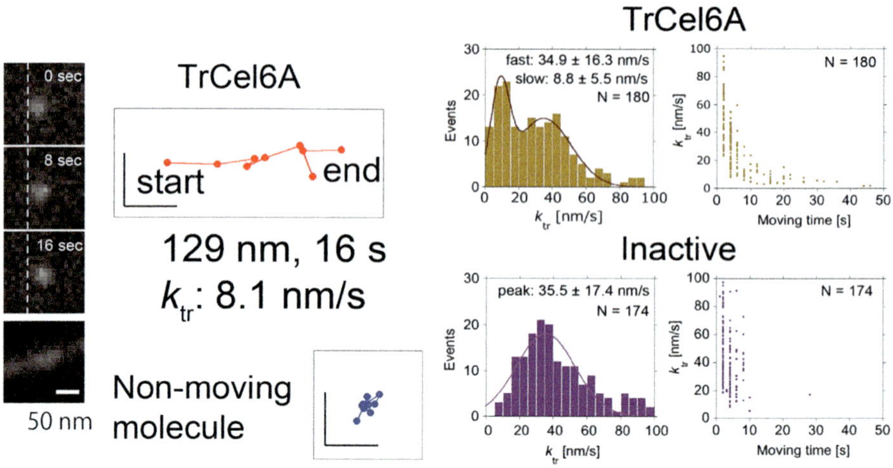

Fig. 10.7 Trajectory and velocity of TrCel6A measured by single-molecule fluorescence microscopy

slowly-moving TrCel6A was longer than 10 s. Distributions of moving times of long movement were fitted with a single exponential decay with a time constant of 7.7 s. When the step size was estimated as 1 nm, the processivity was calculated as 68 ± 42. Furthermore, the distribution of velocity of the CD was fitted with a Gaussian distribution with peak at 39.2 ± 28.3 nm/s and moving times lower than 10 s, even though it was still active. From these results, it was shown that TrCel6A moves processively at 8.8 nm/s for 68 times, and the CBD-linker increases the processivity in the interaction with cellulose.

10.5 Conclusions and Future Directions

In this chapter, visualization of protonation states and the proton pathway of cellulase by neutron crystallography and visualization of kinetic elementary steps of processive cellulases by HS-AFM and single-molecule fluorescence microscopy were introduced. These methods are very useful to directly analyze the functional structure and dynamic motion of a cellulase on the surface of cellulose.

One of the good targets for the future study of structural analysis of cellulases is the catalytic mechanism of GH6 enzymes. TrCel6A is the most historical cellulase, but its catalytic mechanism is still under discussion, because the possible catalytic base is far from the putative catalytic acid residue. To determine the catalytic base, protonation state analysis by neutron crystallography is helpful. Another useful method is the characterization of processive cellobiohydrolases, as linear molecular motors. In the case of other linear motors, like kinesin-1 and myosin, distribution of step sizes and dwell times during pauses have been reported (Svoboda et al. 1993;

Kitamura et al. 1999), and their moving mechanisms have been discussed. HS-AFM could not resolve the single steps of TrCel7A, because the frame rate is still not fast enough. The expected step size of cellobiohydrolases is 1 nm, and the step size of TrCel7A was reported as 1.3 ± 0.7 nm, using optical tweezers with 1.25 μm polystyrene beads (Brady et al. 2015). The moving power of TrCel7A was shown as more than 20 pN, but the observed velocities of TrCel7A were less than one tenth of that in HS-AFM observation. Thus, it is difficult to discuss kinetics and moving mechanisms. For further analysis of the moving mechanism of cellulases, single-molecule observation with higher precision and time resolution and lower inhibition to movement, e.g., using a gold nanoparticle as a probe (Isojima et al. 2016), is needed. Furthermore, the difference between cellulase and chitinase is interesting. Chitinase A from a bacteria *Serratia marcescens* (SmChiA) hydrolyzes and moves on chitin crystals that have a similar structure to cellulose. The k_{on} and k_{off} of SmChiA were both about 3 times larger than those of TrCel7A, indicating a similar (K_d) (Nakamura et al. 2018). However, a moving velocity of 70.5 nm/s was observed in SmChiA using HS-AFM that was approximately ten times faster than that of TrCel7A (Igarashi et al. 2014). When the moving mechanism of cellulase is revealed, the big difference in moving velocities will be explained. At last, the synergistic effect between GH7 and GH6 cellobiohydrolases that move in opposite directions of each other is the most mysterious topic. Localization and movement of each enzyme should be analyzed at the same time using multicolor imaging techniques.

Acknowledgments This study was supported by Grants-in-Aid for Scientific Research from the Ministry of Education, Culture, Sports, Science, and Technology, Japan [grant numbers JP17K19213, JP16H00789, JP16H00858, and JP15H04366 to R. I., JP15H06898, JP17K18429, and JP17H05899 to A. N.], Advanced Technology Institute Research Grants 2015 (RG2709 to A.N.).

References

Ando T (2013) High-speed atomic force microscopy (AFM). In: Encyclopedia of biophysics. vol Chapter 478. Springer, Berlin/Heidelberg/Berlin/Heidelberg, pp 984–987. https://doi.org/10.1007/978-3-642-16712-6_478

Ando T (2017) Directly watching biomolecules in action by high-speed atomic force microscopy. Biophys Rev 9(4):421–429. https://doi.org/10.1007/s12551-017-0281-7

Brady SK, Sreelatha S, Feng Y, Chundawat SPS, Lang MJ (2015) Cellobiohydrolase 1 from Trichoderma reesei degrades cellulose in single cellobiose steps. Nat Commun 6:10149. https://doi.org/10.1038/ncomms10149

Chanzy H, Henrissat B (1985) Undirectional degradation of valonia cellulose microcrystals subjected to cellulase action. FEBS Lett 184(2):285–288. https://doi.org/10.1016/0014-5793(85)80623-2

Davies G, Henrissat B (1995) Structures and mechanisms of glycosyl hydrolases. Structure 3(9):853–859. https://doi.org/10.1016/S0969-2126(01)00220-9

Igarashi K, Koivula A, Wada M, Kimura S, Penttila M, Samejima M (2009) High speed atomic force microscopy visualizes processive movement of Trichoderma reesei cellobiohydrolase I on crystalline cellulose. J Biol Chem 284(52):36186–36190. https://doi.org/10.1074/jbc.M109.034611

Igarashi K, Uchihashi T, Koivula A, Wada M, Kimura S, Okamoto T, Penttila M, Ando T, Samejima M (2011) Traffic jams reduce hydrolytic efficiency of cellulase on cellulose surface. Science 333(6047):1279–1282. https://doi.org/10.1126/science.1208386

Igarashi K, Uchihashi T, Uchiyama T, Sugimoto H, Wada M, Suzuki K, Sakuda S, Ando T, Watanabe T, Samejima M (2014) Two-way traffic of glycoside hydrolase family 18 processive chitinases on crystalline chitin. Nat Commun 5:1–7. https://doi.org/10.1038/ncomms4975

Iino R, Iida T, Nakamura A, E-i S, You H, Sako Y (2017) Single-molecule imaging and manipulation of biomolecular machines and systems. Biochim Biophys Acta S0304-4165(17):30253–30252. https://doi.org/10.1016/j.bbagen.2017.08.008

Imai T, Boisset C, Samejima M, Igarashi K, Sugiyama J (1998) Unidirectional processive action of cellobiohydrolase Cel7A on Valonia cellulose microcrystals. FEBS Lett 432(3):113–116. https://doi.org/10.1016/S0014-5793(98)00845-X

Isojima H, Iino R, Niitani Y, Noji H, Tomishige M (2016) Direct observation of intermediate states during the stepping motion of kinesin-1. Nat Chem Biol 12(4):290–297. https://doi.org/10.1038/nchembio.2028

Jalak J, Kurasin M, Teugjas H, Valjamae P (2012) Endo-exo synergism in cellulose hydrolysis revisited. J Biol Chem 287(34):28802–28815. https://doi.org/10.1074/jbc.M112.381624

Julliand V, de Vaux A, Millet L, Fonty G (1999) Identification of Ruminococcus flavefaciens as the predominant cellulolytic bacterial species of the equine cecum. Appl Environ Microbiol 65(8):3738–3741

Kitamura K, Tokunaga M, Iwane AH, Yanagida T (1999) A single myosin head moves along an actin filament with regular steps of 5.3 nanometres. Nature 397(6715):129–134. https://doi.org/10.1038/16403

Kodera N, Uchida K, Ando T, Aizawa S-I (2015) Two-ball structure of the flagellar hook-length control protein FliK as revealed by high-speed atomic force microscopy. J Mol Biol 427(2):406–414. https://doi.org/10.1016/j.jmb.2014.11.007

Kodera N, Yamamoto D, Ishikawa R, Ando T (2010) Video imaging of walking myosin V by high-speed atomic force microscopy. Nature 468(7320):72–76. https://doi.org/10.1038/nature09450

Kusaka K, Hosoya T, Yamada T, Tomoyori K, Ohhara T, Katagiri M, Kurihara K, Tanaka I, Niimura N (2013) Evaluation of performance for IBARAKI biological crystal diffractometer iBIX with new detectors. J Synchrotron Radiat 20(6):994–998. https://doi.org/10.1107/S0909049513021845

Leschine SB (1995) Cellulose degradation in anaerobic environments. Annu Rev Microbiol 49(1):399–426. https://doi.org/10.1146/annurev.mi.49.100195.002151

Lombard V, Ramulu HG, Drula E, Coutinho PM, Henrissat B (2014) The carbohydrate-active enzymes database (CAZy) in 2013. Nucleic Acids Res 42(D1):D490–D495. https://doi.org/10.1093/nar/gkt1178

McCarter JD, Stephen Withers G (1994) Mechanisms of enzymatic glycoside hydrolysis. Curr Opin Struct Biol 4(6):885–892. https://doi.org/10.1016/0959-440X(94)90271-2

McMaster WH, Del Grande NK, Mallett JH, Hubbell JH (1969) Compilation of X-ray cross sections. Lawrence Livermore National Laboratory Report UCRL-50174

Nakamura A, Ishida T, Fushinobu S, Kusaka K, Tanaka I, Inaka K, Higuchi Y, Masaki M, Ohta K, Kaneko S, Niimura N, Igarashi K, Samejima M (2013a) Phase-diagram-guided method for growth of a large crystal of glycoside hydrolase family 45 inverting cellulase suitable for neutron structural analysis. J Synchrotron Radiat 20(6):859–863. https://doi.org/10.1107/S0909049513020943

Nakamura A, Ishida T, Kusaka K, Yamada T, Fushinobu S, Tanaka I, Kaneko S, Ohta K, Tanaka H, Inaka K, Higuchi Y, Niimura N, Samejima M, Igarashi K (2015) "Newton's cradle" proton relay with amide-imidic acid tautomerization in inverting cellulase visualized by neutron crystallography. Sci Adv 1(7):e1500263–e1500263. https://doi.org/10.1126/sciadv.1500263

Nakamura A, Tasaki T, Ishiwata D, Yamamoto M, Okuni Y, Visootsat A, Maximilien M, Noji H, Uchiyama T, Samejima M, Igarashi K, Iino R (2016) Single-molecule imaging analysis of binding, processive Movement, and dissociation of cellobiohydrolase Trichoderma reesei

Cel6A and its domains on crystalline cellulose. J Biol Chem 291(43):22404–22413. https://doi.org/10.1074/jbc.M116.752048

Nakamura A, Tasaki T, Okuni Y, Song C, Murata K, Kozai T, Hara M, Sugimoto H, Suzuki K, Watanabe T, Uchihashi T, Noji H, Iino R (2018) Rate constants, processivity, and productive binding ratio of chitinase A revealed by single-molecule analysis. Phys Chem Chem Phys 20(5):3010–3018. https://doi.org/10.1039/c7cp04606e

Nakamura A, Tsukada T, Auer S, Furuta T, Wada M, Koivula A, Igarashi K, Samejima M (2013b) The tryptophan residue at the active site tunnel entrance of Trichoderma reesei cellobiohydrolase Cel7A is important for initiation of degradation of crystalline cellulose. J Biol Chem 288(19):13503–13510. https://doi.org/10.1074/jbc.M113.452623

Nakamura A, Watanabe H, Ishida T, Uchihashi T, Wada M, Ando T, Igarashi K, Samejima M (2014) Trade-off between processivity and hydrolytic velocity of cellobiohydrolases at the surface of crystalline cellulose. J Am Chem Soc 136(12):4584–4592. https://doi.org/10.1021/ja4119994

Ogata H, Nishikawa K, Lubitz W (2015) Hydrogens detected by subatomic resolution protein crystallography in a [NiFe] hydrogenase. Nature 520(7548):571–574. https://doi.org/10.1038/nature14110

Payne CM, Resch MG, Chen L, Crowley MF, Himmel ME, Taylor LE II, Sandgren M, Ståhlberg J, Stals I, Tan Z, Beckham GT (2013) Glycosylated linkers in multimodular lignocellulose-degrading enzymes dynamically bind to cellulose. Proc Natl Acad Sci U S A 110(36):14646–14651. https://doi.org/10.1073/pnas.1309106110

Shibafuji Y, Nakamura A, Uchihashi T, Sugimoto N, Fukuda S, Watanabe H, Masahiro S, Toshio A, Hiroyuki N, Koivula A, Kiyohiko I, Ryota I (2014) Single-molecule imaging analysis of elementary reaction steps of Trichoderma Reesei cellobiohydrolase I (Cel7A) hydrolyzing crystalline cellulose Iα and IIII. J Biol Chem 289(20):14056–14065. https://doi.org/10.1074/jbc.M113.546085

Shibata M, Yamashita H, Uchihashi T, Kandori H, Ando T (2010) High-speed atomic force microscopy shows dynamic molecular processes in photoactivated bacteriorhodopsin. Nat Nanotechnol 5(3):208–212. https://doi.org/10.1038/nnano.2010.7

Štursová M, Žifčáková L, Leigh MB, Burgess R, Baldrian P (2012) Cellulose utilization in forest litter and soil: identification of bacterial and fungal decomposers. FEMS Microbiol Ecol 80(3):735–746. https://doi.org/10.1111/j.1574-6941.2012.01343.x

Svoboda K, Schmidt CF, Schnapp BJ, Block SM (1993) Direct observation of kinesin stepping by optical trapping interferometry. Nature 365(6448):721–727. https://doi.org/10.1038/365721a0

Uchihashi T, Iino R, Ando T, Noji H (2011) High-speed atomic force microscopy reveals rotary catalysis of rotorless F1-ATPase. Science 333(6043):755–758. https://doi.org/10.1063/1.115365

Varley FS (1992) Neutron scattering lengths and cross section. Neutron News 3(3):26–37. https://doi.org/10.1080/10448639208218770

Wood RKS (1960) Pectic and cellulolytic enzymes in plant disease. Annu Rev Plant Physiol 11(1):299–322. https://doi.org/10.1146/annurev.pp.11.060160.001503

Chapter 11
Structure and Dynamics of Immunoglobulin G Glycoproteins

Hirokazu Yagi, Saeko Yanaka, and Koichi Kato

Abstract Immunoglobulin G (IgG) is a major serum glycoprotein that exerts the role of antibody in the immune system. This multifunctional glycoprotein couples antigen recognition with a variety of effector functions promoted *via* interactions with various IgG-binding proteins. Given its versatile functionality, IgG has recently been used for therapeutic interventions. Evidence indicates that the carbohydrate moieties of IgG glycoproteins critically affect their antibody functions, particularly the effector functions mediated by the interactions with Fcγ receptors (FcγRs). *N*-glycans at specific positions of FcγR also contribute both positively and negatively to the interactions with IgG. The integration of multilateral biophysical approaches, including X-ray crystallography, nuclear magnetic resonance spectroscopy, and molecular dynamics simulations, has provided structural insights into the mechanisms underlying the glycofunctions of this interacting system. The *N*-glycans of IgG and FcγR mediate their interactions by either strengthening or weakening the affinity on the basis of their glycoforms. Moreover, the *N*-glycosylation of IgG-Fc is a prerequisite to maintain the integrity of the quaternary structure of the sites interacting with the effector molecules and can also control functionally relevant local conformations. The biopharmaceutical significance of these glycan functions is discussed from a structural point of view.

H. Yagi
Graduate School of Pharmaceutical Sciences, Nagoya City University, Nagoya, Japan

S. Yanaka
Exploratory Research Center on Life and Living Systems (ExCELLS) and Institute for Molecular Science, National Institutes of Natural Sciences, Okazaki, Aichi, Japan

K. Kato (✉)
Exploratory Research Center on Life and Living Systems,
National Institutes of Natural Sciences, Okazaki, Aichi, Japan
e-mail: kkato@phar.nagoya-cu.ac.jp

Y. Yamaguchi, K. Kato (eds.), *Glycobiophysics*, Advances in Experimental Medicine and Biology 1104, https://doi.org/10.1007/978-981-13-2158-0_11

Keywords Immunoglobulin G · Therapeutic antibody · Fcγ receptor · *N*-linked oligosaccharide · X-ray crystallography · Nuclear magnetic resonance spectroscopy · Stable isotope labeling

11.1 Introduction

Many biopharmaceuticals, including therapeutic antibodies, are modified with glycans, which exert an important contribution to their physicochemical properties, such as water solubility and thermal stability, and govern their therapeutic efficacy by influencing bioactivity, pharmacokinetics, and allergenicity (Solá and Griebenow 2011; Strohl 2015; Lalonde and Durocher 2017). Therefore, detailed insights into the glycoprotein structure, including the carbohydrate moieties, will allow the development of biopharmaceutical glycoproteins. The carbohydrate moieties of glycoproteins present the microheterogeneities that is determined by the presence or absence of terminal sugar units and exhibit the conformational fluctuations generated by the motional freedom of glycosidic linkages (Kato et al. 2018). Although such structural complexity, diversity, and fluctuation hamper a structural biology approach, detailed structural data have recently been accumulated using X-ray crystallography, nuclear magnetic resonance (NMR) spectroscopy, and other biophysical methods, thus allowing us to understand the functional mechanisms of immunoglobulin G (IgG) glycoproteins. This article outlines the structural biophysics approaches used to characterize the structure, dynamics, and interactions of IgG as a model biopharmaceutical glycoprotein.

11.2 The Structure of the IgG Glycoprotein

IgG forms a "Y"-shaped structure composed of two identical heavy chains and two identical light chains, each of which comprises two distinct regions: variable (V) and constant (C) regions (Fig. 11.1A). The light chain consists of two domains, namely, V_L and C_L, whereas the heavy chain is divided into four domains: V_H, C_H1, C_H2, and C_H3. The antigen recognition of IgG is performed by its Fab regions, specifically the V_L and V_H domains. These correspond to the top ends of the fork of the Y-shaped structure, whereas the Fc region comprises the C_H2, and C_H3 domains corresponding to the stem of the Y shape. The latter interacts with the complement component C1q and Fcγ receptors (FcγRs) of the immune cells to express its effector functions. Upon interaction with the hexavalent C1q, IgG forms a hexameric structure, whereas 1:1 binding to FcγR causes Fc to undergo an asymmetric conformational change (Sondermann et al. 2000; Kato et al. 2000; Yogo et al. 2017; Ugurlar et al. 2018).

Although IgG is classified into several subclasses (IgG1–4 in human and IgG1, IgG2a, IgG2b, and IgG3 in mouse), the Fc regions of each subclass share one conserved *N*-glycosylation site (Asn297) in each C_H2 domain, which displays a

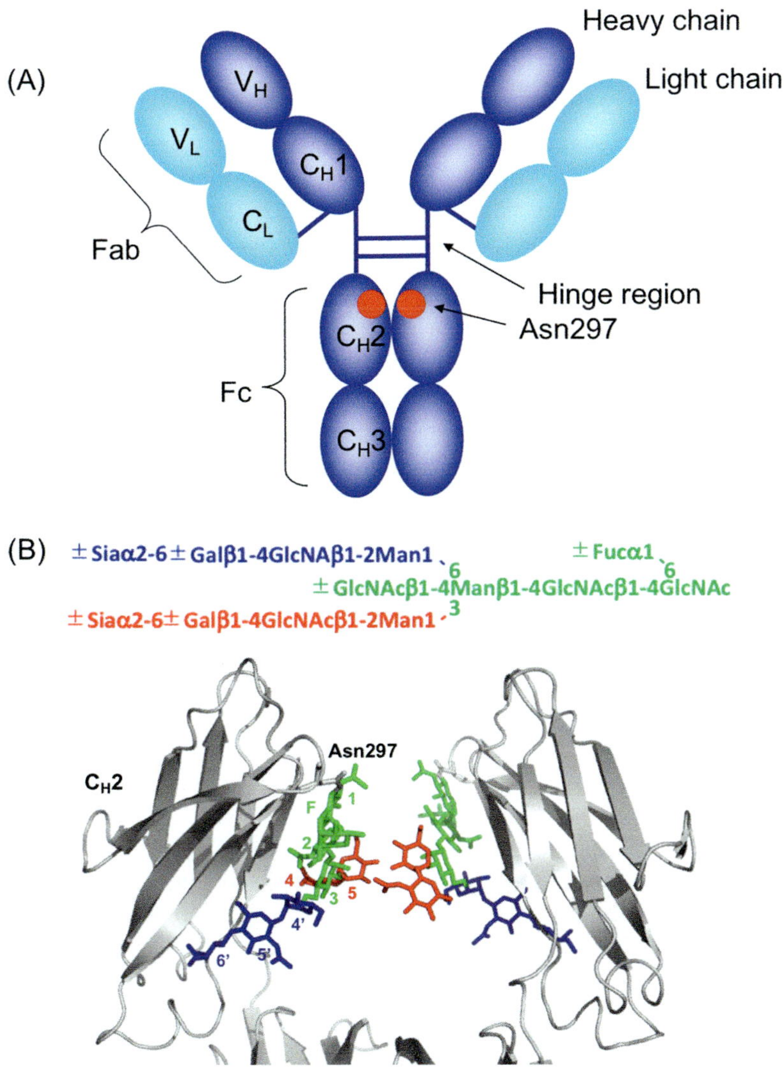

Fig. 11.1 Basic structure of IgG. (**A**) Schematic representation of IgG, which consists of two heavy chains (blue) and two light chains (cyan). *N*-glycosylation sites are indicated by a red circle. (**B**) Crystal structure of human IgG1-Fc. The oligosaccharides attached to Asn297 are colored in green (core), blue (Manα1-6 arm), and red (Manα1-3 arm). Key: galactose, Gal; mannose, Man; fucose, Fuc; *N*-acetylglucosamine, GlcNAc; Sialic acid, Sia

biantennary complex-type oligosaccharide. The pair of *N*-glycans is packed between the two C_H2 domains (Deisenhofer 1981) and exhibits a microheterogeneity that results from the presence or absence of the nonreducing terminal fucose, galactose, and sialic acid residues (Fig. 11.1B) (Mizuochi et al. 1982). The glycosylation profiles of IgG-Fc depend on the species, age, pathological status, and culture conditions of the IgG-producing cells. The profiles of IgG glycosylation can be altered by the occurrence and progression of many inflammatory diseases, such as rheumatoid arthritis and Wegener's granulomatosis (Parekh et al. 1985; Mizuochi et al. 1990; Holland et al. 2006; Shade and Anthony 2013; Ercan et al. 2017). The Manα1-6 branch contacts the inner surface of the C_H2 domain, whereas the Manα1-3 branch projects into the cavity space and maintains motional freedom (Fig. 11.1B) (Wright and Morrison 1993; Yamaguchi et al. 1998; Jo et al. 2018).

By contrast, the Fab V region occasionally expresses nonconserved *N*-glycans (carried by ~20% of human IgG), which are more exposed to solvent; therefore, this region undergoes higher degrees of sialylation than Fc glycans (Holland et al. 2006). Fab glycosylation can affect antigen binding according to glycan positions (van de Bovenkamp et al. 2016).

The Fab and Fc regions are separated by a hinge region, which harbors a disulfide-linked core part, and flanked by flexible segments (Kim et al. 1994a). *O*-glycans have been identified in the Fab-proximal flexible segments in the hinge region of human IgG3 (Plomp et al. 2015), mouse IgG2b (Kim et al. 1994b), and a mouse IgG2a variant that lacks the entire C_H1 domain (Masuda et al. 1999). The remaining classes of antibodies (IgA, IgD, IgE, and IgM) share the conserved *N*-glycosylation sites (corresponding to Asn297 in IgG) and additional *N*- and *O*-glycosylation sites (Yamaguchi et al. 2007; Shade and Anthony 2013).

The crystallographic data of the entire IgG molecule remain limited because the flexible nature of the hinge region hampers crystallization, unlike the cumulative crystal structures of the Fab and Fc fragments in complexes with their ligands (Yamaguchi et al. 2007). To date, only four crystal structures have provided entire views of IgG structure: the PDB codes 1igt, 1igy, 1hzh, and 5dk3 for mouse IgG1, mouse IgG2a, human IgG1, and humanized IgG4 structures, respectively (Harris et al. 1997, 1998; Saphire et al. 2001; Scapin et al. 2015). In all these cases, IgG molecules are packed into the crystal lattice by extensive intermolecular interactions that exhibit distinct orientations of the Fc stem and Fab arms. In the humanized IgG4 structure, one C_H2 domain flips, thus exposing the glycan (Scapin et al. 2015). This unique conformational snapshot may be driven by the shorter hinge. The dynamic structure of a full-length IgG has been characterized by biophysical techniques, including NMR spectroscopy, electron microscopy, and solution scattering (Kim et al. 1994a; Roux et al. 1997; Eryilmaz et al. 2013; Zhang et al. 2015). All these studies have highlighted the flexible nature of the hinge regions as a source of the internal motion of the IgG; therefore, these regions yield varying orientations of the Fab arms as well as the Fc stem.

11.3 Impact of Glycosylation on the IgG Structure

It has long been recognized that the *N*-glycosylation of IgG influences its Fc-mediated effector functions (Nose and Wigzell 1983; Leatherbarrow et al. 1985; Burton 1985; Rademacher et al. 1988; Tao and Morrison 1989; Kiyoshi and Akira

1991; Yamaguchi et al. 2007; Dekkers et al. 2017). Aglycosyl IgG exhibits impaired complement activation because of a reduced affinity for C1q. Similarly, IgG-Fc deglycosylation results in a loss of affinity for FcγRs. The primary binding sites of these effector molecules are provided by the pair of C_H2 domains, which surround the *N*-glycans and exhibit higher motion amplitude than the C_H3 domains, as indicated by molecular dynamics (MD) simulation (Frank et al. 2014). The crystal structures of human IgG1-Fc, which contain a series of truncated glycoforms, have indicated that glycan trimming affects the overall quaternary structure of Fc *via* a conformational alteration of the C_H2–C_H3 interdomain angles (Fig. 11.2A) (Krapp et al. 2003), which correspond to the maximum C_H2–C_H2 distance in the digalactosyl glycoform and the minimum C_H2–C_H2 distance in a truncated glycoform, with a single mannose residue. However, a crystal structure of deglycosyl IgG1-Fc (3dkn) has been determined, and this structure exhibits an open conformation compared with the fully glycosylated IgG1-Fc (3ave) (Fig. 11.2B) (Matsumiya et al. 2007; Kiyoshi et al. 2017). Small-angle X-ray scattering and single-molecule Förster resonance energy transfer data have also shown that deglycosylated Fc exhibits an open-form structure in terms of C_H2–C_H2 distance compared with the closed conformation presented by glycosylated Fc (Borrok et al. 2012; Ju et al. 2015). In the hinge-proximal C_H2 domains, deglycosylation-coupled conformational alterations have been identified by stable isotope-assisted NMR methods (*vide infra*) (Yamaguchi et al. 2006). Considering that C_H2 domains provide C1q and FcγRs bindings sites, the deglycosylation-coupled dislocation of the C_H2 domains explains the affinity reduction for those effector molecules (Leatherbarrow et al. 1985; Tao and Morrison 1989; Jung et al. 2010; Ju et al. 2015).Therefore, the *N*-glycans of Fc contribute to the structural integrity of the functional sites of IgG.

The interaction between IgE and its high-affinity Fc receptor, namely, FcεRI, also requires the *N*-glycans at Asn 394 of IgE-Fc (corresponding to Asn297 of IgG) (Nettleton and Kochan 1995; Plomp et al. 2014). This suggests the conserved role of *N*-glycans in maintaining the structural integrity of the receptor-binding sites despite the high-mannose type of the *N*-glycans at this IgE site (Plomp et al. 2014). In chicken IgG (often called IgY), the major glycan displayed at the conserved *N*-glycosylation site is a monoglucosyl high-mannose-type oligosaccharide irrespective of the production vehicles (Suzuki and Lee 2004; Kondo et al. 2012). Although the mechanisms behind the formation of such specific glycoform remain largely unknown, some structural properties of glycoproteins are thought to comprise critical determining factors. Recently, aglycosyl IgG-Fc mutants exhibiting higher binding affinity for FcγRIIIa have been developed by high-throughput screening, and these mutants might exhibit a C_H2–C_H2 close conformer (Jo et al. 2018).

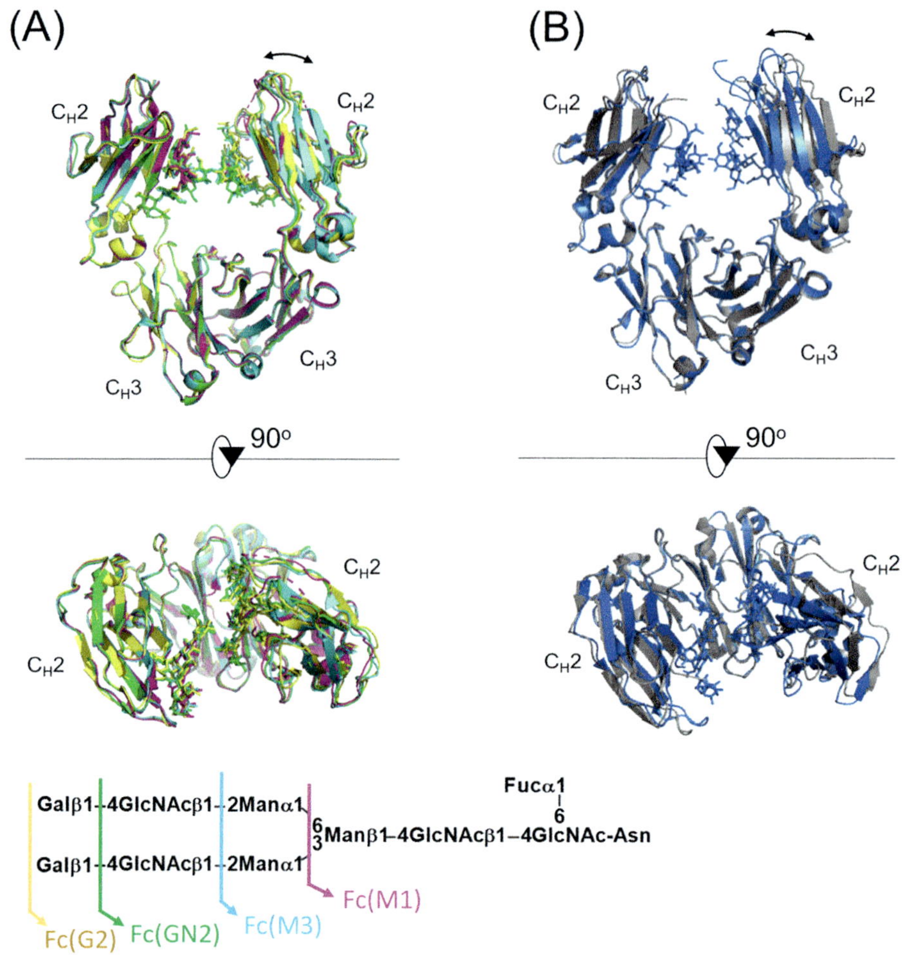

Fig. 11.2 Effects of the *N*-glycan truncation on IgG1-Fc crystal structure. (**A**) Superposition of the four crystal structures of IgG1-Fc fragments with different glycoforms (G2, 1h3v, yellow; GN2, 1h3x, green; M3, 1h3u, cyan; M1, 1h3t, magenta). The enzymatic trimming giving rise to the truncated Fc glycoforms used in the crystal graphic study is shown at the bottom. (**B**) Superposition of the crystal structures of glycosylated (3ave, blue) and nonglycosylated (3dkn, gray) IgG1-Fc fragments

11.4 NMR Characterization of Glycoform-Dependent IgG Structures

NMR offers a sensitive tool for characterizing the structure of IgG-Fc with various glycoforms under varying pH and temperature conditions. Stable isotope-assisted observation is required for the detailed NMR analyses of proteins in a solution, and it can be obtained by the labeling of target proteins with ^{13}C and ^{15}N. Typically, this is achieved by using *Escherichia coli* and cell-free expression systems (Ikeya et al.

2018). However, bacterial expression systems cannot produce proteins with functional glycosylation. This limitation has been overcome by a recently developed methodology consisting of the metabolic isotope labeling of glycoproteins by using mammalian cells (Kato et al. 2010, 2018). This method has enabled the preparation of uniformly $^{13}C/^{15}N$-labeled mouse IgG2b and recombinant mouse/human chimeric IgG1 glycoproteins by cell cultivation in synthetic media containing a mixture of $^{15}N/^{13}C$-labeled algal amino acids supplemented with isotope-labeled metabolic precursors.

The assignments of NMR signals originating from the backbone of human IgG1 Fc fragment with a molecular mass of 53 kDa have been almost completely accomplished (Fig. 11.3A) (Yagi et al. 2015c), thus providing useful probes for the site-specific detection of structural disturbances by molecular interactions, amino acid substitutions, and glycan modification. Intriguingly, a cluster of amino acid residues (Gln295–Thr299) near the Asn297-linked *N*-glycan exhibits significant differences in terms of chemical shift compared with the nonglycosylated Fc protein produced by *E. coli* (Liu et al. 2007). This reflects the conformational alteration of the C_H2 domains resulting from an impaired effector function of IgG (*vide supra*). By using NMR peaks as spectroscopic probes, the conformational disturbances of the systematic *N*-glycan truncation were mapped on the IgG-Fc structure (Yamaguchi et al. 2006; Kato et al. 2010).

Furthermore, NMR spectroscopy can be used to validate the structural integrity of therapeutic antibodies produced by heterologous vehicles (Loos and Steinkellner 2012; Yagi et al. 2015a, b). The ^1H-^{15}N HSQC spectra of the Fc fragment of human IgG1 expressed by tobacco and silkworm as production vehicles are shown in Fig. 11.3B and D. The preparation of stable isotope-labeled antibodies produced by tobacco plants was performed by culturing the plants for 1.5 months with the isotopically labeled Murashige and Skoog medium containing potassium nitrate ($K^{15}NO_3$) and ammonium nitrate ($^{15}NH_4{}^{15}NO_3$) as nitrogen sources. The production of stable isotope-labeled antibodies using silkworms was performed by rearing larvae infected with baculovirus by using an artificial diet containing yeast-derived ^{15}N-labeled protein and then harvesting the labeled antibodies from the body fluids. The antibody's expression efficiency was enhanced by the coexpression of molecular chaperones. By comparing the NMR spectrum of human IgG1-Fc produced by CHO cells with those produced by tobacco and silkworms, chemical shift differences were observed for the peaks originating from the amino acid residues in the spatial vicinity of the sugar chains (Fig. 11.3C and E). Unlike CHO cell-expressed IgG-Fc, the plant-derived IgG1-Fc exhibits α1,3-linked fucose and β1,2-linked xylose residues on the $Man_3GlcNAc_2$ core, whereas the insect cell-derived IgG-Fc expresses paucimannose-type glycans as the major components. Therefore, the observed spectral differences could be attributed to the microenvironmental changes caused by the distinct glycosylation profiles.

A stable isotope-assisted NMR approach is useful for selectively observing specific component(s) in an interaction system. Therefore, the NMR technique has been used to detect the semispecific interactions of IgG with serum proteins, including polyclonal IgG (Yanaka et al. 2017, 2018).

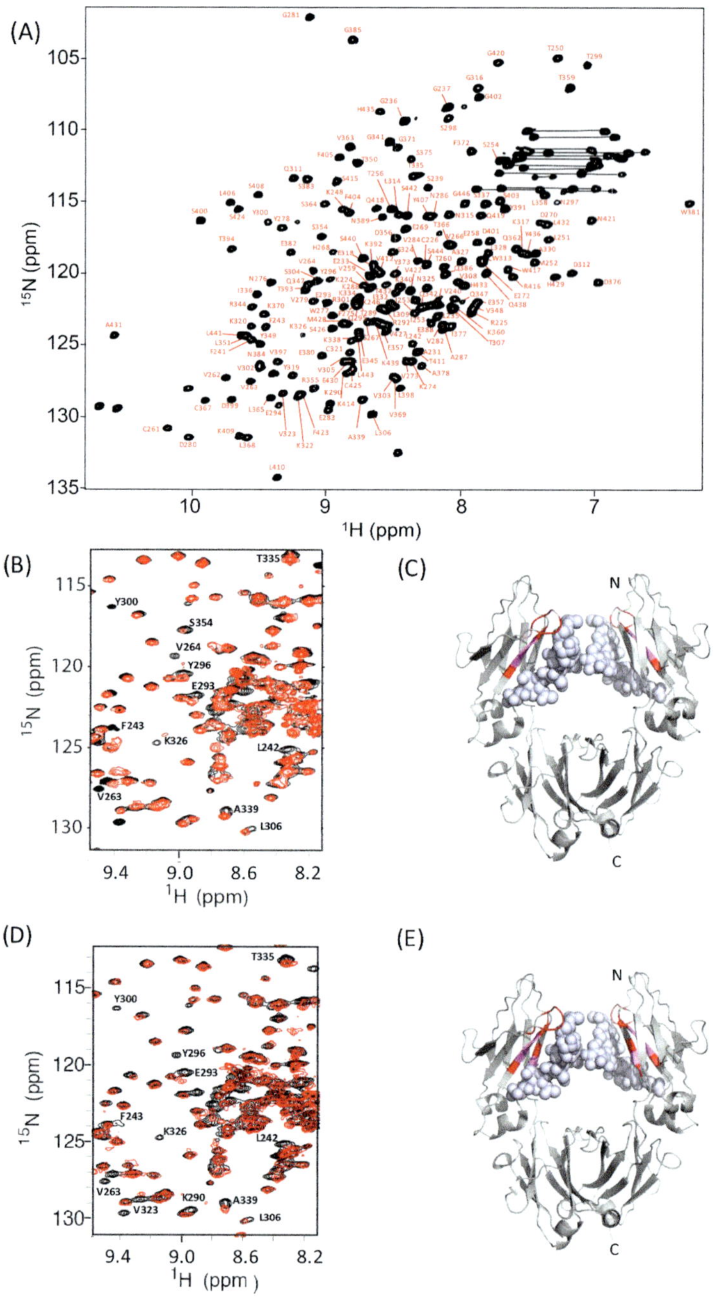

Fig. 11.3 ^{1}H-^{15}N HSQC spectra of isotope-labeled IgG1-Fc obtained from (**A**) CHO cells, (**B**) transgenic tobacco plants, and (**D**) silkworms. In **B** and **D**, the NMR spectrum of IgG-Fc glycoprotein expressed by tobacco or silkworm (red) was superposed on that of the Fc derived from CHO cells (black). In **C** and **E**, mapping on the crystal structure of human IgG1-Fc (PDB code: 3ave) of

11.5 Functional Roles of the *N*-Glycans of Fc Receptors in Their Interactions with IgG

FcγRs possess an extracellular region composed of Ig-like domains and can be categorized into three classes: FcγRI, FcγRII, and FcγRIII. The three classes are differentially expressed by various immune cells with different binding affinities for each IgG subclass (Hayes et al. 2014, 2016). In humans, each FcγR class shows structural variations resulting from multigenes, alternative splicing, and genetic polymorphisms (Fridman et al. 1993). Although FcγRIIb acts as an inhibitory receptor, FcγRI, FcγRIIa, FcγRIIIa, and FcγRIIIb act as activating receptors and induce effector activities on innate effector cells, such as macrophages and natural killer (NK) cells. Neonatal Fc receptor (FcRn) is another class of Fc receptor, which assumes an MHC-like structure and is expressed by the epithelial cells of the human mammary gland and regulates the serum half-life of IgGs (Kacskovics et al. 2006). All these receptor molecules are also *N*-glycosylated, and their interactions with IgG-Fc are influenced by the *N*-glycosylations of either or both glycoproteins (Hayes et al. 2014, 2016).

FcγRIII possess four to six *N*-glycosylation sites depending on the isoforms and genetic polymorphisms. The functional roles of individual *N*-glycans have been investigated using the soluble form of FcγRIIIa (sFcγRIIIa) produced in HEK293T and CHO cells (Shibata-Koyama et al. 2009; Zeck et al. 2011). The results indicate that Asn45 and Asn162 display a biantennary complex-type glycans with partial sialylation, which, respectively, negatively and positively affect the affinity for human IgG1. However, an earlier study indicated that the endogenous FcγRIII exhibits cell-type-specific glycoforms judging from its distinct lectin-binding properties (Edberg and Kimberly 1997). A recently reported glycosylation profile of the endogenous FcγRIIIa on NK cells has shown to express a large proportion of hybrid and oligomannose *N*-glycans (45%), in contrast to those of recombinant FcγRIIIa (Patel et al. 2018). Furthermore, the site-specific *N*-glycosylation profiling of sFcγRIIIb isolated from human serum has indicated that among the six *N*-glycosylation sites, Asn45 was exclusively occupied by high-mannose-type oligosaccharides; the remaining sites were modified by the sialyl complex-type oligosaccharides with fucosylation (Yagi et al. 2018). In this receptor, the major glycoforms at Asn162 are partially sialylated tri-antennary glycans. These results suggest differences in the functional *N*-glycosylation between endogenous and recombinant FcγRIII glycoproteins.

Fig. 11.3 (continued) the observed chemical shift changes between the IgG-Fc produced by CHO cells and those of (**D**) transgenic tobacco plants or (**E**) silkworms. The amino acid residues with observable chemical shift changes ($(0.2\delta_N^2+\delta_H^2)^{1/2}$ > 0.1 ppm and >0.2 ppm) are colored in pink and red. The molecular graphics were generated with PyMOL. In **A**, the backbone assignments are annotated by the resonance peaks with one-letter amino acid codes and sequence numbers. Side-chain resonances corresponding to NH_2 amides are connected by horizontal lines. Adapted and modified from the references (Yagi et al. 2015a, b, c)

The functional significance of the FcγRIIIa glycans as well as the IgG1-Fc glycans has been highlighted in the context of antibody-dependent cell-mediated cytotoxicity (ADCC), which depends on the glycoforms of these glycoproteins (Shields et al. 2002; Shinkawa et al. 2003; Niwa et al. 2004; Yamaguchi et al. 2006; Kato et al. 2010; Mizushima et al. 2011; Ferrara et al. 2011; Dekkers et al. 2017; Bernado et al. 2018). IgG1 having the Fc *N*-glycans without core fucosylation exhibits higher affinity for FcγRIIIa on NK cells, thus evoking a dramatically improved ADCC. Such defucosylation-dependent enhancement of IgG1-FcγRIIIa interaction was not observed for nonglycosylated FcγRIIIa (Shibata-Koyama et al. 2009), thus indicating a significant involvement of the receptor *N*-glycosylation in this potentiation. Furthermore, site-specific analysis revealed that the Asn162 glycan of FcγRIIIa is the key behind the ADCC improvement observed on the IgG1-Fc defucosylation (Ferrara et al. 2006; Shibata-Koyama et al. 2009).

The structural impact of IgG1-Fc defucosylation on its interaction with FcγRIIIa has been delineated on the basis of a multilateral biophysical approach. The X-ray crystallographic analyses of human IgG1-Fc in complex with sFcγRIIIa having two *N*-glycans at Asn45 and Asn162 provide the structural basis for the functional roles of these two glycans (Mizushima et al. 2011; Ferrara et al. 2011). In the complex made with nonfucosyl Fc, the Asn162 glycan of sFcγRIIIa is involved in the intermolecular interaction, providing unambiguous electron density for nine sugar residues (Fig. 11.4A and C). This interaction is mediated by van der Waals contacts and by hydrogen bonds between an Fc *N*-glycan and the Asn162 glycan, thus suggesting that the carbohydrate–carbohydrate interactions reinforce the protein–protein interactions in the complex. By contrast, the crystal structure of the complex formed by fucosylated IgG1-Fc presents the electron density of the Asn162 glycan only on the two residues of the reducing terminal (i.e., Fucα1-6GlcNAc) (Sakae et al. 2017). MD simulation confirmed that the Asn162 glycan of sFcγRIIIa exhibits a higher degree of conformational fluctuations in the complex formed with the fucosyl Fc in comparison with the nonfucosyl Fc complex (Sakae et al. 2017). All of these data indicate that the core fucosylation of the Fc glycan disrupts the optimum intermolecular carbohydrate–carbohydrate interactions, thus rendering the Asn162 glycan more mobile (Fig. 11.4B and D). Additionally, crystallographic data suggest that the Asn45 glycan of sFcγRIIIa cause steric hindrance against Fc, thereby negatively affecting IgG-binding affinity.

The possible involvement of an Fc receptor glycan in IgG interaction has been suggested by the early crystal structure of rat FcRn ectodomain in complex with the

Fig. 11.4 (continued) are blue and pink, respectively; sFcγRIIIa is yellow. Carbohydrates are represented as spheres. Fucose residue is colored in red. (**C** and **D**) Schematic representation of the carbohydrate–carbohydrate interactions of sFcγRIIIa complexed with (**C**) nonfucosylated and (**D**) fucosylated Fc. Principal component analysis of the simulation trajectory of the complexes between sFcγRIIIa and (**E**) nonfucosylated or (**F**) fucosylated IgG-Fc. Free energy landscape along the first two principal components estimated from the trajectory of both complexes at 300 K. The Asn162 glycan on sFcγRIIIa in the crystal structure (purple) is superimposed onto the representative simulated structure (yellow) derived from the major conformational state in the simulation trajectory. Adapted and modified from the references (Mizushima et al. 2011; Sakae et al. 2017)

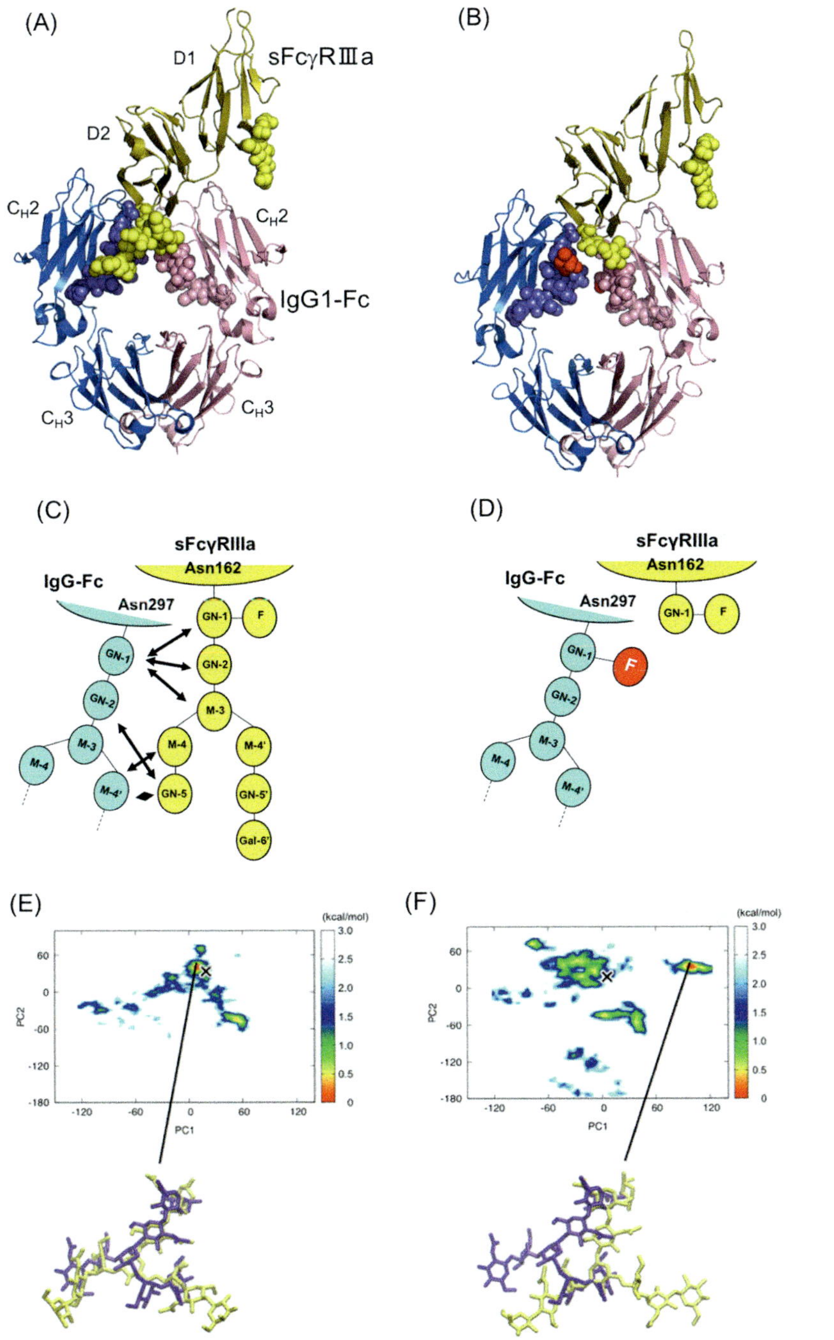

Fig. 11.4 Structure of (**A**) nonfucosylated and (**B**) fucosylated Fc complexed with bis-glycosylated soluble form of Fcγ receptor IIIa (sFcγRIIIa). Overall view of the complex. Chains A and B of Fc

Fc fragment of rat IgG2a (Burmeister et al. 1994). In this crystal structure, the Asn128 glycan of FcRn contributes 10%–15% of the total buried surface area at the interaction interface. In fact, the interaction stoichiometry between mouse Fc and FcRn depends on the glycosylation of mouse FcRn (Schuck et al. 1999).

Crystallographic data have also shown that the Asn162 glycan of sFcγRIIIa interacts with the Tyr296 of Fc, which is proximal to the glycosylation site (i.e., Asn297) (Mizushima et al. 2011). Mutational studies have underscored the functional importance of this tyrosine residue in FcγRIIIa binding (Isoda et al. 2015). In the free form of Fc in solution, this tyrosine residue is involved in the intramolecular interaction with the core fucose residue, as revealed by the nuclear Overhauser effect (NOE) experiment (Fig. 11.5A) (Kato et al. 2018). HSQC spectral comparison between fucosylated and nonfucosylated glycoforms of human IgG1-Fc has shown the significant line broadening of the Tyr296 peak (Fig. 11.5B) (Matsumiya et al. 2007), thus suggesting that Tyr296 exhibits conformational multiplicity in the absence of the core fucose residue. These data have implied that the core fucose residue of the Fc glycan comes into contact with Tyr296, thereby restricting the motional freedom of Tyr296 and hindering the involvement of Tyr296 in the intermolecular interaction with the receptor.

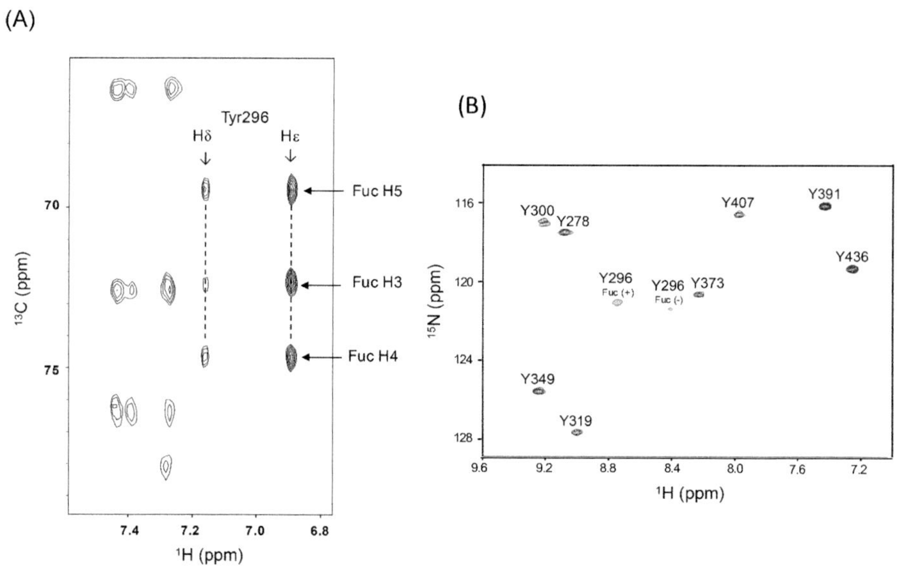

Fig. 11.5 (**A**) Parts of the 2D HSQC-NOESY spectrum of ^{13}C-labeled human IgG1-Fc exhibiting intramolecular NOE connectivities between Tyr296 and the core fucose residue. (**B**) ^{1}H-^{15}N HSQC spectral comparison between the fucosylated (red) and Fuc nonfucosylated (black) glycoforms of ^{15}N-Tyr-labeled human IgG1-Fc. Tyr296 of the nonfucosylated Fc exhibits a significant broadening of the HSQC peak. Adapted from the reference with permission (Matsumiya et al. 2007; Kato et al. 2018)

The sialylation of the Fc *N*-glycans influences the biological functions of IgG. Although α2,3 sialylation compromises ADCC activity (Scallon et al. 2007), α2,6 sialylation promotes anti-inflammatory activity (Kaneko et al. 2006). The α2,3 sialylation disturbs the functional C_H2 arrangement and interferes with FcγRIII binding (Crispin et al. 2013). The mechanism underlying the α2,6 sialylation-dependent anti-inflammatory activity remains to be fully understood. However, it has been suggested that the potential ability of the sialyl IgG to interact with C-type lectins (DC-SIGN and SIGN-RI) may be involved, thus resulting in the increased expression level of the inhibitory receptor FcγRIIb (Anthony et al. 2011).

11.6 Concluding Remarks

Glycosylation is currently considered a crucial factor in improving the efficacy of biopharmaceutical glycoproteins, including therapeutic antibodies. This article has highlighted the functional significance of the *N*-glycosylation of IgG-Fc and that of Fcγ receptors. In this system, *N*-glycans play versatile roles by positively and negatively controlling the intermolecular interactions between those glycoproteins. First, glycans are directly involved in the interaction, thereby enhancing their affinity or weakening their affinity by steric hindrance. This interaction can be finely tuned depending on their glycoforms, as best exemplified by the core fucosylation of IgG1-Fc interacting with FcγRIII.

The glycosylation also contributes to the formation of the protein's functionally active conformations. The *N*-glycosylation of IgG-Fc is a prerequisite for maintaining the integrity of the quaternary structure of the sites interacting with the effector molecules. Glycosylation can also determine the local conformations of functional relevance for IgG-Fc, as exemplified by the motional restriction of Tyr296 by the core fucose residue.

Structural insights into the mechanisms of these glycofunctions have been provided by integrating multilateral biophysics approaches including X-ray crystallography, NMR spectroscopy, and MD simulation. Given the potential to improve the efficacy of glycoprotein drugs, the structural optimization of their glycoforms will become more important in the design and development of biopharmaceuticals. Glycobiophysics approaches comprise valuable tools for the detailed characterization of glycan functions toward tailored creation of optimum glycoforms for specific pharmaceutical contexts.

Acknowledgments We would like to thank Ms. Kiyomi Senda and Ms. Kumiko Hattori (Nagoya City University) for their help with the protein purification. This work was partly supported by the Nanotechnology Platform Program (Molecule and Material Synthesis) of MEXT and MEXT/JSPS Grants in Aid for Scientific Research (JP25102008, JP15K07935, JP17H06414, JP17H05893, and JP18K14892).

References

Anthony RM, Kobayashi T, Wermeling F, Ravetch JV (2011) Intravenous gammaglobulin suppresses inflammation through a novel T_H2 pathway. Nature 475:110–113. https://doi.org/10.1038/nature10134
Bernado P, Shimizu N, Zaccai G et al (2018) Solution scattering approaches to dynamical ordering in biomolecular systems. Biochim Biophys Acta 1862:253–274. https://doi.org/10.1016/j.bbagen.2017.10.015
Borrok MJ, Jung ST, Kang TH et al (2012) Revisiting the role of glycosylation in the structure of human IgG Fc. ACS Chem Biol 7:1596–1602. https://doi.org/10.1021/cb300130k
Burmeister WP, Huber AH, Bjorkman PJ (1994) Crystal structure of the complex of rat neonatal Fc receptor with Fc. Nature 372:379–383. https://doi.org/10.1038/372379a0
Burton DR (1985) Immunoglobulin G: functional sites. Mol Immunol 22:161–206. https://doi.org/10.1016/0161-5890(85)90151-8
Crispin M, Yu X, Bowden TA (2013) Crystal structure of sialylated IgG Fc: implications for the mechanism of intravenous immunoglobulin therapy. Proc Natl Acad Sci U S A 110:E3544–E3546. https://doi.org/10.1073/pnas.1310657110
Deisenhofer J (1981) Crystallographic refinement and atomic models of a human Fc fragment and its complex with fragment B of protein A from Staphylococcus aureus at 2.9- and 2.8-Å resolution. Biochemistry 20:2361–2370. https://doi.org/10.1021/bi00512a001
Dekkers G, Treffers L, Plomp R et al (2017) Decoding the human immunoglobulin G-glycan repertoire reveals a spectrum of Fc-receptor- and complement-mediated-effector activities. Front Immunol 8. https://doi.org/10.3389/fimmu.2017.00877
Edberg JC, Kimberly RP (1997) Cell type-specific glycoforms of FcγRIIIa (CD16): differential ligand binding. J Immunol 159:3849–3857
Ercan A, Kohrt WM, Cui J et al (2017) Estrogens regulate glycosylation of IgG in women and men. JCI Insight 2. https://doi.org/10.1172/jci.insight.89703
Eryilmaz E, Janda A, Kim J et al (2013) Global structures of IgG isotypes expressing identical variable regions. Mol Immunol 56:588–598. https://doi.org/10.1016/j.molimm.2013.06.006
Ferrara C, Stuart F, Sondermann P et al (2006) The carbohydrate at FcγRIIIa Asn-162. An element required for high affinity binding to non-fucosylated IgG glycoforms. J Biol Chem 281:5032–5036. https://doi.org/10.1074/jbc.M510171200
Ferrara C, Grau S, Jäger C et al (2011) Unique carbohydrate-carbohydrate interactions are required for high affinity binding between FcγRIII and antibodies lacking core fucose. Proc Natl Acad Sci U S A:1–6. https://doi.org/10.1073/pnas.1108455108
Frank M, Walker RC, Lanzilotta WN et al (2014) Immunoglobulin G1 Fc domain motions: implications for Fc engineering. J Mol Biol 426:1799–1811. https://doi.org/10.1016/j.jmb.2014.01.011
Fridman WH, Teillaud J-L, Bouchard C et al (1993) Soluble Fcγ receptors. J Leukoc Biol 54:504–512. https://doi.org/10.1002/jlb.54.5.504
Harris LJ, Larson SB, Hasel KW, McPherson A (1997) Refined structure of an intact IgG2a monoclonal antibody. Biochemistry 36:1581–1597. https://doi.org/10.1021/bi962514+
Harris LJ, Skaletsky E, McPherson A (1998) Crystallographic structure of an intact IgG1 monoclonal antibody. J Mol Biol 275:861–872. https://doi.org/10.1006/jmbi.1997.1508
Hayes JM, Cosgrave EFJ, Struwe WB et al (2014) Glycosylation and Fc receptors. Curr Top Microbiol Immunol 382:165–199
Hayes JM, Wormald MR, Rudd PM, Davey GP (2016) Fcγ receptors: glycobiology and therapeutic prospects. J Inflamm Res 9:209–219. https://doi.org/10.2147/JIR.S121233
Holland M, Yagi H, Takahashi N et al (2006) Differential glycosylation of polyclonal IgG, IgG-Fc and IgG-Fab isolated from the sera of patients with ANCA-associated systemic vasculitis. Biochim Biophys Acta–Gen Subj 1760:669–677. https://doi.org/10.1016/j.bbagen.2005.11.021
Ikeya T, Ban D, Lee D et al (2018) Solution NMR views of dynamical ordering of biomacromolecules. Biochim Biophys Acta–Gen Subj 1862:287–306

Isoda Y, Yagi H, Satoh T et al (2015) Importance of the side chain at position 296 of antibody Fc in interactions with FcγRIIIa and other Fcγ receptors. PLoS One 10. https://doi.org/10.1371/journal.pone.0140120

Jo M, Kwon HS, Lee KH et al (2018) Engineered aglycosylated full-length IgG Fc variants exhibiting improved FcγRIIIa binding and tumor cell clearance. MAbs 10:278–289. https://doi.org/10.1080/19420862.2017.1402995

Ju MS, Na JH, Yu YG et al (2015) Structural consequences of aglycosylated IgG Fc variants evolved for FcγRI binding. Mol Immunol 67:350–356. https://doi.org/10.1016/j.molimm.2015.06.020

Jung ST, Reddy ST, Kang TH et al (2010) Aglycosylated IgG variants expressed in bacteria that selectively bind FcγRI potentiate tumor cell killing by monocyte-dendritic cells. Proc Natl Acad Sci 107:604–609. https://doi.org/10.1073/pnas.0908590107

Kacskovics I, Kis Z, Mayer B et al (2006) FcRn mediates elongated serum half-life of human IgG in cattle. Int Immunol 18:525–536. https://doi.org/10.1093/intimm/dxh393

Kaneko Y, Nimmerjahn F, Ravetch JV (2006) Anti-inflammatory activity of immunoglobulin G resulting from Fc sialylation. Science 313:670–673. https://doi.org/10.1126/science.1129594

Kato K, Sautès-Fridman C, Yamada W et al (2000) Structural basis of the interaction between IgG and Fcγ receptors. J Mol Biol 295:213–224. https://doi.org/10.1006/jmbi.1999.3351

Kato K, Yamaguchi Y, Arata Y (2010) Stable-isotope-assisted NMR approaches to glycoproteins using immunoglobulin G as a model system. Prog Nucl Magn Reson Spectrosc 56:346–359. https://doi.org/10.1016/j.pnmrs.2010.03.001

Kato K, Yanaka S, Yagi H (2018) Technical basis for nuclear magnetic resonance approach for glycoproteins. In: Naito Asakura T, Shimada I, Takegoshi K, Yamamoto YA (eds) Experimental approaches of NMR spectroscopy. Springer, Singapore, pp 415–438

Kim HH, Matsunaga C, Yoshino A et al (1994a) Dynamical structure of the hinge region of immunoglobulin G as studied by ^{13}C nuclear magnetic resonance spectroscopy. J Mol Biol 236:300–309. https://doi.org/10.1006/jmbi.1994.1136

Kim HH, Yamaguchi Y, Masuda K et al (1994b) *O*-glycosylation in hinge region of mouse immunoglobulin G2b. J Biol Chem 269:12345–12350

Kiyoshi F, Akira K (1991) IgG galactosylation-its biological significance and pathology. Mol Immunol 28:1333–1340. https://doi.org/10.1016/0161-5890(91)90035-I

Kiyoshi M, Tsumoto K, Ishii-Watabe A, Caaveiro JMM (2017) Glycosylation of IgG-Fc: a molecular perspective. Int Immunol 29:311–317. https://doi.org/10.1093/intimm/dxx038

Kondo S, Yagi H, Kamiya Y et al (2012) *N*-glycosylation profiles of chicken immunoglobulin Y glycoproteins expressed by different production vehicles. J Glycomics Lipidomics S5-002. https://doi.org/10.4172/2153-0637.S5-002

Krapp S, Mimura Y, Jefferis R et al (2003) Structural analysis of human IgG-Fc glycoforms reveals a correlation between glycosylation and structural integrity. J Mol Biol 325:979–989. https://doi.org/10.1016/S0022-2836(02)01250-0

Lalonde ME, Durocher Y (2017) Therapeutic glycoprotein production in mammalian cells. J Biotechnol 251:128–140. https://doi.org/10.1016/j.jbiotec.2017.04.028

Leatherbarrow RJ, Rademacher TW, Dwek RA et al (1985) Effector functions of a monoclonal aglycosylated mouse IgG2a: binding and activation of complement component C1 and interaction with human monocyte Fc receptor. Mol Immunol 22:407–415

Liu D, Cocco MJ, Rosenfied R et al (2007) Assignment of backbone ^{1}H, ^{13}C and ^{15}N resonances of human IgG1 Fc (51.4 kDa). Biomol NMR Assign 1:233–235. https://doi.org/10.1007/s12104-007-9065-5

Loos A, Steinkellner H (2012) IgG-Fc glycoengineering in non-mammalian expression hosts. Arch Biochem Biophys 526:167–173

Masuda K, Yamaguchi Y, Kato K et al (1999) Post-translational modifications of immunoglobulin G: a mouse IgG variant that lacks the entire C_H1 domain. Mol Immunol 36:993–1003. https://doi.org/10.1016/S0161-5890(99)00131-5

Matsumiya S, Yamaguchi Y, Saito J et al (2007) Structural comparison of fucosylated and nonfucosylated Fc fragments of human immunoglobulin G1. J Mol Biol 368:767–779. https://doi.org/10.1016/j.jmb.2007.02.034

Mizuochi T, Taniguchi T, Shimizu A, Kobata A (1982) Structural and numerical variations of the carbohydrate moiety of immunoglobulin G. J Immunol 129:2016–2020
Mizuochi T, Hamako J, Nose M, Titani K (1990) Structural changes in the oligosaccharide chains of IgG in autoimmune MRL/Mp-lpr/lpr mice. J Immunol 145:1794–1798
Mizushima T, Yagi H, Takemoto E et al (2011) Structural basis for improved efficacy of therapeutic antibodies on defucosylation of their Fc glycans. Genes to Cells 16:1071–1080. https://doi.org/10.1111/j.1365-2443.2011.01552.x
Nettleton MY, Kochan JP (1995) Role of glycosylation sites in the IgE Fc molecule. Int Arch Allergy Immunol 107:328–329. https://doi.org/10.1159/000237017
Niwa R, Hatanaka S, Shoji-Hosaka E et al (2004) Enhancement of the antibody-dependent cellular cytotoxicity of low-fucose IgG1 is independent of FcγRIIIa functional polymorphism. Clin Cancer Res 10:6248–6255. https://doi.org/10.1158/1078-0432.CCR-04-0850
Nose M, Wigzell H (1983) Biological significance of carbohydrate chains on monoclonal antibodies. Proc Natl Acad Sci U S A 80:6632–6636. https://doi.org/10.1073/pnas.80.21.6632
Parekh RB, Dwek RA, Sutton BJ et al (1985) Association of rheumatoid arthritis and primary osteoarthritis with changes in the glycosylation pattern of total serum IgG. Nature 316:452–457. https://doi.org/10.1038/316452a0
Patel KR, Roberts JT, Subedi GP, Barb AW (2018) Restricted processing of CD16a/Fcγ receptor IIIa *N*-glycans from primary human NK cells impacts structure and function. J Biol Chem 293:3477–3489. https://doi.org/10.1074/jbc.RA117.001207
Plomp R, Hensbergen PJ, Rombouts Y et al (2014) Site-specific *N*-glycosylation analysis of human immunoglobulin E. J Proteome Res 13:536–546. https://doi.org/10.1021/pr400714w
Plomp R, Dekkers G, Rombouts Y et al (2015) Hinge-region *O*-glycosylation of human immunoglobulin G3 (IgG3). Mol Cell Proteomics 14:1373–1384. https://doi.org/10.1074/mcp.M114.047381
Rademacher TW, Parekh RB, Dwek RA (1988) Glycobiology. Annu Rev Biochem 57:785–838. https://doi.org/10.1146/annurev.bi.57.070188.004033
Roux KH, Strelets L, Michaelsen TE (1997) Flexibility of human IgG subclasses. J Immunol 159:3372–3382
Sakae Y, Satoh T, Yagi H et al (2017) Conformational effects of *N*-glycan core fucosylation of immunoglobulin G Fc region on its interaction with Fcγ receptor IIIa. Sci Rep 7:13780. https://doi.org/10.1038/s41598-017-13845-8
Saphire EO, Parren PW, Pantophlet R, et al (2001) Crystal structure of a neutralizing human IgG against HIV-1: a template for vaccine design. Science 293:1155–1159. https://doi.org/10.1126/science.1061692
Scallon BJ, Tam SH, McCarthy SG et al (2007) Higher levels of sialylated Fc glycans in immunoglobulin G molecules can adversely impact functionality. Mol Immunol 44:1524–1534. https://doi.org/10.1016/j.molimm.2006.09.005
Scapin G, Yang X, Prosise WW et al (2015) Structure of full-length human anti-PD1 therapeutic IgG4 antibody pembrolizumab. Nat Struct Mol Biol 22:953–958. https://doi.org/10.1038/nsmb.3129
Schuck P, Radu CG, Ward ES (1999) Sedimentation equilibrium analysis of recombinant mouse FcRn with murine IgG1. Mol Immunol 36:1117–1125. https://doi.org/10.1016/S0161-5890(99)00093-0
Shade K-T, Anthony R (2013) Antibody glycosylation and inflammation. Antibodies 2:392–414. https://doi.org/10.3390/antib2030392
Shibata-Koyama M, Iida S, Okazaki A et al (2009) The *N*-linked oligosaccharide at FcγRIIIa Asn-45: an inhibitory element for high FcγRIIIa binding affinity to IgG glycoforms lacking core fucosylation. Glycobiology 19:126–134. https://doi.org/10.1093/glycob/cwn110
Shields RL, Lai J, Keck R et al (2002) Lack of fucose on human IgG1 *N*-linked oligosaccharide improves binding to human FcγRIII and antibody-dependent cellular toxicity. J Biol Chem 277:26733–26740. https://doi.org/10.1074/jbc.M202069200
Shinkawa T, Nakamura K, Yamane N et al (2003) The absence of fucose but not the presence of galactose or bisecting *N*-acetylglucosamine of human IgG1 complex-type oligosaccharides

shows the critical role of enhancing antibody-dependent cellular cytotoxicity. J Biol Chem 278:3466–3473. https://doi.org/10.1074/jbc.M210665200

Solá RJ, Griebenow K (2011) Glycosylation of therapeutic proteins: an effective strategy to optimiza efficacy. BioDrugs 24:9–21. https://doi.org/10.2165/11530550-000000000-00000

Sondermann P, Huber R, Oosthuizen V, Jacob U (2000) The 3.2-A crystal structure of the human IgG1 Fc fragment-FcγRIII complex. Nature 406:267–273. https://doi.org/10.1038/35018508

Strohl WR (2015) Fusion proteins for half-life extension of biologics as a strategy to make biobetters. BioDrugs 29:215–239. https://doi.org/10.1007/s40259-015-0133-6

Suzuki N, Lee YC (2004) Site-specific *N*-glycosylation of chicken serum IgG. Glycobiology 14:275–292. https://doi.org/10.1093/glycob/cwh031

Tao MH, Morrison SL (1989) Studies of aglycosylated chimeric mouse-human IgG. Role of carbohydrate in the structure and effector functions mediated by the human IgG constant region. J Immunol 143:2595–2601

Ugurlar D, Howes SC, de Kreuk BJ, et al (2018) Structures of C1-IgG1 provide insights into how danger pattern recognition activates complement. Science 359:794–797. https://doi.org/10.1126/science.aao4988

van de Bovenkamp FS, Hafkenscheid L, Rispens T, Rombouts Y (2016) The emerging importance of IgG fab glycosylation in immunity. J Immunol 196:1435–1441. https://doi.org/10.4049/jimmunol.1502136

Wright A, Morrison SL (1993) Antibody variable region glycosylation: biochemical and clinical effects. Springer Semin Immunopathol 15:259–273. https://doi.org/10.1007/BF00201106

Yagi H, Fukuzawa N, Tasaka Y et al (2015a) NMR-based structural validation of therapeutic antibody produced in Nicotiana benthamiana. Plant Cell Rep 34:959–968. https://doi.org/10.1007/s00299-015-1757-1

Yagi H, Nakamura M, Yokoyama J et al (2015b) Stable isotope labeling of glycoprotein expressed in silkworms using immunoglobulin G as a test molecule. J Biomol NMR 62:157–167. https://doi.org/10.1007/s10858-015-9930-y

Yagi H, Zhang Y, Yagi-Utsumi M et al (2015c) Backbone ^{1}H, ^{13}C, and ^{15}N resonance assignments of the Fc fragment of human immunoglobulin G glycoprotein. Biomol NMR Assign 9:257–260. https://doi.org/10.1007/s12104-014-9586-7

Yagi H, Takakura D, Roumenina LT et al (2018) Site-specific *N*-glycosylation analysis of soluble Fcγ receptor IIIb in human serum. Sci Rep 8:2719. https://doi.org/10.1038/s41598-018-21145-y

Yamaguchi Y, Kato K, Shindo M et al (1998) Dynamics of the carbohydrate chains attached to the Fc portion of immunoglobulin G as studied by NMR spectroscopy assisted by selective ^{13}C labeling of the glycans. J Biomol NMR 12:385–394

Yamaguchi Y, Nishimura M, Nagano M et al (2006) Glycoform-dependent conformational alteration of the Fc region of human immunoglobulin G1 as revealed by NMR spectroscopy. Biochim Biophys Acta 1760:693–700

Yamaguchi Y, Takahashi N, Kato K (2007) Antibody structures. In: Comprehensive glycoscience: from chemistry to systems biology. Elsevier, Oxford, pp 745–763

Yanaka S, Yamazaki T, Yogo R et al (2017) NMR detection of semi-specific antibody interactions in serum environments. Molecules 22:1619. https://doi.org/10.3390/molecules22101619

Yanaka S, Yagi H, Yogo R et al (2018) Stable isotope labeling approaches for NMR characterization of glycoproteins using eukaryotic expression systems. J Biomol NMR:1–10. https://doi.org/10.1007/s10858-018-0169-2

Yogo R, Yanaka S, Yagi H et al (2017) Characterization of conformational deformation-coupled interaction between immunoglobulin G1 Fc glycoprotein and a low-affinity Fcγ receptor by deuteration-assisted small-angle neutron scattering. Biochem Biophys Rep 12:1–4. https://doi.org/10.1016/j.bbrep.2017.08.004

Zeck A, Pohlentz G, Schlothauer T et al (2011) Cell type-specific and site directed *N*-glycosylation pattern of FcγRIIIa. J Proteome Res 10:3031–3039. https://doi.org/10.1021/pr1012653

Zhang X, Zhang L, Tong H et al (2015) 3D structural fluctuation of IgG1 antibody revealed by individual particle electron tomography. Sci Rep 5:9803. https://doi.org/10.1038/srep09803

Chapter 12
Biophysical Approaches to Solve the Structures of the Complex Glycan Shield of Chloroviruses

Cristina De Castro, Garry A. Duncan, Domenico Garozzo, Antonio Molinaro, Luisa Sturiale, Michela Tonetti, and James L. Van Etten

Abstract The capsid of *Paramecium bursaria* chlorella virus (PBCV-1) contains a heavily glycosylated major capsid protein, Vp54. The capsid protein contains four glycans, each N-linked to Asn. The glycan structures are unusual in many aspects: (1) they are attached by a β-glucose linkage, which is rare in nature; (2) they are highly branched and consist of 8–10 neutral monosaccharides; (3) all four glycoforms contain a dimethylated rhamnose as the capping residue of the main chain, a hyper-branched fucose residue and two rhamnose residues "with opposite absolute configurations; (4) the four glycoforms differ by the nonstoichiometric presence of

C. De Castro (✉)
Department of Agricultural Sciences, University of Napoli, Portici, NA, Italy
e-mail: decastro@unina.it

G. A. Duncan
Department of Biology, Nebraska Wesleyan University, Lincoln, NE, USA
e-mail: gduncan@nebrwesleyan.edu

D. Garozzo · L. Sturiale
CNR, Institute for Polymers, Composites and Biomaterials, Catania, Italy
e-mail: domenico.garozzo@cnr.it; luisella.sturiale@cnr.it

A. Molinaro
Department of Chemical Sciences, University of Napoli, Napoli, Italy
e-mail: molinaro@unina.it

M. Tonetti
Department of Experimental Medicine and Center of Excellence for Biomedical Research, University of Genova, Genova, Italy
e-mail: tonetti@unige.it

J. L. Van Etten
Department of Plant Pathology and Nebraska Center for Virology, University of Nebraska, Lincoln, NE, USA
e-mail: jvanetten1@unl.edu

Y. Yamaguchi, K. Kato (eds.), *Glycobiophysics*, Advances in Experimental Medicine and Biology 1104, https://doi.org/10.1007/978-981-13-2158-0_12

two monosaccharides, L-arabinose and D-mannose; (5) the N-glycans from all of the chloroviruses have a strictly conserved core structure; and (6) these glycans do not resemble any structures previously reported in the three domains of life.

The structures of these N-glycoforms remained elusive for years because initial attempts to solve their structures used tools developed for eukaryotic-like systems, which we now know are not suitable for this noncanonical glycosylation pattern. This chapter summarizes the methods used to solve the chlorovirus complex glycan structures with the hope that these methodologies can be used by scientists facing similar problems.

Keywords Giant viruses · GC-MS · NMR · MALDI · N-glycosylation

Abbreviation List

AA	acetylated alditol
AMG	acetylated methyl glycosides
AOG	acetylated octyl glycoside
COSY	correlation spectroscopy
GC-MS	gas chromatography-mass spectrometry
HMBC	heteronuclear multiple bond coherence
HSQC	heteronuclear single quantum coherence
NOESY	nuclear Overhauser effect spectroscopy
PBCV-1	*Paramecium bursaria* chlorella virus type 1
PMAA	partially methylated and acetylated alditol
ROESY	rotating frame Overhauser effect spectroscopy
TOCSY	total correlation spectroscopy

12.1 Introduction

In considering enzymes involved in manipulating carbohydrates, one usually does not think about viruses playing a role in this important activity; however, exceptions to this common belief are beginning to emerge from viruses often referred to as giant viruses. Giant viruses are a rapidly expanding group of viruses (their taxonomy is still being resolved) characterized by a large particle size and very large dsDNA genomes (typically from >300 kb to ~2.5 Mb) that encode many proteins involved in functions not normally found in typical viruses such as HIV or Ebola (Colson et al. 2013).

Paramecium bursaria chlorella virus 1 (PBCV-1, genus, *Chlorovirus*), the first member of the giant virus group, was discovered more than 35 years ago (Van Etten et al. 1982). PBCV-1 is an icosahedron with a diameter of 190 nm, and it has an ~331 kb genome that encodes ~416 proteins and 11 tRNAs (Dunigan et al. 2012).

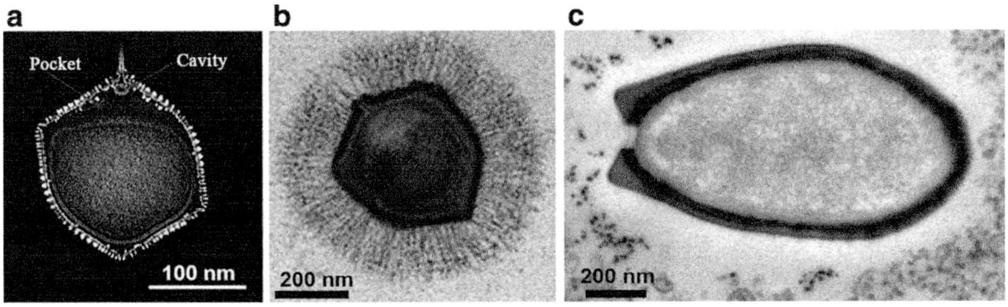

Fig. 12.1 Electron microscopy of (**a**) PBCV-1. (Adapted from Zhang et al. 2011). (**b**) *Mimivirus*. (Adapted from Ameobal Pathogen Mimivirus Infects Macrophages through Phagocytosis, Ghigo E et al. PLOS Pathogens 2008, 4(6): e1000087, doi: 10.1371/journal.ppat.1000087), (**c**) *Pandoravirus*. (Adapted by permission from Macmillan Publishers Ltd: Nature, Ed Young, Giant Viruses open Pandora's box, doi: 10.1038/nature.2013.13410, copyright 2013)

The virus infects the unicellular, eukaryotic, and symbiotic microalga *Chlorella variabilis* NC64A (Fig. 12.1a). Since then many new giant viruses have been discovered, with a burst of activity in the last 15 years (Abergel et al. 2015).

For example, *Mimivirus* was first described in 2004 (Raoult et al. 2004, Fig. 12.1b). It is more complex than PBCV-1 in terms of physical and genome size. The capsid has an overall size of 700 nm in diameter and a genome of 1.18 Mb, with ~1000 protein-encoding genes, and its capsid is covered with a thick layer of long fibers (~200 nm). Recently, even larger viruses have been isolated. Pandoraviruses (Fig. 12.1c) have a size reaching 1 μm with a shape reminiscent of some types of bacteria (Philippe et al. 2013). The *Pandoravirus salinus* genome is 2.5 Mb and contains 2,556 putative protein-encoding genes, of which only 7% have recognizable relationships with genes from other known organisms. *Pithovirus sibericus* was isolated from a 30,000-year-old permafrost sample harvested from Siberia (Legendre et al. 2014). This virus is most closely related to Marseilleviridae, another group of giant viruses. Its maximum size can reach 1.5 μm, and it has an oval shape similar to Pandoraviruses; however, its genome only has ~467 protein-encoding genes.

Thus, giant viruses differ in shape, genome size and number of encoded proteins. New members are continually being discovered, and on-going genomic analyses indicate that some of them encode many glyco-related genes, including those for the biosynthesis of nucleotide-sugar precursors and glycosyltransferases. However, information on the role of these genes and on the structure of the glycans associated with them are still fragmentary. Furthermore, such viral glycans have been overlooked because of the lack of the appropriate tools to investigate them.

Currently, the most studied genus is Chloroviruses; accordingly, this chapter focuses on PBCV-1 as a representative chlorovirus member. We provide a brief description of its features and then describe the methods that allowed us to determine its new and unusual glycan structure.

12.2 Chloroviruses and *Paramecium bursaria* Chlorella Virus (PBCV-1)

The plaque-forming chloroviruses (family *Phycodnaviridae*) are ubiquitous in freshwater throughout the world with titers occasionally reaching thousands of plague-forming units (PFU) per ml of indigenous water. Chlorovirus hosts are unicellular microalgae, often referred to as zoochlorellae and mostly found in nature as endosymbionts of protists (Karakashian and Karakashian 1965).

Chloroviruses are divided in four groups depending on the host selectivity, and the genomes of 43 chloroviruses have been sequenced, assembled and annotated (Jeanniard et al. 2013; Quispe et al. 2017). Collectively, the viruses encode genes from 643 predicted protein families, and at least 17 of them are predicted to be involved in manipulating carbohydrates including (1) enzymes involved in making extracellular polysaccharides such as hyaluronan and chitin; (2) enzymes that make nucleotide sugars such as GDP-L-fucose and GDP-D-rhamnose; and (3) enzymes involved in the synthesis of glycans attached to the virus major capsid proteins (Van Etten et al. 2010, 2017).

The capsid of the prototype chlorovirus PBCV-1 is an icosahedron with a spike-like structure at one vertex, which makes the first contact with the wall of the host cell (Zhang et al. 2011; Van Etten and Dunigan 2012), and a few external fibers that extend from some of the viral capsomeres (Cherrier et al. 2009). The PBCV-1 major capsid protein (named Vp54) is a glycoprotein, and the glycosylation pattern of this protein, which is the subject of this manuscript, differs from that of all other organisms known to date.

Typically, viruses, such as HIV or Ebola, use host-encoded glycosyltransferases and glycosidases located in the endoplasmic reticulum (ER) and Golgi apparatus to add and remove N-linked sugar residues from virus glycoproteins (Vigerust and Shepherd 2007). Consequently, the glycan portion of the glycoproteins of these viruses is host-specific and resembles that of the host.

However, glycosylation of the PBCV-1 major capsid protein differs from the scenario described above because the virus encodes most, if not all, of the machinery for the process (Van Etten et al. 2010, 2017). Solving the structure of the four Vp54 N-linked glycans (De Castro et al. 2013) proved the uniqueness of PBCV-1 glycosylation: (1) Vp54 has four Asn-linked glycans, and none of the Asn are located in an Asn-X-(Thr/Ser) sequon characteristic of ER-located glycosyltransferases (Nandhagopal et al. 2002; De Castro et al. 2018); (2) the glycans are attached to Asn by a β-glucose linkage, which is rare in nature; (3) the glycans are highly branched and consist of eight to ten neutral monosaccharides (Fig. 12.2); (4) the four glycoforms contain a dimethylated rhamnose as the capping residue of the main chain, a hyper-branched fucose residue, and two rhamnose residues with opposite absolute configurations; (5) the four glycoforms differ by the nonstoichiometric presence of two monosaccharides, L-arabinose and D-mannose; and (6) there is a core region (Fig. 12.2) strictly conserved in the N-glycans of all the chloroviruses studied to date (De Castro et al. 2016; Quispe et al. 2017; Speciale et al. 2017).

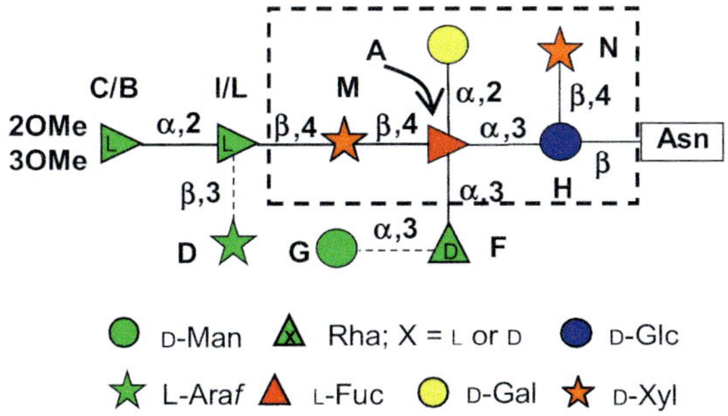

Fig. 12.2 Structures of glycoforms of the major capsid protein Vp54 from chlorovirus PBCV-1. Residues are labeled with the letter used during the NMR assignment (Fig. 12.5). Glycoform 1 lacks residue **D** and the two L-rhamnose units are labeled **C** and **I**. Glycoform 2 has the arabinose residue **D**, and the two L-rhamnose residues are labeled **B** and **L**. (Adapted from Van Etten et al. 2017). Residues within the box are conserved among all chloroviruses; rhamnose **F** is a semiconserved element because its absolute configuration is virus-dependent. Additional decorations occur on this core N-glycan and represent a molecular signature for each chlorovirus

The glycan structures do not resemble any structure previously reported in the three domains of life.

Solving the structure of the Vp54 glycans was a challenging issue that was accomplished by a combination of gas chromatography-mass spectrometry (GC-MS), NMR spectroscopy, and MALDI spectrometry techniques as described in the following sections.

12.3 GC-MS Approach

GC-MS is a powerful technique. Its detection limit is very low, but it has one prerequisite: the glycan (or the glycoprotein) needs to be depolymerized and transformed into a proper volatile derivative. Depending on the type of derivative, GC-MS helps to elucidate the monosaccharides, their absolute configuration and substitution pattern. This information is of paramount importance because it supports the interpretation of NMR or MALDI spectra.

12.3.1 Determination of the Glycan Composition

To detect the type of sugars, monosaccharides are transformed into the corresponding acetylated alditols (AA) or acetylated methyl glycosides (AMG), depending on the type of depolymerization procedure used, hydrolysis or methanolysis, respectively. Both procedures can be applied directly to the glycoprotein with no significant interference from the amino acids, and they are complementary in terms of disadvantages and advantages. The AA procedure is limited to neutral or basic monosaccharides, either aldose or ketose, but it cannot detect acidic monosaccharides unless the reduction of the carboxylic group is performed first. The advantage of this approach is that aldoses are transformed into an acyclic molecule, which gives only one peak in the chromatogram, a feature that is desirable for accurate quantification; in addition, if a monosaccharide is partially methylated, the position of the substituent is easily inferred by applying the fragmentation rules that exist for the partially methylated and acetylated alditol derivatives. The main disadvantage is that enantiomeric monosaccharides, such as L- and D-rhamnose, cannot be distinguished, while in some cases, such as arabinose and lyxose, the two alditols are identical once the aldehydic function is reduced. In addition, ketoses produce two different alditols upon the reduction of their carbonyl function.

To solve some of the pitfalls of the AA procedure, it is good practice to combine the results with those from other types of derivatives; acetylated methylglycoside is the best alternative. This procedure (De Castro et al. 2010) is less laborious than AA and detects neutral, basic, and acidic monosaccharides but fails with neutral ketoses, which are not detected. Solvolysis of the glycoprotein or of the glycopeptides with hydrochloric methanol depolymerizes the glycan and transforms the residues in the corresponding methylglycosides, which are later acetylated. The main difference with the AA method is that the AMG approach produces cyclic sugar derivatives, either in pyranose or furanose form and with both configurations at the anomeric center; in other words, one monosaccharide can present up to four peaks in the chromatogram, with one usually more abundant than the others. Detection of the monosaccharide as cyclic derivatives removes the problem related to the production of identical derivatives, as mentioned earlier for the alditols of arabinose and lyxose.

Both AA and AMG approaches were performed on PBCV-1 glycopeptides. The AA profiles (Fig. 12.3a) contained eight monosaccharides in different proportions, of which only seven could be assigned with confidence by comparing their electron-impact (EI) MS spectrum and their retention times with that of authentic standards. The species that eluted at 12 min had an EI-MS spectrum (Fig. 12.3b) interpreted on the basis of the rules valid for partially methylated and acetylated alditols (Lönngren and Svensson 1974). In brief, the EI-MS spectrum contains cations that originate from the rupture of the carbon-carbon linkage of the molecule (Fig. 12.3b). Importantly, the carbon bearing the methoxyl, rather than an acetyl group, retains a positive charge and is revealed. Therefore, the ions displayed in the EI-MS spectrum result from the position of the methyl group(s) of the alditol, and the spectrum of species at 12 min was consistent with a derivative of a 6-deoxyhexose methylated at both 2 and 3 positions (Fig. 12.3b). The *manno*-stereochemistry of the residue was inferred later on the basis of the PMAA derivatives and NMR studies.

The GC-MS chromatogram of the acetylated methylglycosides (Fig. 12.3c) provided information similar to that from the acetylated alditols procedure; it confirmed the identity of arabinose as distinguished from all the other residues. Notably, direct methanolysis of the glycopeptide did not cleave the linkage of the N-linked glucose, which can be accomplished by adding a hydrolytic step prior to methanoly-

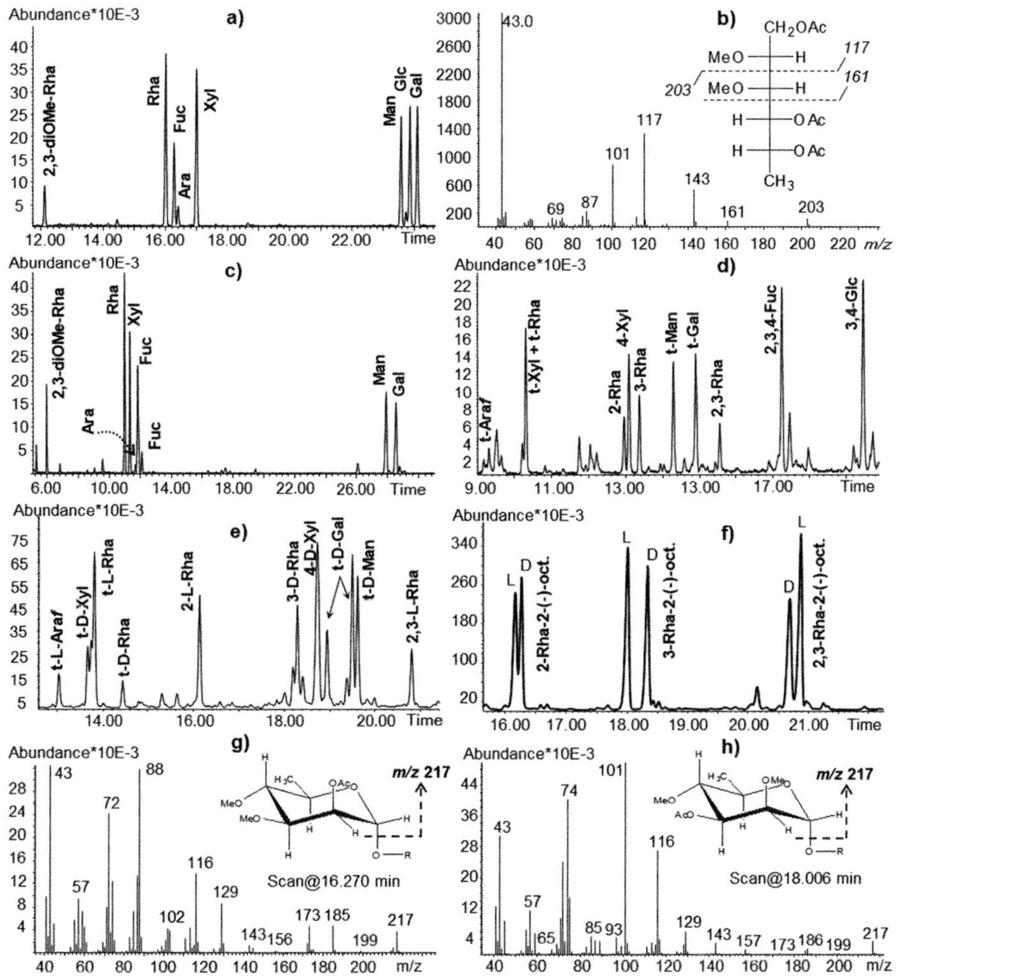

Fig. 12.3 GC-MS chromatograms from PBCV-1 glycopeptide Vp54, each analysis was performed by derivatizing 0.2–0.3 mg of sample. (**a**) Profile of the acetylated alditol derivatives. (**b**) EI-MS spectrum of the species at 12 min. of chromatogram 3A, together with the structure of the alditol and of the fragmentation pathway. (**c**) Profile of the acetylated methylglycosides, (**d**) of the PMAA, and (**e**) expansion of the chromatogram of the partially methylated and acetylated octyl glycosides reporting the different rhamnose species. (**f**) Chromatogram of the standard of the partially methylated and acetylated octyl glycosides of rhamnose 2-, 3-, and 2,3-linked. (**g** and **h**) EI-MS spectra of the peaks at 16.3 and 18.0 min. of panel (**f**), representing the derivatives of 2- and 3-linked octyl rhamnosides, respectively. (Fig. 12.3**f**–**h** is adapted from De Castro et al. (2013))

sis. Thus, by comparing results from these two derivatization procedures, it was possible to deduce which was the first monosaccharide N-linked to the protein, information that is not trivial in a system with unusual N-linked glycans.

12.3.2 Determination of the Glycan Substitution Pattern

Evaluation of the substitution pattern of the residues in a polysaccharide is also called linkage analysis (De Castro et al. 2010). The key step is the transformation of the glycan into its methylated derivative (Fig. 12.4a); during this procedure the available free hydroxyl functions are converted into methyl ethers, *N*-acyl amino sugars are *N*-methylated, and uronic acids are esterified. At this stage, if a uronic acid is present, the methyl ester function is reduced; otherwise the residue is lost during successive steps. Next, the hydrolysis step (Fig. 12.4b) cleaves all the glycosidic linkages and releases the partially methylated monosaccharides; the free hydroxyl groups that now appear are those originally involved in the glycosidic linkages. Successively, the monosaccharide is transformed into an alditol by reduction with a hydride, usually $NaBD_4$, to label the anomeric carbon; finally, acetylation yields the so-called partially methylated and acetylated alditols (PMAA, Fig. 12.4c).

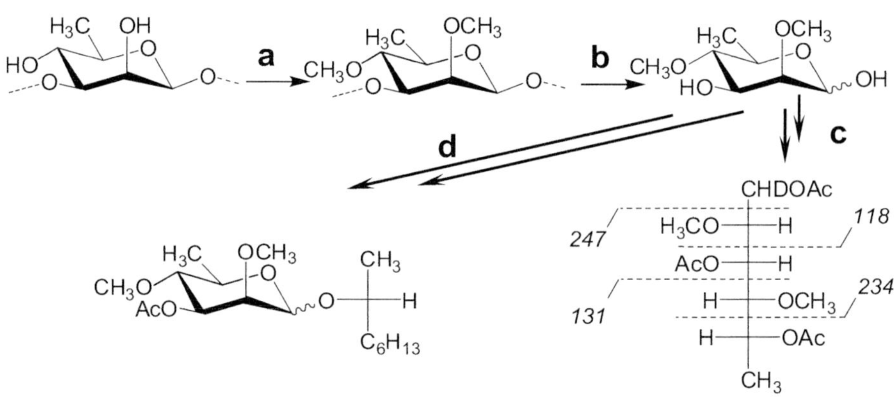

Fig. 12.4 Reaction scheme for the production of the partially methylated and acetylated alditol (PMAA) derivative of a 3-linked rhamnose. (**a**) Methylation of the native glycan: the free hydroxyl groups are transformed in methyl ethers. (**b**) The methylated glycan is hydrolyzed to give a monosaccharide partially methylated. (**c**) Carbon 1 is labeled by reduction with a deuterated hydride and acetylated to produce the final partially methylated and acetylated derivative (PMAA), reported with its fragmentation scheme. Acetyl groups at carbons 1 and 5 indicate the ring closure of the monosaccharide; acetyl group at position 3 indicates a previously linked position of the residue. (**d**) Partially methylated and acetylated octyl rhamnoside obtained after methylation, hydrolysis, and octanolysis of the glycopeptide. The acetyl group is indicative of a previously linked position of the residue as in the PMAA in Fig. 12.4c

Table 12.1 Linkage analysis of PBCV-1 glycopeptide. Unless indicated otherwise, each PMAA derivative denotes a residue in the pyranose form and the number, the alditol stereochemistry is indicated with the acronym of the corresponding monosaccharide, and the number indicates the position substituted, while "t" means that the monosaccharide is terminal

Component	Retention time (min.)	Diagnostic ions[a]
t-Araf	9.30	101[a], 102[a], 118, 129[a], 161, 162
t-Xyl	10.32	101[a], 102[a], 118, 161, 162
t-Rha	10.32	89[a], 102[a], 118, 131, 162, 175
2-Rha	12.95	89[a], 130[a], 131, 190
4-Xyl	13.08	102[a], 118, 129[a], 162, 189
3-Rha	13.37	89[a], 118, 131, 202[a], 234
t-Man	14.28	102[a], 118, 129[a], 145[a], 161, 162, 205
t-Gal	14.88	102[a], 118, 129[a], 145[a], 161, 162, 205
2,3-Rha	15.57	89[a], 131, 202[a], 262
2,3,4-Fuc	17.22	87, 129[a], 146, 159, 171[a], 218, 231, 290
3,4-Glc	19.48	118, 185[a], 305

Ions generated from the primary fragments by loss of a neutral molecule: acetic acid (*m/z* 60), and/or acetic anhydride (*m/z* 102), and/or methanol (*m/z* 32)

This approach was applied to the glycopeptides of PBCV-1, omitting the reduction of the methyl ester function because no uronic acid was detected in the compositional analysis.

The GC-MS chromatogram presented a large array of peaks (Fig. 12.3d), and the interpretation of their EI-MS spectra (Table 12.1), by the rules mentioned in Sect. 12.3.1, identified the type of monosaccharide (pentose, hexose, or deoxyhexose) present along with their substitution pattern. Finally, determination of the stereochemistry of each species was accomplished by injecting the PMAA standards available in the laboratory collection.

Using these procedures, 11 different PMAAs were found (Fig. 12.3d), indicating a very complex glycan. Notably, rhamnose existed in four different forms (terminal, 2-linked, 3-linked and 2,3-linked), xylose in two forms (terminal and 4-linked), while fucose, arabinose and the three hexoses only formed one PMAA derivative.

These results established the nature of the 2,3-diOMe-6-deoxyhexose found in the acetylated alditol analysis (Sect. 12.3.1). Depending on its location in the glycopeptide, the methylation protocol could transform this residue into two different PMAAs, a 4-linked deoxyhexose if substituted at the only available position, or into a fully methylated derivative if it is located in a terminal position. Accordingly, the only PMAA derivative compatible with one of the two options was a terminal rhamnose unit.

12.3.3 Determination of the Monosaccharide Absolute Configuration

The acetylated methyl glycosides of enantiomeric sugars, as D- and L-rhamnose, are equivalent during GC-MS analysis, because the GC column used can separate diastereoisomers but not enantiomers. The identification of the absolute configuration is therefore possible by replacing the methanol with a chiral alcohol, usually 2-octanol, to prepare the acetylated glycoside (acetylated octyl glycoside, AOG). The advantage of this approach over the traditional one is that the isolation of mg amounts of the target monosaccharide to determine its chirality with the polarimeter is not necessary. In addition, the AOG method enables the contemporaneous determination of the stereochemistry of a mixture of monosaccharides, if the appropriate standards are available.

Preparation of the acetylated octyl glycoside standard is a procedure that consists of two steps: (1) reaction of the monosaccharide with racemic 2-(±)-octanol and (2) reaction of the monosaccharide with one optically pure enantiomer of the alcohol, for instance, 2-(-)-octanol. The first step, using L-rhamnose (L-Rha) as an example, produces two diastereoisomers, L-Rha-(+)-oct and L-Rha-(-)-oct, each equivalent to its enantiomer, D-Rha-(-)-oct and D-Rha-(+)-oct, respectively, during GC-MS analysis. The second step gives only one diastereomer, i.e., L-Rha-(-)-oct whose GC-MS retention time is also equivalent to that of its enantiomer D-Rha-(+)-oct. Thus, the two separate reactions allow the determination of the retention time of L-Rha-(-)-oct and D-Rha-(-)-oct. This information is used to assign the configuration of a monosaccharide, prepared by treating the sample with enantiomeric pure 2-(-)-octanol, followed by acetylation.

This procedure applied to PBCV-1 glycopeptides revealed that fucose and arabinose are L-configured, the three hexoses and xylose are D, while rhamnose exists in both L and D configurations. Collectively, PMAA and AOG analyses indicated that different types of rhamnose residues were in the glycopeptide and that they differed in the type of substitution and in their absolute configuration. This finding prompted an improvement in the octyl glycoside approach that produced a derivative able to provide the absolute configuration of the residue and its substitution pattern. Accordingly, the glycopeptide was methylated and hydrolyzed as in Fig. 12.4b, and then the mixture of the partially methylated monosaccharide was directly converted into the corresponding partially methylated and acetylated octyl glycosides (Fig. 12.4d).

The standards needed for the identification of rhamnose in its variously acetylated and methylated forms were obtained by applying the same approach to the polysaccharide from *Kaistella flava* (Gargiulo et al. 2008), because it had a L-rhamnose linked at either 2 or 3 and at both 2 and 3 positions. The standard for the terminal rhamnose was prepared by methylation of the octyl rhamnoside.

Attribution of the GC-MS profile (Fig. 12.3e) was possible by extension of the fragmentation rules of the methyl glycosides to the octyl glycosides (Lönngren and Svensson 1974). In general, the most diagnostic ion is the oxonium ion that originates from the loss of the aglycon of the anomeric carbon. For the two peaks at ca. 21 min (Fig. 12.3f), it was *m/z* 245 that indicated a rhamnose derivative with two acetyl groups and one methyl group, as expected from the transformation of the 2,3-linked rhamnose unit of the glycan. As for the two peaks at 16 and 18 min (Fig. 12.3f), they had different EI-MS spectra (Fig. 12.3g, h) but the same oxonium ion at *m/z* 217, pointing to a rhamnose with two methyl groups and one acetyl, as expected for a 2- or 3-linked residue. The 2-linked rhamnose species were identified at 16 min because the EI-MS spectrum had an intense ion at m/z 88 that arose from two vicinal methylated hydroxyl groups, which is only possible for a 2-linked residue. Thus, the peaks at 18 min were assigned to 3-linked rhamnose. Identification of the L and D isomer (Fig. 12.3f) was possible by preparing the AOG from the methylated polysaccharide with pure 2-(-)-octanol.

Finally, the PBCV-1 glycopeptide reaction with pure 2-(-)-octanol and acetylation indicated that D-rhamnose was mainly 3-linked (major form), but also terminal, while L-rhamnose was 2- or 2,3-linked or terminal (Fig. 12.3e). Thus, preparation of partially methylated and acetylated octyl glycosides is useful to solve glycan structures that have monosaccharides with opposite configurations; while these cases are rather rare, they are quite challenging when they exist.

12.4 NMR Spectroscopy in Glycan Analysis, Determination of Glycan Sequence, and Anomeric Configuration

NMR spectroscopy is without doubt the best technique for providing information about several features of a glycan, including the configuration of the anomeric carbon, the stereochemistry of the residues and the way the residues are connected. This approach is well developed for samples in solution and combines the information from a certain range of experiments, viz., the homonuclear ^{1}H-^{1}H 2D sequences COSY, TOCSY, and ROESY (or NOESY) and the heteronuclear ^{1}H-^{13}C 2D sequences HSQC, HMBC, and HSQC-TOCSY. The major limitation of this technique is related to the amount of sample available, a limit that nowadays is less severe because of the availability of instruments at high field, equipped with cold probes, and gradient-selected sequences that shorten the acquisition time. Thus, the following comments describe the major shortcuts for quickly obtaining the main structural information of a glycan.

12.4.1 Configuration of the Anomeric Carbon

The regions of interest are those at 5.5–4.3 ppm and 110–90 ppm for proton and carbon spectra, respectively. Proton signals found above 5 ppm are generally considered diagnostic of a residue in the α-pyranosidic form, while those below 5 ppm

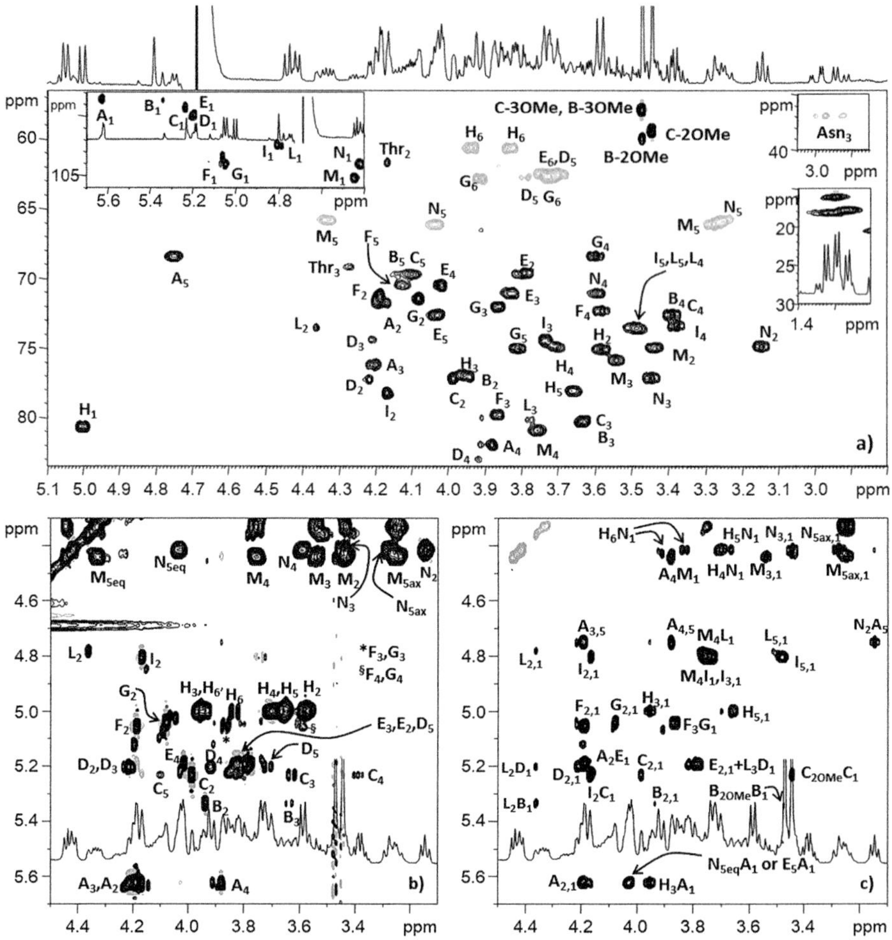

Fig. 12.5 NMR spectra acquired for PBCV-1 glycopeptide structural studies (refer to Fig. 12.2 for labeling system, glycopeptide amount 0.5 mg). (**a**) HSQC spectrum. (**b**) Expansion of the TOCSY spectrum. (**c**) Expansion of T-ROESY spectrum. (Adapted from De Castro et al. 2013)

are diagnostic of β-configured pyranose sugars. In regard to the carbon signals, the anomeric carbon of an α-configured pyranose residue has a chemical shift lower than a β-configured unit, and it is generally found at less than 102 ppm. This generalization does not consider residues in the furanose form or having a N-glycosidic linkage; for these cases, the determination of the anomeric configuration needs information on all of the proton and carbon signals.

In the case of the PBCV-1 glycan, the HSQC spectrum (Fig. 12.5a) had about 11 anomeric signals. Of these 11 signals, 6 were assigned to α-configured monosaccharides (**A**–**C**, **E**–**G**) because they were found at proton signals above 5 ppm. Unit **D** represented an arabinose in the β-furanose form, in agreement with linkage analysis and with the carbon four (C-4) chemical shifts at 83.0 ppm (Bock and Pedersen

1983). All the other signals belonged to β-configured pyranose residues, with glucose **H** having an N-type glycosidic linkage as inferred from the characteristic signal at 80 ppm.

12.4.2 Stereochemistry of the Residues

For most of the residues, this feature can be quickly determined by analyzing the anomeric region of the TOCSY spectrum (Fig. 12.5b). In this experiment, the transfer of the magnetization from one proton to the next is controlled by the value of the coupling constants ($^2J_{H,H}$ and $^3J_{H,H}$), which in turn depends on the stereochemistry of the monosaccharides investigated. In the case of a pyranose ring, small $^3J_{H,H}$ values are characteristic of axial/equatorial- or equatorial/equatorial-arranged protons and hinder the magnetization transfer, thus decreasing the intensity of the successive correlations in the spectrum or, in some cases, preventing them completely. For example, *manno*- and *galacto*-configured residues in the pyranose form have small $^3J_{H1,H2}$ and $^3J_{H3,H4}$ values, respectively. Accordingly, the anomeric proton (H-1) of an α-*manno*-configured monosaccharide correlates with H-2 and has very poor correlations, if at all, with the other protons in the residue; the H-1 of the β-isomer correlates only with H-2. A *galacto*-configured residue instead has only three correlations, from H-2 to H-4, independent of the configuration at the anomeric center. In contrast, residues with large coupling values such as *gluco*-configured forms have the complete correlation pattern, up to both H-6 protons (or up to H-5 protons in xylose residues). This approach is less straightforward for recognizing other isomers, such as those with *altro*- or *talo*-stereochemistry, or residues in the furanose form. Therefore, it is always good practice to confirm the identity of a monosaccharide by comparing its NMR data with that of a reference compound (Bock and Pedersen 1983) and with information from chemical analysis.

Regarding the PBCV-1 glycopeptides, the TOCSY spectrum in Fig. 12.5b identified the residues **I**, **L**, **F**, **G**, **C**, and **B** as *manno*-configured, and successive integration of TOCSY with COSY and HSQC spectra determined that all the residues were 6-deoxy-mannose or rhamnose units except **G**, which was a mannose. Similarly, **A** and **E** had a *galacto*-configuration, but **A** was a fucose, while **E** was a galactose. The residues **H**, **M**, and **N** had the complete correlation pattern (Fig. 12.5b), and further analysis disclosed that **H** was a glucose, while **M** and **N** were two units. The arabinose residue escaped from this scheme because it was in the furanose form; thus, its identity was determined by assignment of all the chemical shift values as reported in the previous section.

12.4.3 Sequence of the Residues

This aspect is pursued by two main approaches. Homonuclear ^{1}H-^{1}H NOESY and ROESY (or its variation T-ROESY) sequences exploit the nuclear Overhauser effect (NOE) that relates nuclei at less than 4 Å. Thus, NOE effects occur within the same residue (intra-residue) as well as between different sugar units (inter-residues). Both types of NOEs are useful during structural determination, but only inter-residue NOEs define the sequence of the residues in the polymer and, in most of the cases, also the substitution point.

As for the PBCV-1 glycopeptide, T-ROESY (Fig. 12.5c) displayed several correlations. For instance, the anomeric signal of **M** (a xylose unit) had three correlations: those with protons H-3 and H-5 ($M_{3.1}$ and $M_{5ax.1}$, respectively) confirmed the β configuration of the anomeric center, while the last, A_4M_1, placed this residue unequivocally at position 4 (C-4) of **A**, the fucose residue. A similar pattern of intra- and inter-residue NOEs was detected for the other residues β-configured at the anomeric position, namely, **I** and **L**.

The anomeric proton of an α-configured residue displays the inter-residue correlation with the proton at the position of the glycosidic linkage, while the intra-residue correlations occur with H-2. However, NOE effects report spatial proximity, meaning that care is necessary during interpretation. For instance, arabinose **D** is linked at C-3 of **L**, but it correlates with both H-2 and H-3 of this unit (correlations L_2D_1 and L_3D_1 in Fig. 12.5c). Similarly, xylose **N** is linked at C-4 of H, but it correlates with both H-4 and H-6 of the glucose unit (correlations H_4N_1 and H_6N_1 in Fig. 12.5c). In addition, other NOE effects connect protons of residues not even directly linked, such as H-2 of **N** with H-5 of **A** (N_2A_5 in Fig. 12.5c), which means that the glycan adopts a conformation that brings these two parts close to each other.

The second approach consists of recording the heteronuclear ^{1}H-^{13}C HMBC spectrum that relates proton and carbon nuclei separated from two (^{2}J) or three (^{3}J) chemical bonds and that are scalar (or J) coupled. Indeed, as for NOE-based techniques, HMBC detects intra- (^{2}J and ^{3}J) and inter-residue correlations (^{3}J), and the intensity associated depends on the J value. The absence of a correlation does not mean that two residues are not linked but simply that the ^{3}J value is close to null or far from the value for which the HMBC acquisition is optimized. It is good practice to record this type of spectrum to confirm the results from the NOESY (or T-ROESY) experiments as was done for PBCV-1.

12.5 Glycoproteomic Analysis of MCP from Chloroviruses by MALDI-TOF MS and MALDI-TOF-TOF MS/MS

Complete characterization of glycoproteins requires the identification of the glycosylation sites, as well as the site-specific analysis of the glycan components, which often have a broad number of glycoforms (microheterogeneity) that enhances the overall molecular complexity. Due to the many structural components to be examined, conventional MS-based methods typically rely on a combination of two different types of analyses: (1) a basic glycomic strategy that analyzes the enzymatically or chemically released glycan mixture and (2) a glycoproteomic approach which also provides information on the glycosylation sites by analyzing the glycopeptides resulting from protein digestion. We adopted this last procedure to analyze the N-glycosylation pattern of the PBCV-1 Vp54 protein because the first approach was hampered by the lack of N-glycan releasing enzymes with the appropriate specificity. Moreover, chemical deglycosylation methods would have led to the loss of the protein counterpart and therefore loss of the structural information at the glycosylation sites. Protease digested samples often contain a complex mixture of peptides and glycopeptides, which are poorly detected because they can be heterogeneous in composition and because their ionization efficiency is less than that of the unmodified peptide. Accordingly, a number of chromatographic methods based on physical and chemical properties of the glycosylated species have been developed for the enrichment/separation of the glycopeptide fraction (Wuhrer et al. 2007; Ongay et al. 2012).

As a first step, our experimental protocol involved the enzymatic digestion of the viral Vp54 glycoprotein with two different proteases, proteinase K and thermolysin, to generate a mixture of properly sized N-linked glycopeptides (Table 12.2). With its broad specificity, proteinase K produced fractions enriched in glycopeptide fragments that were separated via size-exclusion chromatography (SEC) on a Bio-Gel P-10 column. The more selective protease, thermolysin, in association with alternative glycopeptide purification procedures, such as reverse-phase (RP) fractionation on C18 columns, hydrophilic interaction liquid chromatography (HILIC) on microcrystalline cellulose, and solid-phase extraction (SPE) on graphitized columns, produced a different set of glycopeptides. Because of its low specificity, proteinase K produced glycans with either a single amino acid or attached to short peptides, whereas thermolysin produced glycopeptides with larger peptides. Trypsin digestion, despite having more targeted cleavages of the protein backbone, was not used because it was expected to generate glycosylated peptides too large for MALDI MS and MS/MS analyses.

By using a combination of the two proteases and the various enrichment columns, we were able to isolate several fractions enriched in glycopeptides (Table 12.2). In the majority of the cases, the samples were still complex mixtures, containing some peptides and mainly glycopeptides whose heterogeneity depended on both peptide and glycan components. All the fractions were analyzed off-line by MALDI-TOF MS and MALDI-TOF-TOF MS/MS in order to identify protein glycosylation sites and glycan sequences.

First-line analysis employed MALDI MS identification of glycopeptides, which always detected the positive ionization mode as adducts with sodium $[M+Na]^+$ and potassium $[M+K]^+$. The spacing pattern ($\Delta m/z=16$) between these two adducts proved useful to distinguish glycopeptide from peptide ions. Moreover, the mass differences between the glycopeptides disclosed that some monosaccharides were present as nonstoichiometric substituents.

Table 12.2 Structures of Vp54 glycopeptides identified by MALDI-TOF MS and MALDI-TOF-TOF MS/MS. Individual species are reported together with the relative glycosylation site(s) assignment and the use of proteolytic enzymes and enrichment methods

m/z [M+Na]⁺ observed (theoretical)	Structure	Glycosylation site	Protease digestion glycopeptide enrichment	*m/z* [M+Na]⁺ observed (theoretical)	Structure	Glycosylation site	Protease digestion glycopeptide enrichment
1649.51 (1649.59)	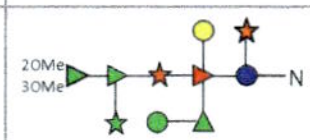	[280]Asn [302]Asn [399]Asn [406]Asn	Proteinase K Bio-Gel P10	1686.51 (1686.63)	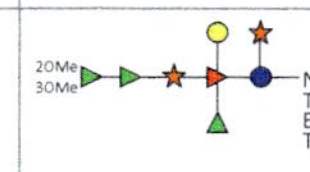	[399]Asn	Proteinase K Bio-Gel P10
1750.54 (1750.64)	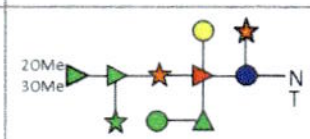	[302]Asn [399]Asn [406]Asn	Proteinase K Bio-Gel P10	1754.64 (1754.69)	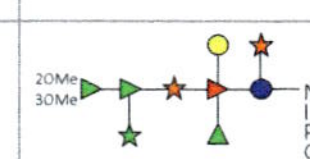	[280]Asn	Proteinase K Bio-Gel P10
1784.60 (1784.70)	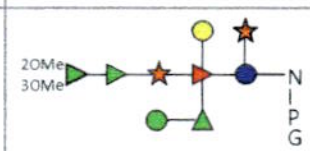	[280]Asn	Proteinase K Bio-Gel P10	1818.55 (1818.67)	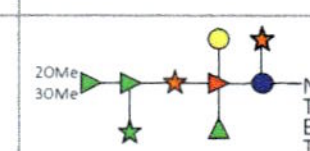	[399]Asn	Proteinase K Bio-Gel P10
1848.58 (1848.68)	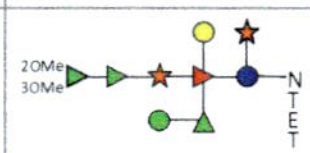	[399]Asn	Proteinase K Bio-Gel P10	1899.60 (1899.78)	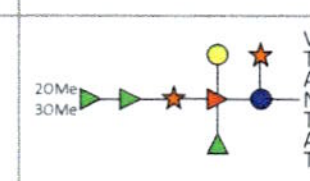	[406]Asn	Thermolysin RP-C18 Proteinase K Bio-Gel P10
1916.66 (1816.75)	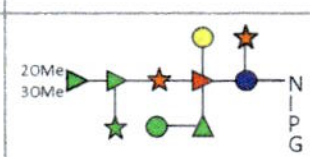	[280]Asn	Proteinase K Bio-Gel P10	1955.67 (1955.81)	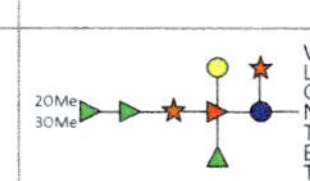	[399]Asn	Thermolysin RP-C18
1980.62 (18980.73)	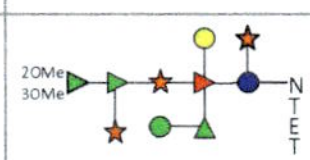	[399]Asn	Proteinase K Bio-Gel P10	2061.74 (2061.83)	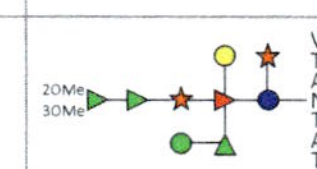	[406]Asn	Thermolysin RP-C18/SPE

2117.91 (2117.86)	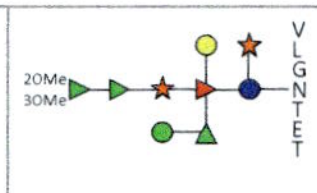	³⁹⁹Asn	Thermolysin RP-C18	2174.82 (2174.92)	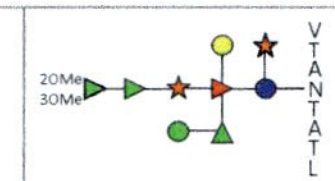	⁴⁰⁶Asn	Thermolysin RP-C18/SPE
2249.96 (2249.90)	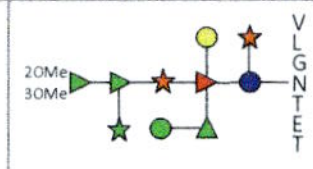	³⁹⁹Asn	Thermolysin RP-C18	2282.16 (2282.15)	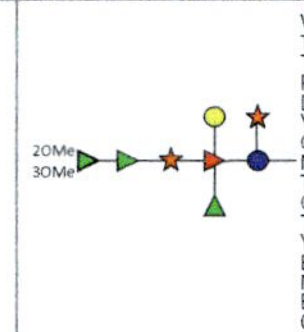	³⁰²Asn	Thermolysin RP-C18/HILIC
3044.14 (3044.20)	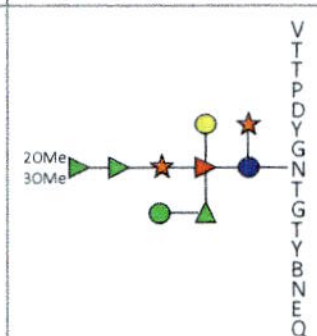	³⁰²Asn	Thermolysin RP-C18/HILIC	3176.17 (3176.24)	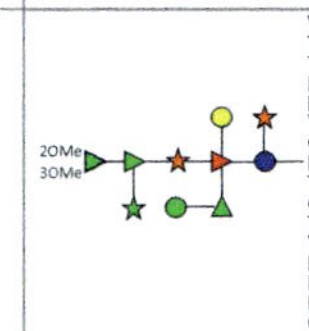	³⁰²Asn	Thermolysin RP-C18

Complete assignment of the more significant glycopeptide MS signals was based on the accurate measurement of the molecular mass, which inferred both peptide and glycan composition (accuracy was more than 75 ppm for Vp54 main glycopeptides), and on tandem MS analysis which provided evidence for the glycan sequence. MALDI-TOF-TOF spectra gave primarily low-energy fragments consisting of the cleavage of the glycoside bonds between adjacent sugar rings – the B-, C-, and Y-type ions (rarely Z-type ions) according to Domon and Costello nomenclature (Domon and Costello 1988). B and C ions comprise the nonreducing sugar terminal, the first arising from a glycosyl bond β-elimination implying the further loss of a water molecule, $[B]^+ = [C - H_2O]^+$, whereas, analogously, Y and Z ions include the reducing end, with $[Z]^+ = [Y - H_2O]^+$. Glycan Y-type fragments were found as the predominant cleavage, which generated highly informative ions retaining the peptide glycosylation sites. On the other hand, B and C ions, typically occuring at or near the nonreducing end side, provided valuable information on the molecular mass of the whole glycan moiety and on the substitution of the innermost glucose residue. Notably, species with larger peptide portions also led to y- and b-type cleavage (Biemann 1992) of the amino acid backbone linkages, allowing further characterization of the peptide side chain.

By combining all the MS data with peptide sequence information, as well as those provided by the FindPept bioinformatic tool (www.expasy.org) and the compositional analysis, we were able to define the PBCV-1 N-glycosylation pattern. MS and MS/MS analyses helped to identify individual amino acid sequences associated with Vp54 glycosylation (at ^{280}NIPG, ^{302}NTGT, ^{399}NTET, ^{406}NTAT), as well as the characterization and relative evaluation of their attached glycoforms. Of note, none of the asparagine is in a typical consensus sequence; Fig. 12.6 shows MS/MS analysis of the four glycopeptides that were more representative and reveal the complex pattern of Vp54 N-glycosylation.

12.6 Conclusions

The structure of the glycans of the PBCV-1 major capsid protein Vp54 had remained elusive because the tools commonly used to study eukaryotic-like glycosylation patterns failed. This glycosylation pattern has been solved only recently (De Castro et al. 2013) by combining three techniques, GC-MS, NMR and MALDI, and this same approach was effective when applied to other chlorovirus encoded glycoproteins (De Castro et al. 2016; Quispe et al. 2017; Speciale et al. 2017). As a result, the structures of other chlorovirus N-glycans are now available, confirming that glycosylation is a trait common to these viruses. The results also revealed that chlorovirus glycans share the same core oligosaccharide structure that is unique (Fig. 12.2) and not shared by any other form of life.

Chloroviruses are members of a rapidly expanding group of viruses characterized by large particle size with huge genomes and are often referred to as giant

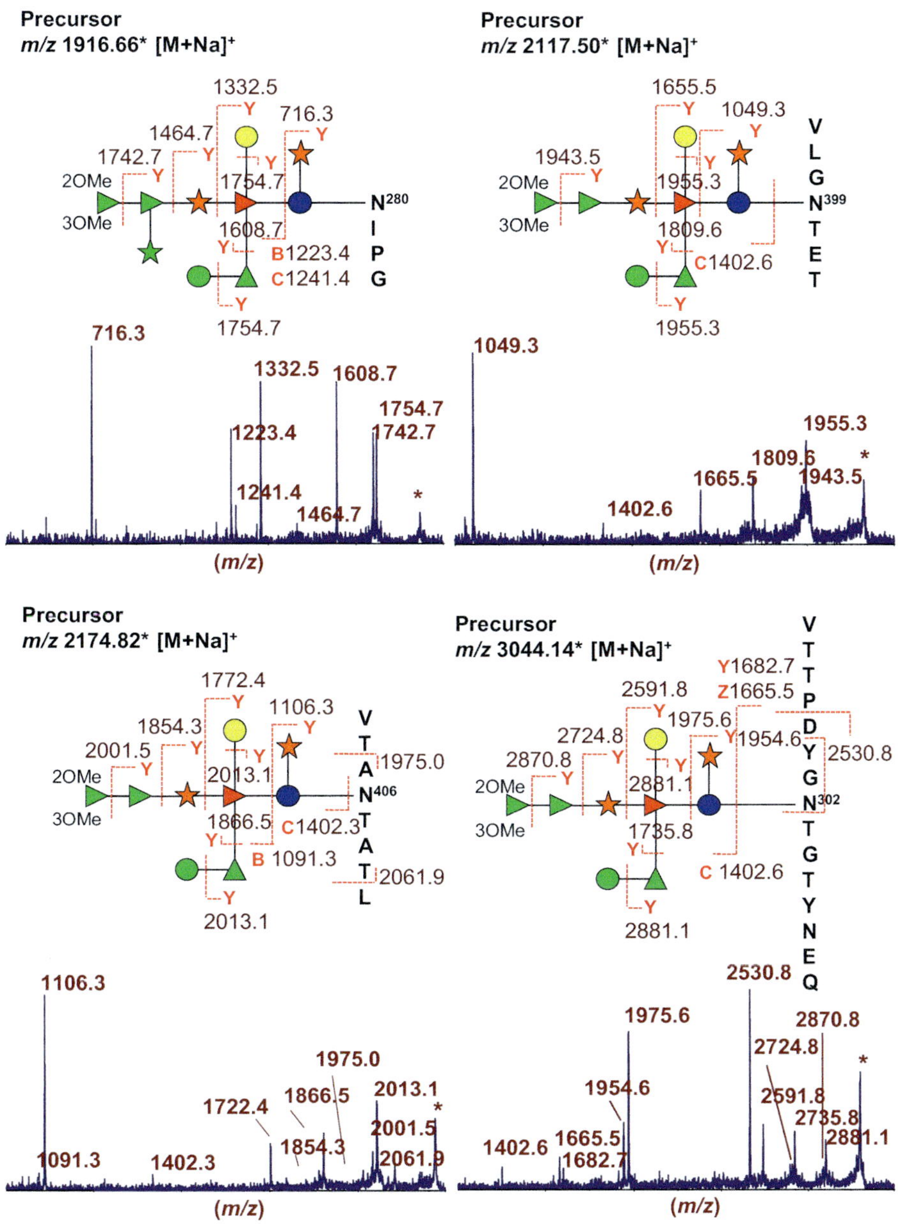

Fig. 12.6 MALDI-TOF-TOF MS/MS spectra of selected glycopeptides from Vp54. Each precursor ion, as detected by MALDI-TOF MS, is associated with one of the four distinct Vp54 glycosylation sites. (Adapted from De Castro et al. 2013). The horizontal axis indicates the mass/charge ratio (*m/z*) of the ions, and numbers are omitted to avoid crowding. Most relevant peaks are labeled with the mass value

viruses. Many giant viruses have putative genes involved in carbohydrate metabolism, such as *Mimivirus* and *Megavirus* (both in the *Mimiviridae* family) for which we have direct evidence that a thick layer of polysaccharides coats their surface (De Castro and Tonetti, unpublished results). The take-home message is that viral autonomous glycosylation probably occurs in a wide range of viruses. The insightful use of the available methods has now added a new topic to the flourishing field of glycobiology.

References

Abergel C, Legendre M, Claverie JM (2015) The rapidly expanding universe of giant viruses: Mimivirus, Pandoravirus, Pithovirus and Mollivirus. FEMS Microbiol Rev 39:779–796

Biemann K (1992) Mass spectrometry of peptides and proteins. Annu Rev Biochem 61:977–1010

Bock K, Pedersen C (1983) Carbon-13 nuclear magnetic resonance spectroscopy of monosaccharides. Adv Carbohydr Chem Biochem 41:27–66

Cherrier MV, Kostyuchenko VA, Xiao C et al (2009) An icosahedral algal virus has a complex unique vertex decorated by a spike. PNAS 106:11085–11089

Colson P, De Lamballerie X, Yutin N et al (2013) "Megavirales", a proposed new order for eukaryotic nucleocytoplasmic large DNA viruses. Arch Virol 158:2617–2521

De Castro C, Parrilli M, Holst O et al (2010) Microbe-associated molecular patterns in innate immunity: extraction and chemical analysis of gram-negative bacterial lipopolysaccharides. Methods Enzymol 480:89–115

De Castro C, Molinaro A, Piacente F et al (2013) Structure of N-linked oligosaccharides attached to chlorovirus PBCV-1 major capsid protein reveals unusual class of complex N-glycans. PNAS 110:13956–13960

De Castro C, Speciale I, Duncan G et al (2016) N-linked glycans of chloroviruses sharing a core architecture without precedent. Angew Chem Int Ed 55:654–658

De Castro C, Klose T, Speciale I et al (2018) Structure of the chlorovirus PBCV-1 major capsid glycoprotein determined by combining crystallographic and carbohydrate molecular modeling approaches. PNAS 115:E44–E52

Domon B, Costello CE (1988) A systematic nomenclature for carbohydrate fragmentations in FAB-MS/MS spectra of glycoconjugates. Glycoconj J 5:397–409

Dunigan DD, Cerny RL, Bauman AT et al (2012) Paramecium bursaria chlorella virus 1 proteome reveals novel architectural and regulatory features of a giant virus. J Virol 86:8821–8834

Gargiulo V, De Castro C, Lanzetta R et al (2008) Structural elucidation of the capsular polysaccharide isolated from *Kaistella flava*. *Carbohydr Res* 343:2401–2405

Jeanniard A, Dunigan DD, Gurnon JR et al (2013) Towards defining the chloroviruses: a genomic journey through a genus of large DNA viruses. BMC Genomics 14:158

Karakashian SJ, Karakashian MW (1965) Evolution and symbiosis in the genus Chlorella and related algae. Evolution 19:368–377

Legendre M, Bartolia J, Shmakov L et al (2014) Thirty-thousand-year-old distant relative of giant icosahedral DNA viruses with a pandoravirus morphology. PNAS 111:4274–4279

Lönngren J, Svensson S (1974) Mass spectrometry in structural analysis of natural carbohydrates. *Adv Carbohydr Chem Biochem* 29:41–106

Nandhagopal N, Simpson AA, Gurnon JR et al (2002) The structure and evolution of the major capsid protein of a large, lipid-containing DNA virus. PNAS 99:14758–14763

Ongay S, Boichenko A, Govorukhina N et al (2012) Glycopeptide enrichment and separation for protein glycosylation analysis. J Sep Sci 35:2341–2372

Philippe N, Legendre M, Doutre G et al (2013) Pandoraviruses: Amoeba viruses with genomes up to 2.5 Mb reaching that of parasitic eukaryotes. Science 341:281–286

Quispe CF, Esmael A, Sonderman O et al (2017) Characterization of a new chlorovirus type with permissive and non-permissive features on phylogenetically related algal strains. Virology 500:103–113

Raoult D, Audic S, Robert C et al (2004) A huge virus that infects amoebae contains genes that are not usually part of the viral repertoire and defines a family of ancient nucleocytoplasmic DNA viruses. Science 306:1344–1350

Speciale I, Agarkova I, Duncan GA et al (2017) Structure of the N-glycans from the chlorovirus NE-JV-1. Anton van Leeuw 110:1391–1399

Van Etten JL, Dunigan DD (2012) Chloroviruses: not your everyday plant virus. Trends Plant Sci 17:1–8

Van Etten JL, Meints RH, Kuczmarski D et al (1982) Viruses of symbiotic Chlorella-like algae isolated from *Paramecium bursaria* and *Hydra viridis*. PNAS 79:3867–3871

Van Etten JL, Gurnon JR, Yanai-Balser GM et al (2010) Chlorella viruses encode most, if not all, of the machinery to glycosylate their glycoproteins independent of the endoplasmic reticulum and Golgi. Biochim Biophys Acta 1800:152–159

Van Etten JL, Agarkova I, Dunigan DD et al (2017) Chloroviruses have a sweet tooth. Viruses 9:E88

Vigerust DJ, Shepherd VL (2007) Virus glycosylation: role in virulence and immune interactions. Trends Microbiol 15:211–218

Wuhrer M, Catalina MI, Deelder AM et al (2007) Glycoproteomics based on tandem mass spectrometry of glycopeptides. J Chromatogr B 849:115–128

Zhang X, Xiang Y, Dunigan DD, Klose T et al (2011) Three-dimensional structure and function of the Paramecium bursaria chlorella virus capsid. PNAS 108:14837–14842

Chapter 13
Quantifying Weak Glycan-Protein Interactions Using a Biolayer Interferometry Competition Assay: Applications to ECL Lectin and X-31 Influenza Hemagglutinin

Ye Ji and Robert J. Woods

Abstract This chapter introduces two formats using bio-layer interferometry competition assays to determine the solution K_D values of weak glycan-protein interactions. This approach overcomes the challenge of determining weak interactions while minimizing the amount of reagents required. Accurate solution K_D values aid in understanding the complex relationships between monomeric versus multimeric interactions and affinity versus avidity. The assays have been applied to a well-studied lectin (*Erythrina crista-galli* lectin) and influenza hemagglutinin (X-31). The solution K_D values determined from this approach are in good agreement with previous reported literature values from isothermal titration calorimetry and NMR. Additionally, this approach appears robust and precise.

Keywords Bio-layer interferometry · Weak interaction · Glycan-protein interaction · BLI-based inhibition assay · ECL · X-31

Abbreviations

BLI	Bio-layer interferometry
CFG	Consortium for Functional Glycomics
ECL	*Erythrina crista-galli* lectin
ELISA	Enzyme-linked immunosorbent assay
HA	Hemagglutinin
ITC	Isothermal titration calorimetry

Y. Ji · R. J. Woods (✉)
Complex Carbohydrate Research Center, University of Georgia, Athens, GA, USA
e-mail: rwoods@ccrc.uga.edu

Y. Yamaguchi, K. Kato (eds.), *Glycobiophysics*, Advances in Experimental Medicine and Biology 1104, https://doi.org/10.1007/978-981-13-2158-0_13

MST	Microscale thermophoresis
SPR	Surface plasmon resonance
NA	Neuraminidase
NMR	Nuclear magnetic resonance spectroscopy
3D	Three-dimensional

13.1 Introduction

Bio-layer interferometry (BLI) is a relatively new label-free technique to study the interactions between an immobilized receptor and soluble analytes in high-throughput, label-free, real-time molecular interaction analysis (Rich and Myszka 2007). This optical technique analyzes the change in the interference pattern of white light reflected from a biosensor surface carrying a chemically immobilized ligand (such as a protein) that occurs as the analyte in solution (such as a carbohydrate) binds to the ligand (Fig. 13.1).

The BLI approach shares some conceptual similarities with surface plasmon resonance (SPR), in the sense that the ligand is immobilized on a surface. Both methods use similar immobilization chemistry, and the binding is detected in real time. However, there are differences. Whereas SPR employs a microfluidic flow cell, BLI employs a multi-well plate (typically 96 or 384), with a row of eight sensors that are dipped into the wells. The association phase of the binding is thus primarily dependent on the concentration of the analyte in the wells and to the time the probe is exposed to the analyte. To measure analyte dissociation, the probes are automatically raised and transferred to wells containing buffer (Fig. 13.2). The times for association and dissociation phases can be adjusted during the experiment, and more complex assays may be programmed. To assist in minimizing mass-transport effects, the analyte tray oscillates at up to 1000 rpm.

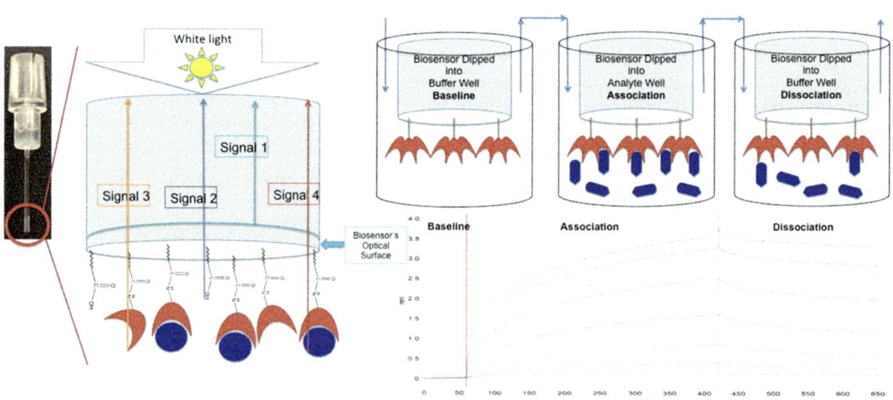

Fig. 13.1 Mechanism of ForteBio bio-layer interferometry

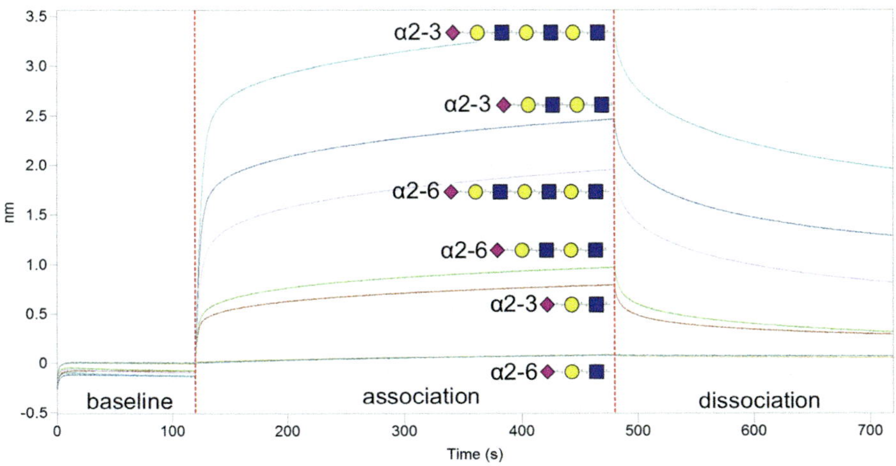

Fig. 13.2 An example in which BLI is used to screen the binding of six oligosaccharides to the lectin wheat germ agglutinin (WGA). Baseline (sensor soaking in buffer) 0~120 s, association phase 120–480 s, and dissociation phase 480–720 s

The performance of SPR and BLI has been reviewed recently (Abdiche et al. 2008, 2009; Yang et al. 2016), and while SPR may be slightly more sensitive (Abdiche et al. 2008), K_D values from BLI are generally within a factor of 2 (larger) than those from SPR (Abdiche et al. 2008, 2009). A significant advantage of BLI over SPR is the lack of any limit on measurement times for association or dissociation, which are limited in SPR by the volume of the microfluidics. On the other hand, long exposures in the analyte wells can lead to solvent evaporation degrading the BLI measurements (Yang et al. 2016). Other benefits of BLI include the ability to measure interactions on disposable tips that do not need to be regenerated and to not necessarily consume precious samples. This makes it a versatile complement to Biacore (Abdiche et al. 2008). Overall the high throughput of the BLI platform appears to be its most significant benefit over SPR (Yang et al. 2016), and the popularity of the BLI method is steadily increasing (Fig. 13.3). BLI provides a relatively high-throughput, cost-effective method for quantifying the strengths of carbohydrate-protein interactions. This helps bridge the gap between qualitative affinity data from high-throughput glycan array screening, and quantitative data from lower-throughput methods such as SPR, ELISA, ITC, MST, and NMR. Additionally, like SPR, BLI provides the ability to determine kinetic binding rate constants (kon, koff), which are critical to a complete understanding of molecular recognition processes, such as between antibodies or lectins and carbohydrates (Lin et al. 2012; Lian et al. 2013; Schieffelin et al. 2010).

BLI is recommended for characterizing interactions with affinities between 1 mM and 10 pM, and for analytes with molecular weights of at least 1500 Da, but preferably closer to 4 kDa (Rich and Myszka 2007). A convenient feature of the ForteBio implementation of BLI is the ability to alter the times allocated for the association or dissociation phases in real time, in response to the observed data.

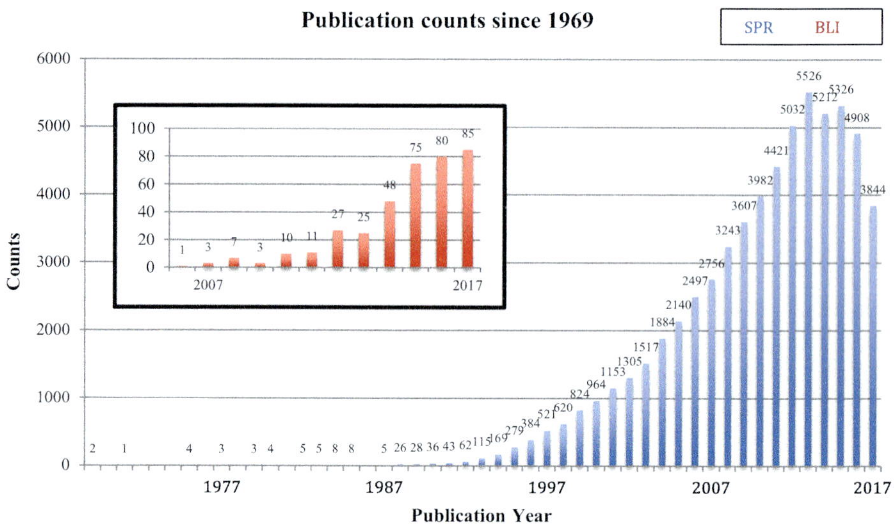

Fig. 13.3 Number of publications citing SPR and or BLI reported per year. Data obtained from SciFinder searched with the keyword "surface plasmon resonance" or "bio-layer interferometry." The first publication of SPR (Stanford and Bennett 1969) was reported in 1969. Then the first publication for BLI appeared over 35 years later in 2007 (Peng et al. 2007)

However, as is the case with all techniques for determining affinities that require immobilization of the receptor, multimeric analytes may form multivalent interactions. Such interactions boost the observed affinity through avidity, and hence do not result in monomeric affinity values (Mammen et al. 1998; Gelinsky-Wersing et al. 2017). Additional issues, which may be particularly problematic in multivalent binding (Liang et al. 2007), include artificially slow dissociation rates due to rapid analyte rebinding or "mass transport" (Schuck and Zhao 2010). Slow off rates ultimately lead to an overestimation of the affinity when K_D is derived from k_{off}/k_{on}.

Protein-carbohydrate interactions mediate countless biological events, from common ligand-analyte binding to cancer metastasis, and viral and bacterial infection (Solís et al. 2015). To overcome the notoriously low affinity (mM) of these interactions, biological glycan interactions frequently take advantage of multimeric or multivalent binding (Mammen et al. 1998). However, the structures of these multimeric complexes are generally unknown, while what is are most often available instead are the three-dimensional (3D) structures of monomeric complexes, generated by x-ray diffraction. To develop robust structure-function relationships for glycans, a first step is therefore to be able to measure monomeric binding affinities that can be interpreted with regard to the available structural data. The ability to quantify monomeric binding affinities for glycans is essential for understanding the effects of protein mutations on biological function, such as, in defining the origin of species specificity of influenza infections (Ji et al. 2017). This information is equally important for the development of inhibitors that target glycan receptors (Ernst and Magnani 2009; Magnani and Ernst 2009).

The low affinity interactions associated with proteins and carbohydrates are challenging to measure by direct binding techniques such as NMR, MST, and ITC (Gupta et al. 1996; Notenboom et al. 2002; Sauter et al. 1992; Xiong et al. 2013), often because of the limited availability of sufficient quantities of large biologically relevant glycans. A competition assay overcomes this limitation by employing the minimum-binding determinant (or other small carbohydrate structure) as the inhibitor while requiring only a minimal amount of the large glycan as the direct binding analyte.

Given that many carbohydrate-binding proteins are multimeric, as are most biologically relevant glycans, surface immobilization techniques are not ideal for quantifying carbohydrate-protein interactions. However, the extreme complexity of many glycans limits their availability; thus we sought to develop an approach to measuring monomeric solution K_D values for carbohydrate-protein interactions that exploited the economies and convenience gained from the BLI approach. To do this, we created a competition-based protocol that reproduces K_D values from solution-phase equilibrium binding measurements (Cheng and Prusoff 1973). We will demonstrate the use of this assay in two common formats; with the lectin from *Erythrina crista-galli* (ECL), we will determine solution K_D with the glycan immobilized, while with influenza A hemagglutinin (HA), we will reverse these conditions.

Given that BLI is not designed to measure interactions weaker than mM, and this is the range of many carbohydrate-protein affinities, the proposed BLI-based inhibition assay was developed to overcome this limitation.

$$K_{D,solution} = IC_{50} / \left(1 + [\text{protein}] / K_{D,surface}\right) \qquad (13.1)$$

As illustrated in Fig. 13.4, a small oligosaccharide is used to inhibit binding of an amplified oligosaccharide to the immobilized protein. By analogy to enzyme inhibition, when the oligosaccharides in both cases are exactly the same glycan, the inhibition constant K_i of small oligosaccharide is equivalent to its solution K_D (Cheng and Prusoff 1973). The solution K_D is obtained from Eq. 13.1 and requires measurement of an IC_{50} for the analyte (oligosaccharide) and a surface K_D ($K_{D,surface}$).

13.2 Assay Format 1: Immobilized Glycan, Protein Analyte

Lectins such as ECL have long been used as carbohydrate detection reagents, and considerable data are available regarding their affinities. Thus ECL was chosen as a standard for development of the present protocol. With the protocol established, we then examined the biologically significant interaction of influenza hemagglutinin with glycans associated with human and avian infection.

Example: *Erythrina crista-galli* Llectin (ECL)-glycan interactions

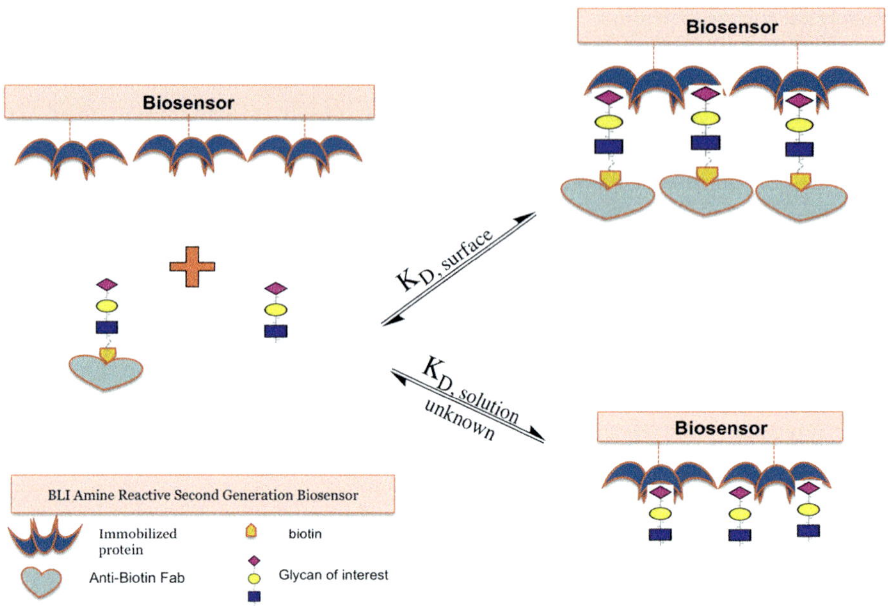

Fig. 13.4 Schematic representation of the BLI-based competition assay. The ability of a small carbohydrate (typically the minimal binding determinant) to inhibit (IC_{50}) the direct binding of the analyte of interest ($K_{D,\,surface}$) is determined in two sets of experiments. From these values the solution K_D ($K_{D,\,solution}$) of the minimal binding determinant may be determined

ECL is widely used as a reagent for the detection of terminal galactopyranose (Gal) residues in glycans (its canonical specificity is for Gal), yet it also binds to *N*-acetylgalactosamine (GalNAc) and fucosylated galactose (L-Fucα1-2Gal). Although its function in the legume is unknown, understanding the complex specificity of lectins, such as ECL, is fundamental to the interpretation of specificity data, as in histology studies (Tong et al. 2001) and in the rational design of diagnostic and therapeutic agents (Farhadi and Hudalla 2016) that target specific glycans. It has been reported from ITC measurement that ECL binds to lactosamine (LacNAc) at 0.26 mM (Gupta et al. 1996) and 0.32 mM (Notenboom et al. 2002). Such weak binding affinities are on the edge of or below the detection limit of SPR and BLI. In this example, the biotinylated oligosaccharides will be immobilized on a streptavidin (SA) biosensor surface, and the ECL will be the analyte for direct binding as $K_{D,\,surface}$. As the second step in the assay, the ability of eight oligosaccharides to inhibit the ECL binding to the LacNAc biosensor surface was quantified in terms of their IC_{50} values (Table 13.1). Subsequently, solution K_D values were derived from Eq. 13.2 (Table 13.2). Like many lectins, ECL is dimeric, so this assay format might be biased by multimeric (avidity) interactions between the lectin and the immobilized oligosaccharides. However, the agreements (Table 13.2) between the reported monomeric solution K_D values from ITC measurements and ours suggest that the dimeric ECL does not form multiple simultaneous interactions with the surface in

Table 13.1 IC_{50}, of all carbohydrate candidates, experiments reported in replicates of three

Analyte	Average (mM)	STDEV (mM)
Lactose (Lac)	0.66	0.044
epi-Lactose	0.44	0.012
LacNAc	0.17	0.011
2′-FucosylLacNAc-sp[a]-N_3	0.07	0.014
2′-FucosylLactose	0.49	0.050
2′-FucosylLactose-N_3	0.47	0.027
Maltose	ND[b]	ND
Cellobiose	ND	ND

[a]sp = -CH_2CH_2-
[b]ND not detected

Table 13.2 Solution K_D (mM) values for several analytes binding to ECL

Oligosaccharide	BLI	ITC (Gupta et al. 1996)	ITC (Notenboom et al. 2002)
Lactose (Lac)	0.32	0.26	0.32
epi-Lactose	0.21	NR[a]	NR
LacNAc	0.08	0.09	NR
2′-FucosylLacNAc-sp[b]-N_3	0.03	NR	NR
2′-FucosylLactose-sp-N_3	0.22	NR	NR
2′-FucosylLactose	0.23	NR	0.31
Maltose	ND[c]	NR	NR
Cellobiose	ND	NR	NR

[a]NR not reported
[b]sp = -CH_2CH_2-
[c]ND not detected

this particular format. The surface K_D and representative sensor plots for direct binding are presented in Fig. 13.5.

$$K_{D,solution} = IC_{50} / \left(1 + [ECL] / K_{D,surface}\right) \tag{13.2}$$

Figure 13.5 shows that the surface K_D determined by fitting to a 1:1 binding model (0.92 μM ± 0.02 (stdev from triplicates)), which compared well to the value from the Scatchard analysis: 0.91 μM ± 0.02 (stdev from triplicates).

13.3 Assay Format 2: Immobilized Protein and Glycan Analyte

Many studies have applied direct ligand binding assays to measure the surface/apparent K_D of ligand-receptor interactions (Khurana et al. 2014; Matsumoto et al. 2010; Makeneni et al. 2014). However, effects from mass transfer, steric effects

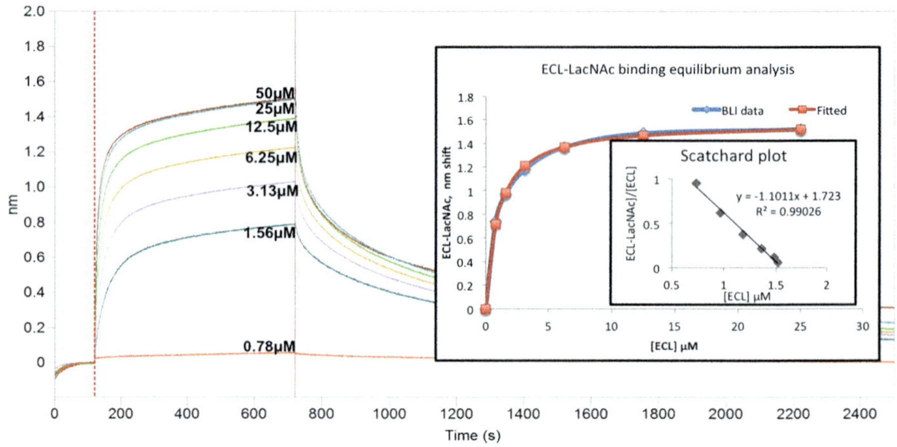

Fig. 13.5 Left, BLI sensorgram of ECL direct binding to LacNAc on SA biosensors. Inset, the $K_{D,surface}$ resulting from equilibrium binding analysis and associated Scatchard plot

(Schuck and Zhao 2010), and avidity (Liao et al. 2010) can lead to a considerable variation in the K_D values (Gupta et al. 1996; Notenboom et al. 2002; Sauter et al. 1992; Xiong et al. 2013). Additionally, when the ligands are low molecular weight, they may be below the confident detection limit of the assay. Nevertheless, the surface K_D values in many cases are useful for defining analyte specificities if not solution affinities (de Vries et al. 2017). However, to understand of monomeric molecular interactions, it would be extremely beneficial to compare to K_D values that do not have avidity effects (Ji et al. 2017).

It was fortuitous in the case of ECL that the protein did not appear to form multimeric interactions with the immobilized oligosaccharides. However, this would not be expected to be the case with a multimeric protein such as trimeric influenza hemagglutinin, which many have specifically evolved to prefer to bind in a multimeric mode to cell-surface oligosaccharides (de Vries et al. 2017; Peng et al. 2017; Khurana et al. 2010). To evade such multimeric interactions, we investigated the present assay by immobilizing the protein (HA) and detecting the binding to the relevant receptor oligosaccharides.

The canonical view of the relationship between HA receptor specificity and species infectivity is that the HA in human-infective viruses prefers to bind to glycans present on the human cell surface that terminate with the Siaα2-6Gal sequence, whereas the HA in avian-infective viruses prefers to bind to glycans that terminate in Siaα2-3Gal (Ji et al. 2017). The discovery of the α2-6/α2-3 infectivity relationship originated not from quantitative biophysical studies but from more qualitative, yet robust, hemagglutination assays (Miller and Stanley 1944). Turning from array screening, lectin staining, and agglutination assays to biophysical assays in order to quantify the relationship between HA sequence and glycan specificity introduces additional challenges. What complicates the quest for a structural rationalization of HA specificity is that many HAs co-crystalize with sialosides independent of whether the sialic acid is in an α2-3 or α2-6Gal linkage. Additionally, solution binding data confirms that the interactions are weak at the monomeric level (mM) and that there is little measurable difference in affinity for either α2-3 or α2-6 sialosides (Ji et al. 2017).

13.4 The Case Study: Human A/Hong Kong/1/1968 (X-31) H3N2 HA: Glycan Interactions

X-31 is one of the first influenza HAs to have its solution binding K_D characterized (by NMR) (Sauter et al. 1989), thus we chose it for the present study.

The direct binding of α2-3 or α2-6 trisaccharides (3′ or 6′-Sialyl-*N*-acetyllactoseamine, short for 3′/6′SLN) to X-31 resulted in signals that were too weak to interpret quantitatively, and the approximate equilibrium K_D values were not able to be determined (Figs. 13.6 and 13.7).

In order to enhance the signal, the biotinylated glycans (molecular weight approximately 765 Da) were combined with an anti-biotin antibody Fab fragment (Cornell et al. 1997), creating a monomeric neoglycoprotein with a molecular weight of approximately 50 kDa. The BLI sensorgram in Fig. 13.8 shows that utilizing the neoglycoprotein dramatically amplified the maximum BLI signal (6.0 nm) relative to the oligosaccharide alone (0.12 nm).

An initial $K_{D,Surface}$ for 6′SLN-Fab binding to immobilized X-31 was measured as 1.15 μM (see inset in Fig. 13.7 for 1:1 association binding fit). When corrected for non-specific binding using an irrelevant neoglycoprotein (biotinylated mannose-Fab), the final $K_{D,Surface}$ for 6′SLN-Fab was 1.58 μM.

In the second step of the assay, IC_{50} values are measured using soluble (non-biotinylated) oligosaccharides to inhibit the corresponding neoglycoprotein from binding to the immobilized X-31 (Fig. 13.9).

The $K_{D,solution}$ values (Table 13.3) were then simply derived by employing the IC_{50} values and $K_{D,surface}$ values in Equation 13.1. Using the Man-Fab neoglycoconjugate

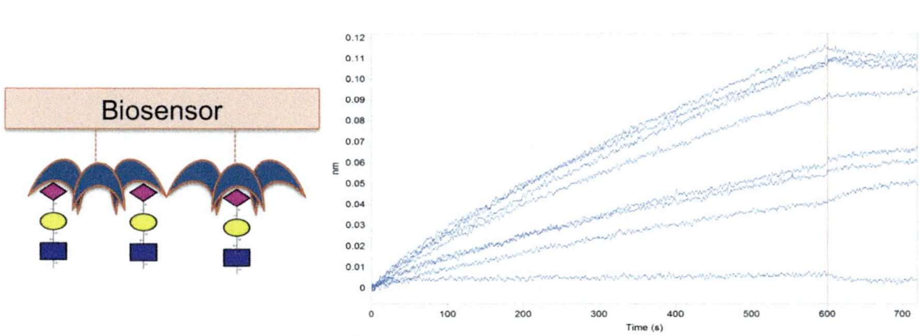

Fig. 13.6 BLI sensorgrams for a serial dilution of the glycan (3′SLN-N_3) direct binding to X-31 H3 HA biosensors. The highest concentration of the glycan is 4mM

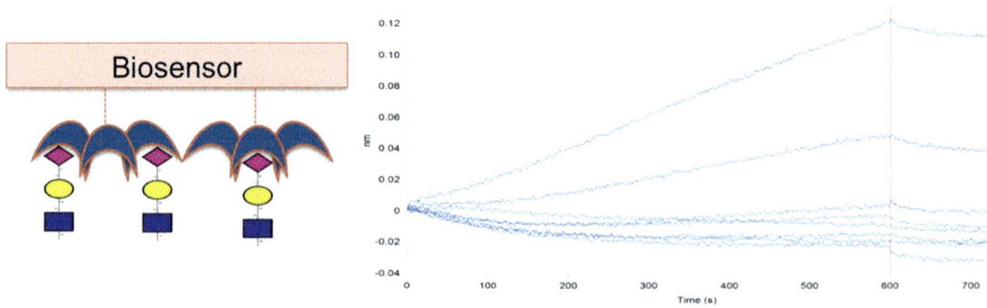

Fig. 13.7 BLI sensorgrams of a serial dilution of the glycan (6′SLN-N_3) direct binding to X-31 H3 HA biosensors. The highest concentration of the glycan is 4mM

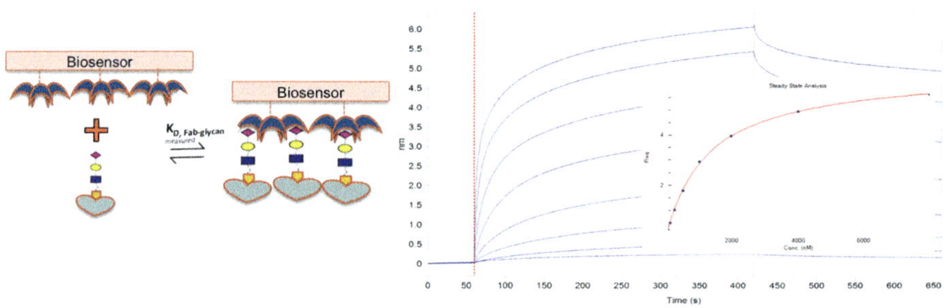

Fig. 13.8 Scheme of Fab-glycan binding directly to HA immobilized biosensor surface

Fig. 13.9 IC50 curve for 6′SLN inhibiting the binding of 6′SLN-Fab (red) and 3′SLN inhibiting the binding of 3′SLN-Fab (blue) to immobilized X-31

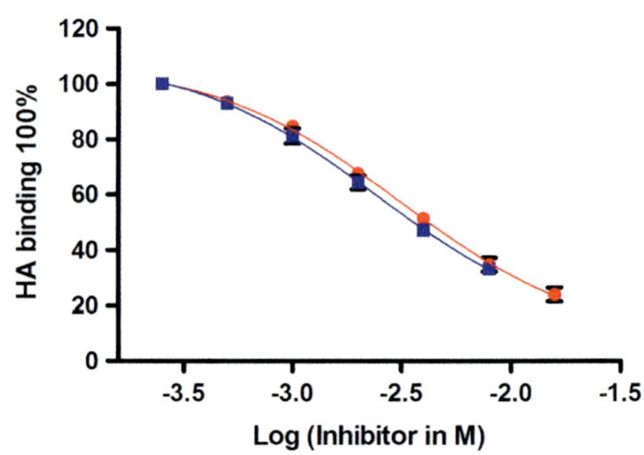

as a correction for non-specific binding modestly altered the final $K_{\mathrm{D,solution}}$ (for 6′SLN) from 2.17 mM (no Man-Fab subtraction) to 2.36 mM (after Man-Fab subtraction).

Thus accounting for the Fab component of the analyte altered the KD values by approximately 10%, but did not alter the relative specificity of 6′SLN versus 3′SLN for X-31 (Table 13.3).

Table 13.3 Solution K_D determined from BLI-based inhibition assay

HA viral strain	Canonical specificity	Ligand	$K_{D,surface}$ (mM)[a]	IC_{50} (mM)	K_D (mM)	Reported K_D (mM)
A/Hong Kong/1/1968	α2-6	3′SLN-N_3	1.7	3.1 ± 0.47	2.4 ± 0.12	NR[b]
(X-31)		3′SLN	1.7	3.6 ± 0.13	2.7 ± 0.01	3.2 ± 0.6 (Sauter et al. 1992) 3.13 ± 0.4 (Xiong et al. 2013)
		6′SLN-N_3	1.2	3.1 ± 0.12	2.2 ± 0.03	NR
		6′SLN	1.2	3.8 ± 0.29	2.7 ± 0.07	2.1 ± 0.3 (Sauter et al. 1992) 2.03 ± 0.2 (Xiong et al. 2013)

[a]At 25 °C
[b]NR not reported

13.5 Concluding Remarks

The BLI competition assay formats appear to be highly suited to measuring weak (mM) solution K_D values for oligosaccharide-protein interactions. Not only does this approach provide accurate values but it also minimized reagent usage, which can be critical, especially for scarce reagents such as complex glycans.

For proteins that cannot form multimeric interactions, Assay Format 1 or 2 may be employed. However to minimize avidity effects from multivalent binding, Assay Format 2 is likely to be preferable. In the case of ECL, it appears that although it is dimeric, it cannot form multimeric interactions with the immobilized oligosaccharides, possibly due to the spacing of the biotinylated oligosaccharides on the SA surface. The fact that the ECL ligands were all short (disaccharides) may also contribute to the inability of ECL to form multimeric interactions in Assay Format 1. Assay Format 1 may be more beneficial when the oligosaccharide is the limiting reagent.

When both the protein and the oligosaccharides are in scarce supply, Assay Format 2 offers the benefit of being able to recover the oligosaccharides used in the IC_{50} step by using a molecular weight cutoff spin column to separate the oligosaccharide from the neoglycoprotien analyte. Oligosaccharide recovery is not as efficient in Assay Format 1 because some amount of the oligosaccharide will remain bound to the protein analyte.

Assay Format 1 Experimental Details

Experimental materials: *Erythrina crista-galli* lectin (ECL, Cat#: L-1140, Vector Lab, Burlingame, CA, USA), N-Acetyl-D-lactosamine (LacNAc, Cat# A7791, Sigma-Aldrich, St. Louis, MO, USA), D-Lactose monohydrate (Lac, Cat#: 61339, Sigma-Aldrich, St. Louis, MO, USA), 4-O-β-Galactopyranosyl-D-mannopyranose (epi-Lac, Cat#: G0886, Sigma-Aldrich, St. Louis, MO, USA),

2-Fucosyllactose (2′FucLac, Cat#: OF06739, Carbosynth Limited, Berkshire, UK), D-(+)-Cellobiose (Cellobiose, Cat# 22150, Sigma-Aldrich, St. Louis, MO, USA), and maltose (Maltose, Cat# 0168-17, DIFCO&BD, Detroit, MI, USA) were purchased from their commercial resources. 2-Fucosyllactose-$OCH_2CH_2CH_2$-N_3 (2′FucLac-N3, CFG#: Tr120) and 2-Fucosyllactosamine-$OCH_2CH_2CH_2$-N_3 (2′FucLacNAc-N3, CFG#: Tr117) were requested from the Consortium for Functional Glycomics (CFG). Biotinylated glycan Galβ1-4GlcNAcβ-$OCH_2CH_2CH_2NH$-biotin (LacNAc-biotin) was received as a gift from Dr. Nicolai Bovin. ECL was weighted and dissolved in the ECL buffer: 10 mM HEPES, 15 mM NaCl, 0.1 mM $CaCl_2$, and 0.1 mM $MnCl_2$ buffered at pH7.4, at 25 °C.

Protein BLI direct binding assay ($K_{D,\ surface}$): Ligand LacNAc-biotin was loaded onto streptavidin biosensors (SA, Cat#: 18-5019, Pall ForteBio Corp., Menlo Park, CA, USA) at 1 μM for 1800s. Then the loaded LacNAc biosensors were dipped into 0.1 μM EZ-link^{TM} Hydrazide-Biocytin (biocytin, Cat#: 28020, Thermo Scientific, Rockford, IL, USA) for 1800s to block the possible unoccupied biotin-SA binding sites for 1800s. The immobilization of ligand onto SA biosensors resulted in ~0.3 nm as loading signal under this condition. ECL direct binding K_D (LacNAc biosensor surface K_D) was measured using a bio-layer interferometer (BLI) Octet Red 96 system (Pall ForteBio Corp., Menlo Park, CA, USA) and data acquired using the ForteBio Data Acquisition 8.2 software (Pall ForteBio Corp., Menlo Park, CA, USA). The protein direct binding experiment was performed at 600 s for association and 1800 s for dissociation in ECL buffer. ECL was prepared in twofold serial dilution in ECL buffer from 0~50 μM, in the replicates of three. Surface K_D ($K_{D,surface\ LacNAc\ biosensor}$) was then calculated by ForteBio Data Analysis 8.2 software (Pall ForteBio Corp., Menlo Park, CA, USA) and Microsoft Office Excel 2011 (Microsoft, USA). Surface K_D ($K_{D,surface\ LacNAc\ biosensor}$) was determined by a 1:1 binding model from both steady state analysis and Scatchard plot (Fig. 13.5) and resulted in 0.92 (STDEV: 0.02) μM of triplicates.

Protein BLI inhibition assay (IC_{50}): ECL protein was prepared at 2 μM in ECL buffer in a large volume for protein inhibition assay. Eight compounds were tested in the inhibition assay including six compounds (inhibitors) and a negative control compound (non-ECL binder glycan). All the compounds were prepared in twofold serial dilution in ECL buffer from 0, 1.25, 2.5, 5, 10, 20, 40, to 80 mM. 100 μL of 2 μM ECL, 20 μL of prepared inhibitor/non-binder at its concentration, and 80μL of ECL buffer were mixed and incubated at room temperature for 1hour. ECL inhibition assay was performed on Octet Red 96 at baseline time 120 s, association time 600 s, and dissociation time 1800 s at shaker speed 1000RPM at room temperature, in replicates of three. IC_{50} was calculated by using the three-parameter dose-response inhibition model in GraphPad Prism 7 (GraphPad, La Jolla, CA, USA). Compounds LacNAc and LacNAc–sp-N_3 resulted in similar IC_{50} values. Therefore, only the final values for LacNAc are reported. This data also shows that the azide group attached to the compound LacNAc–sp-N_3 does not affect binding.

Assay Format 2 Experimental Details

Experimental materials: Influenza A H3N2 (X-31) Hemagglutinin (X-31, Cat#: 40059-V08H-50, Sino Biological, China), 3′-Sialyl-*N*-acetyllactoseamine (3′SLN, Cat#: SLN302, Dextra, UK), and 6′-Sialyl-*N*-acetyllactoseamine (6′SLN, Cat#: SLN306, Dextra, UK) were purchased from their commercial resources. Biotinylated glycan Siaα2-3Galβ1-4GlcNAcβ-$OCH_2CH_2CH_2NH$-biotin (3′SLN-biotin, CFG#B84), Siaα2-6Galβ1-4GlcNAcβ-$OCH_2CH_2CH_2NH$-biotin (6′SLN-biotin, CFG#B87), azido-glycan Siaα2-3Galβ1-4GlcNAcβ-$OCH_2CH_2CH_2$-N_3 (3′SLN-N_3, CFG#Tr33), and Siaα2-6Galβ1-4GlcNAcβ-$OCH_2CH_2CH_2$-N_3 (6′SLN-N_3, CFG#Tr36) were requested from the Consortium for Functional Glycomics (CFG). Anti-biotin-Fab antibody was purchased from Rockland Inc. (Fab, Cat# 800-101-098, Limerick, PA, USA). Fab-glycan was prepared by incubating Fab with biotinylated glycans at 1:1.1 mole ratio in dart at 4 °C overnight. X-31 was weighed and dissolved in the HA buffer: 1X PBS buffer at pH7.4, at 25 °C. Fab-glycans were all prepared in 10mM HEPES buffer at pH7.4 at 25 °C.

Immobilization of HA on streptavidin biosensors: Amine Reactive Second Generation biosensors (AR2G, Cat#: 18-5092, Pall ForteBio Corp., Menlo Park, CA, USA) were activated by dipped into a mixture of 20 mM EDC and 10 mM sulfo-NHS solution for 900 s. Then H3N2 X-31 HA protein was coupled onto activated AR2G biosensors at 20 μg/mL for 1800 s. Then the loaded X-31 H3 HA biosensors were dipped into 1 M ethanolamine (Cat#:110167-100ML, Sigma-Aldrich, St. Louis, MO, USA) pH 8.5 to quench the possible unreacted carboxylic group on the AR2G biosensor surface for 600 s. The immobilization of HA onto AR2G biosensors resulted in ~3 nm as loading signal under this condition.

Protein BLI direct binding assay ($K_{D,\ surface}$): Fab-glycan direct binding K_D (X-31 biosensor surface K_D) was measured using an bio-layer interferometer (BLI) Octet Red 96 system (Pall ForteBio Corp., Menlo Park, CA, USA) and data acquired using ForteBio Data Acquisition 8.2 software (Pall ForteBio Corp., Menlo Park, CA, USA). The protein direct binding experiment was performed at 360 s for association and 240 s for dissociation in HEPES buffer pH 7.4. Fab-glycans were prepared in twofold serial dilution in HEPES buffer from 0~8 μM, in replicates of three. Surface K_D ($K_{D,\text{surface Fab-glycan biosensor}}$) was then calculated by ForteBio Data Analysis 8.2 software (Pall ForteBio Corp., Menlo Park, CA, USA) and Microsoft Office Excel 2011 (Microsoft, USA). Surface K_D ($K_{D,\text{surface Fab-glycan biosensor}}$) was determined by 1:1 binding model from both steady state analysis and Scatchard plot and in triplicates.

Protein BLI inhibition assay (IC_{50}): Fab-glycans were prepared at 1μM in 10mM HEPES buffer pH7.4 in a large volume for protein inhibition assay. 3′SLN(Tr33), 6′SLN(Tr36), 3′SDLN(Te175), and 6′SDLN(Te176) were prepared in twofold serial dilution in water from 0,1.25, 2.5, 5, 10, 20, 40, to 80 mM. 100 μL of 1μM Fab-glycan, 20 μL of prepared azido-glycan at its concentration, and 80μL of 10mM HEPES buffer pH7.4 were mixed and incubated at room temperature for 1 h. X-31 inhibition assay was performed on Octet Red 96 at baseline time 120 s, association time 360 s, and dissociation time 240 s at shaker speed 1000 RPM at room temperature, in replicates of three. IC_{50} was calculated by using three-parameter dose-response inhibition model in GraphPad Prism 7 (GraphPad, La Jolla, CA, USA).

References

Abdiche Y, Malashock D, Pinkerton A, Pons J (2008) Determining kinetics and affinities of protein interactions using a parallel real-time label-free biosensor, the Octet. Anal Biochem 377:209–217

Abdiche YN, Malashock DS, Pinkerton A, Pons J (2009) Exploring blocking assays using Octet, ProteOn, and Biacore biosensors. Anal Biochem 386:172–180

Cheng Y, Prusoff W (1973) Relationship between inhibition constant (K_i) and concentration of inhibitor which causes 50 percent inhibition (I_{50}) of an enzymatic reaction. Biochem Pharm 22:3099–3108

Cornell BA, Braach-Maksvytis VLB, King LG, Osman PDJ, Raguse B, Wieczorek L, Pace RJ (1997) A biosensor that uses ion-channel switches. Nat Lett 387:580–583

de Vries RP, Peng W, Grant OC, Thompson AJ, Zhu X, Bouwman KM, de la Pena ATT, van Breemen MJ, Ambepitiya Wickramasinghe IN, de Haan CAM, Yu W, McBride R, Sanders RW, Woods RJ, Verheije MH, Wilson IA, Paulson JC (2017) Three mutations switch H7N9 influenza to human-type receptor specificity. PLoS Pathog 13:e1006390

Ernst B, Magnani JL (2009) From carbohydrate leads to glycomimetic drugs. Nat Rev Drug Disc 8:661–677

Farhadi SA, Hudalla GA (2016) Engineering galectin-glycan interactions for immunotherapy and immunomodulation. Exp Biol Med (Maywood) 241:1074–1083

Gelinsky-Wersing D, Wersing W, Pompe W (2017) Bivalent kinetic binding model to surface plasmon resonance studies of antigen-antibody displacement reactions. Anal Biochem 518:110–125

Gupta D, Cho M, Cummings RD, Brewer CF (1996) Thermodynamics of carbohydrate binding to galectin-1 from Chinese hamster ovary cells and two mutants. A comparison with four galactose-specific plant lectins. Biochemistry 35:15236–15243

Ji Y, White YJ, Hadden JA, Grant OC, Woods RJ (2017) New insights into influenza A specificity: an evolution of paradigms. Curr Opin Struct Biol 44:219–231

Khurana S, Verma S, Verma N, Crevar CJ, Carter DM, Manischewitz J, King LR, Ross TM, Golding H (2010) Properly folded bacterially expressed H1N1 hemagglutinin globular head and ectodomain vaccines protect ferrets against H1N1 pandemic influenza virus. PLoS One 5:e11548

Khurana S, King LR, Manischewitz J, Coyle EM, Golding H (2014) Novel antibody-independent receptor-binding SPR-based assay for rapid measurement of influenza vaccine potency. Vaccine 32:2188–2197

Lian W, Wu M, Huang N, Gao N, Xiao C, Li Z, Zhang Z, Zheng Y, Peng W, Zhao J (2013) Anti-HIV-1 activity and structure–activity-relationship study of a fucosylated glycosaminoglycan from an echinoderm by targeting the conserved CD4 induced epitope. BBA-Gen Subjects 1830:4681–4691

Liang P-H, Wang S-K, Wong C-H (2007) Quantitative analysis of carbohydrate−protein interactions using glycan microarrays: determination of surface and solution dissociation constants. J Am Chem Soc 129:11177–11184

Liao H-Y, Hsu C-H, Wang S-C, Liang C-H, Yen H-Y, Su C-Y, Chen C-H, Jan J-T, Ren C-T, Chen C-H, Cheng T-JR, Wu C-Y, Wong C-H (2010) Differential receptor binding affinities of influenza hemagglutinins on glycan arrays. J Am Chem Soc 132:14849–14856

Lin YP, Xiong X, Wharton SA, Martin SR, Coombs PJ, Vachieri SG, Christodoulou E, Walker PA, Liu J, Skehel JJ et al (2012) Evolution of the receptor binding properties of the influenza a(H3N2) hemagglutinin. Proc Natl Acad Sci U S A **109**:21474–21479

Magnani JL, Ernst B (2009) Glycomimetic drugs--a new source of therapeutic opportunities. Discov Med 8:247–252

Makeneni S, Ji Y, Watson DC, Young NM, Woods RJ (2014) Predicting the origins of anti-blood group antibody specificity: a case study of the ABO A- and B-antigens. Front Immunol 5:1–9

Mammen M, Choi S-K, Whitesides GM (1998) Polyvalent interactions in biological systems: implications for design and use of multivalent ligands and inhibitors. Angew Chem Int Ed 37:2754–2794

Matsumoto H, Shinzaki S, Narisada M, Kawamoto S, Kuwamoto K, Moriwaki K, Kanke F, Satomura S, Kumada T, Miyoshi E (2010) Clinical application of a lectin-antibody ELISA to measure fucosylated haptoglobin in sera of patients with pancreatic cancer. Clin Chem Lab Med 48:505–512

Miller GL, Stanley WM (1944) Quantitative aspects of the red blood cell agglutination test for influenza virus. J Exp Med 79:185–195

Notenboom V, Boraston AB, Williams SJ, Kilburn DG, Rose DR (2002) High-resolution crystal structures of the lectin-like xylan binding domain from streptomyces liVidans xylanase 10A with bound substrates reveal a novel mode of xylan binding. Biochemisrty 41:4246–4254

Peng L, Varma MM, Cho W, Regnier FE, Nolte DD (2007) Adaptive interferometry of protein on a BioCD. Appl Opt 46:5384–5395

Peng W, de Vries RP, Grant OC, Thompson AJ, McBride R, Tsogtbaatar B, Lee PS, Razi N, Wilson IA, Woods RJ, Paulson JC (2017) Recent H3N2 viruses have evolved specificity for extended, branched human-type receptors, conferring potential for increased avidity. Cell Host Microbe 21:23–34

Rich RL, Myszka DG (2007) Higher-throughput, label-free, real-time molecular interaction analysis. Anal Biochem 361:1–6

Sauter NK, Bednarski MD, Wurzburg BA, Hanson JE, Whitesides GM, Skehel JJ, Wiley DC (1989) Hemagglutinins from two influenza virus variants bind to sialic acid derivatives with millimolar dissociation constants: a 500-MHz proton nuclear magnetic resonance study. Biochemistry 28:8388–8396

Sauter NK, Hanson JE, Glick GD, Brown JH, Crowther RL, Park SJ, Skehel JJ, Wiley DC (1992) Binding of influenza virus hemagglutinin to analogs of its cell-surface receptor, sialic acid: analysis by proton nuclear magnetic resonance spectroscopy and X-ray crystallography. Biochemistry 31:9609–9621

Schieffelin J, Costin J, Nicholson C, Orgeron N, Fontaine K, Isern S, Michael S, Robinson J (2010) Neutralizing and non-neutralizing monoclonal antibodies against dengue virus E protein derived from a naturally infected patient. Virol J 7:1–11

Schuck P, Zhao H (2010) The role of mass transport limitation and surface heterogeneity in the biophysical characterization of macromolecular binding processes by SPR biosensing. Methods Mol Biol 627:15–54

Solís D, Bovin NV, Davis AP, Jiménez-Barbero J, Romero A, Roy R, Smetana K Jr, Gabius H-J (2015) A guide into glycosciences: how chemistry, biochemistry and biology cooperate to crack the sugar code. BBA-Gen Subjects 1850:168–235

Stanford JL, Bennett HE (1969) Enhancement of surface plasma resonance absorption in mirrors by overcoating with dielectrics. Appl Opt 8:2556–2557

Tong HH, James M, Grants I, Liu X, Shi G, DeMaria TF (2001) Comparison of structural changes of cell surface carbohydrates in the eustachian tube epithelium of chinchillas infected with a Streptococcus pneumoniae neuraminidase-deficient mutant or its isogenic parent strain. Microb Pathog 31:309–317

Xiong X, Coombs PJ, Martin SR, Liu J, Xiao H, McCauley JW, Locher K, Walker PA, Collins PJ, Kawaoka Y, Skehel JJ, Gamblin SJ (2013) Receptor binding by a ferret-transmissible H5 avian influenza virus. Nature 497:392–396

Yang DL, Singh A, Wu HL, Kroe-Barrett R (2016) Comparison of biosensor platforms in the evaluation of high affinity antibody-antigen binding kinetics. Anal Biochem 508:78–96

Index

Y. Yamaguchi, K. Kato (eds.), *Glycobiophysics*, Advances in Experimental Medicine and Biology 1104, https://doi.org/10.1007/978-981-13-2158-0

Printed by Printforce, the Netherlands